Elementary
Linear Algebra

Elementary Linear Algebra

Margaret and Erwin Kleinfeld

Nova Science Publishers, Inc.

Art Director: Maria Ester Hawrys
Assistant Director: Elenor Kallberg
Graphics: Eddie Fung, Kerri Pfister,
       Max Dutton and Barbara Minerd
Manuscript Coordinator: Gloria H. Piza
Book Production: Tammy Sauter, Gavin Aghamore
              Christine Mathosian and Joanne Bennette
Circulation: Irene Kwartiroff and Annette Hellinger

Library of Congress Cataloging-in-Publication Data

Kleinfeld, Margaret
    Elementary linear algebra / by Margaret and Erwin
Kleinfeld.
      p. cm.
    Includes bibliographical references (p. –) and index.
    ISBN 1-56072-292-4 (lib. bdg.):
    1. Algebras, Linear. I. Kleinfeld, Erwin. II. Title
QA184.K585 1996   95-49100
512' .5--dc20   CIP

Copyright © 1996
  Nova Science Publishers, Inc.
  6080 Jericho Turnpike, Suite 207
  Commack, New York 11725
  Tele. 516-499-3103 Fax 516-499-3146
  E Mail Novasci1@aol.com

Printed in the United States of America

Contents

Introduction

In this presentation of the fundamentals of linear algebra we have stressed concepts rather than computation. We treat the reduced row echelon form of a matrix as a means rather than an end. Included with this book is a computer program called MATS, designed by our colleague Prof. Eugene Johnson especially for use by students taking a first course in linear algebra. A student can learn to use MATS in only a few minutes. MATS can do elementary row operations, find the reduced row echelon and Hermite forms of a matrix, find determinants, find inverses, etc., and we try to focus on explaining clearly the significance and uses of these things. The reduced row echelon form of a matrix carries a lot of information if you know how to interpret it. We take a strongly matrix oriented approach and try to illuminate and exploit the close connection between matrix operations, linear combinations, subspaces of R^n, and linear equations. In this way we have been able to give clear and easy to understand proofs of all the basic theorems. While some students and instructors may wish to skip some or all of the proofs, we feel that it is good to have them included, especially since they are so accessible. The MATS program will take the computational drudgery out of the subject, and using MATS students can not only do the exercises in the book, but have fun making up problems and examples of their own. Further information on the MATS program is included in Appendix C.

Our approach follows the recommendations of the Linear Algebra Curriculum Study Group in emphasizing matrices rather than linear transformations. We define a linear transformation but immediately show that on R^n every linear transformation T can be realized by matrix multiplication, $T(\mathbf{v}) = A\mathbf{v}$. Our treatment does not require the representation of T with respect to any basis other than the standard basis. This removes a source of great confusion to many students at this level.

Chapter 1 covers vector spaces, subspaces, basis and dimension. We begin with ordered pairs because students are familiar with the XY plane. We generalize to n-tuples and then to an abstract vector space. We do not belabor the axioms, but give a few examples and quickly move on to show that every vector space with a finite spanning set is essentially a copy of the n-tuples. We could restrict ourselves to considering n-tuples, but sometimes we prefer to view vectors more abstractly if that seems to simplify or clarify something. The main reason for putting this material first is so that the idea of linear combination of n-tuples can be related to matrix multiplication, and we can speak of the row space and column space of a matrix as soon as matrices are introduced.

Matrices and matrix operations are introduced next because we want these available so that as soon as we introduce systems of linear equations we can express them in matrix notation. Then we define the reduced row echelon form and elementary row operations and show how these are related to the row space, the column space and the solution of systems of linear equations. We reintroduce the Hermite form and use it to develop a convenient way to get a basis for the null space. We have a new and very elementary proof of the uniqueness of the reduced row echelon form. This is a stronger uniqueness theorem than is usually given, stating that the reduced row echelon form is uniquely determined by the row space of the matrix. This gives a very simple algorithm for determining whether or not two different sets of n-tuples span the same subspace. You simply put them as rows of two matrices, and they span the same subspace if and only if these two matrices have the same nonzero rows in their reduced row echelon forms (affectionately called RREF). We explain in very simple terms that the RREF of A tells you how the columns of A are related, which allows one to see clearly from the RREF why row rank equals column rank.

Chapter 3, determinants, begins with a section on permutations. This has been omitted from most recent elementary texts in an effort to simplify the definition of determinant. We feel that this is a mistake. The material on permutations that we have included is well within the reach of students at this level, and greatly simplifies and clarifies the discussion of determinants. However, this section could be skipped, as it is only used to prove how the determinant changes when elementary row operations are performed on the matrix, so one could just assume these results and go on from there.

In Chapter 4 the students can make good use of everything they have learned up to this point to understand eigenvalues, eigenvectors, and diagonalization. We introduce a method for finding an orthogonal basis for an eigenspace of a symmetric matrix which is more efficient than first finding a non-orthogonal basis and then using the Gram-Schmidt process on it. In simple cases of low dimension, our method allows you to chose orthogonal basis vectors by inspection, in more difficult cases you can use the computer to get the RREFs needed. In order to prove that the eigenvalues of a symmetric matrix must be real, we briefly discuss complex numbers. We give a proof by mathematical induction that a symmetric matrix is orthogonally similar to a diagonal matrix. This proof is very simple if you understand matrix multiplication well, and provides perhaps the first experience a student may have of a non-trivial use of mathematical induction. A similar proof shows that over the complex numbers every matrix A is similar to an upper triangular matrix with the eigenvalues of A on the diagonal (a useful result not usually included). This has several interesting corollaries, and can be used to get an elementary proof of the Cayley-Hamilton Theorem, again based on a good understanding of the various ways of viewing matrix multiplication (as operating on rows or columns). We conclude Chapter 4 with a section on applications of diagonalization, including some of the standard ones, but also some material on nilpotent and idempotent matrices not usually included at this level.

In Chapter 5 we use the RREF to find vectors in the intersection of two subspaces. We discuss the orthogonal complement of a subspace and use this to show that every subspace of R^n can be realized as the null space of a matrix and show how to find this matrix given a basis of the subspace. We discuss projection of a vector on a subspace and several ways to find it. We discuss the relation of projections to the least squares solution. We use Vandermonde matrices to show that there is a unique polynomial of degree less than or equal to n through any $n+1$ points in R^2 with distinct x values, and show how to find this polynomial. We also discuss how to find the least squares fit polynomial of a given degree if you have too many points to find an exact solution.

We include a "How To $\cdots$" chapter recalling all the algorithms that were previously introduced in the book. Problems become easy and fun using these techniques and the MATS program.

In the last chapter we show that vectors can be used to simplify and clarify some topics from analytic geometry. We put the geometry last so that students can take full advantage of the concepts and techniques they have learned in the earlier chapters. However, except for Section 7.6 which involves eigenvalues of orthogonal matrices, an effort was made to make Chapter 7 self-contained so that it could be done earlier if desired.

Finally we include three appendices. Appendix A (Linear Transformations) translates some of the results we have proved for matrices into the language of linear transformations, defines the matrix of a linear transformation on R^n with respect to an arbitrary basis, and shows how this matrix changes when the basis is changed. Appendix B gives answers to most of the exercises (excluding most which begin with "prove" or "show"). Of course some of the answers are not unique, so that the answer given may not be the only correct answer possible. Appendix C discusses the MATS program and gives some instructions for its use and a list of commands.

Chapter 1

Vector Spaces

Section 1.1 Introduction to R^2

Definition 1.1 *The set of ordered pairs of real numbers is called R^2.*

We will refer to the elements of R^2 as "vectors", and to the real numbers as "scalars". We know from analytic geometry that the points in the plane can be represented by pairs of real numbers. It is helpful to visualize the vector (a, b) as the directed line segment from the point $(0, 0)$ to the point (a, b). In the vector (a, b), a is called the first component, and b is called the second component. We have the following definition of equality:

Definition 1.2 $(a, b) = (c, d)$ *if and only if $a = c$ and $b = d$*

We add vectors according to the following definition:

Definition 1.3 $(a, b) + (c, d) = (a + c, b + d)$.

We multiply a vector by a scalar as follows:

Definition 1.4 $c(a, b) = (ca, cb)$.

We call these operations "vector addition" and "scalar multiplication", respectively. As you can see, the addition of pairs is done by

3

adding the components separately, and the scalar multiplication is done by multiplying each component by a real number c. Note that although we use the same plus sign for both vector addition and addition of real numbers, these are different operations. We really should use a different symbol, but the context makes it clear which addition is meant. The same comments apply to scalar multiplication, which we denote by juxtaposition, just as we do regular multiplication of real numbers.

Definition 1.5 *The length of a vector (a, b) is the distance from the origin to the point (a, b), which equals $\sqrt{a^2 + b^2}$.*

To save writing we will frequently denote vectors in R^2 by a single letter, like $\mathbf{u}, \mathbf{v}, \mathbf{w}, \mathbf{x}, \mathbf{y}, \mathbf{z}$. We will write vectors in boldface.

Vector addition and scalar multiplication have geometric interpretations. Figure 1.1 illustrates vector addition. The points $(0, 0)$, (a, b), (c, d), and $(a + c, b + d)$ form a parallelogram. Thus our definition of vector addition agrees with the parallelogram addition used for forces and velocities in physics.

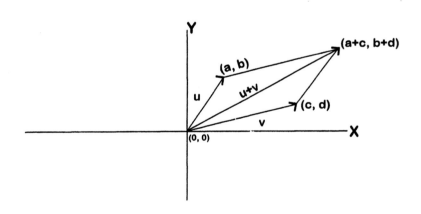

Figure 1.1

Figure 1.2 shows that if a vector **u** is represented by a directed line segment, then a scalar multiple c**u**, $c > 0$ is represented by a directed line segment in the same direction but of length c times the original length. If $c < 0$, c**u** is represented by a directed line segment in the opposite direction, of length $|c|$ times the length of **u**.

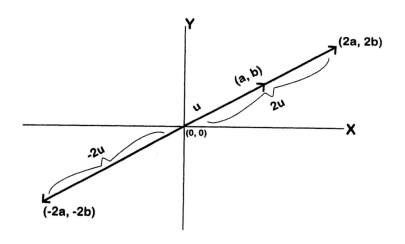

Figure 1.2

Vector addition is a commutative operation. By this we mean that for all **u**, **v** in R^2

$$\mathbf{u} + \mathbf{v} = \mathbf{v} + \mathbf{u}. \tag{1.1}$$

To verify this let $\mathbf{u} = (a, b)$ and $\mathbf{v} = (c, d)$. Then $\mathbf{u} + \mathbf{v} = (a, b) + (c, d) = (a + c, b + d) = (c + a, d + b) = (c, d) + (a, b) = \mathbf{v} + \mathbf{u}$. This constitutes a proof that vector addition in R^2 is commutative. Vector addition is also associative. By this we mean for all **u**, **v**, and **w** in R

$$(\mathbf{u} + \mathbf{v}) + \mathbf{w} = \mathbf{u} + (\mathbf{v} + \mathbf{w}). \tag{1.2}$$

To verify this take $\mathbf{u} = (a, b)$, $\mathbf{v} = (c, d)$, and $\mathbf{w} = (e, f)$. Then $(\mathbf{u} + \mathbf{v}) + \mathbf{w} = ((a, b) + (c, d)) + (e, f) = (a + c, b + d) + (e, f) =$

$(a+c+e, b+d+f) = (a,b)+(c+e, d+f) = (a,b)+((c,d)+(e,f)) =$ $\mathbf{u}+(\mathbf{v}+\mathbf{w})$. You might wonder why it was unnecessary to distinguish between $(a+c)+e$ and $a+(c+e)$. The reason is that we already know that addition of real numbers is associative. This has enabled us to prove that vector addition is associative as well.

The vector $(0,0)$ plays a special role in vector addition because $(0,0)+(a,b) = (a,b)$ for all vectors (a,b) in R^2. This together with equation 1.1 shows that the vector $(0,0)$ under vector addition plays the same role as the number zero under addition of real numbers. For this reason the vector $(0,0)$ is called the zero vector and given the special symbol $\mathbf{O}$. Thus we see that

$$\mathbf{O}+\mathbf{v} = \mathbf{v}, \text{ for all } \mathbf{v} \text{ in } R^2. \tag{1.3}$$

We also note that for any vector $\mathbf{v} = (a,b)$ in R^2 we have $(-a,-b)$ is in R^2 and $(a,b)+(-a,-b) = (0,0) = \mathbf{O}$. Since there is only one such vector $(-a,-b)$ satisfying $\mathbf{v}+\mathbf{w} = \mathbf{O}$, we shall call $(-a,-b)$ the inverse (under vector addition) of $\mathbf{v}$, and denote it $by - \mathbf{v}$. We have shown:

For every $\mathbf{v}$ in R^2 there exists a unique $\mathbf{w}$ in R^2 such that $\mathbf{v}+\mathbf{w} = \mathbf{O}$.
$$\tag{1.4}$$

Properties 1.1 through 1.4 occur frequently in algebraic systems and have a special name. A set S with a binary operation $+$ satisfying these properties is called a commutative group. We have proved our first theorem.

Theorem 1.1 *The vectors in R^2 form a commutative group under addition.*

We also have the following properties relating the scalar multiplication and the vector addition, which hold for all scalars (real numbers) r and s, and for all vectors $\mathbf{u}$ and $\mathbf{v}$ in R^2 :

$$r(\mathbf{u}+\mathbf{v}) = r\mathbf{u}+r\mathbf{v}, \tag{1.5}$$

$$(r+s)\mathbf{u} = r\mathbf{u}+s\mathbf{u}, \tag{1.6}$$

$$(rs)\mathbf{u} = r(s\mathbf{u}), \qquad (1.7)$$

$$1\mathbf{u} = \mathbf{u}. \qquad (1.8)$$

Exercises

1. Find $\mathbf{u} + \mathbf{v}$, $c\mathbf{u}$, and length of $\mathbf{u}$ for the following values of $\mathbf{u}$, $\mathbf{v}$, and c.

   (a) $\mathbf{u} = (2, 3)$, $\mathbf{v} = (-1, 4)$, $c = 3$
   (b) $\mathbf{u} = (-5, 2)$, $\mathbf{v} = (-1, -1)$, $c = -\frac{1}{2}$.

2. Describe algebraically all vectors that lie along a straight line through $(-1, -1)$ and $(1, 1)$.

3. Let $\mathbf{v}$ be a fixed vector. Describe geometrically the set of all scalar multiples $c\mathbf{v}$ when

   (a) $\mathbf{v} = \mathbf{O}$,
   (b) $\mathbf{v} \neq \mathbf{O}$.

4. Describe geometrically all vectors of the form $\mathbf{u} + \mathbf{v}$, where $\mathbf{u} = (1, 2)$ and $\mathbf{v}$ lies along the X axis.

5. Prove that the points $(0, 0)$, (a, b), (c, d), and $(a + c, b + d)$ form a parallelogram. (Hint: Calculate the slopes of all four sides in figure 1.1).

6. Prove that properties 1.5-1.8 hold for all real numbers r, s, and all ordered pairs $\mathbf{u} = (a, b)$, $\mathbf{v} = (c, d)$.

Section 1.2 The Concept of Basis in R^2

Definition 1.6 *If* **u** *and* **v** *are vectors and* c *and* d *are scalars, then every vector of the form* $c\mathbf{u}+d\mathbf{v}$ *is called a linear combination of* **u** *and* **v**.

It is easy to check that $(-1,8)$ is a linear combination of $(1,2)$ and $(2,-1)$, since $3(1,2)+(-2)(2,-1) = (3,6)+(-4,2) = (-1,8)$. A harder problem is to find scalars c and d such that $(2,5) = c(1,2) + d(2,-1)$. This gives rise to a system of two equations in two unknowns, since we must have $(2,5) = (c,2c) + (2d,-d)$, or equivalently $2 = c + 2d$ and $5 = 2c-d$. Multiplying the first equation by 2 and subtracting the result from the second equation gives $1 = -5d$, so that $d = -\frac{1}{5}$. Substituting this in either equation yields $c = \frac{12}{5}$. As a check we compute $\frac{12}{5}(1,2) - \frac{1}{5}(2,-1) = \left(\frac{12}{5},\frac{24}{5}\right) + \left(-\frac{2}{5},\frac{1}{5}\right) = (2,5)$, as desired.

Now we return to the question of whether we can find two vectors **u** and **v** such that every vector **w** can be expressed uniquely in as a linear combination of **u** and **v**. It is very easy to see that by taking $\mathbf{u} = (1,0)$ and $\mathbf{v} = (0,1)$, we have such a pair, since $(a,b) = c(1,0) + d(0,1)$ quickly yields $a = c$ and $b = d$ as the only possible solution, and in fact $a(1,0) + b(0,1) = (a,0) + (0,b) = (a,b)$.

Suppose we take another pair of vectors, say $\mathbf{u} = (1,1)$ and $\mathbf{v} = (2,3)$ and ask the question "Can every vector (a,b) be written uniquely as a linear combination of **u** and **v**?" We must try to solve $c(1,1) + d(2,3) = (a,b)$. Multiplying by the scalars and equating components leads to the equations $c + 2d = a$, and $c + 3d = b$. Subtracting yields $d = b - a$, and substituting this is for d yields $c = 3a - 2b$ as the only possible solution, and we can check that indeed $(3a - 2b)(1,1) + (b - a)(2,3) = (3a - 2b, 3a - 2b) + (2b - 2a, 3b - 3a) = (a,b)$, so we have a unique solution for every choice of a and b.

We try another pair of vectors. Let $\mathbf{u} = (2,5)$ and $\mathbf{v} = (4,10)$. Now try to solve $c(2,5)+d(4,10) = (a,b)$. We get the equations $2c+4d = a$, and $5c+10d = b$. Note that the left hand side of the second equation is $\frac{5}{2}$ the left hand side of the first. Thus no solution will exist unless $b = \frac{5}{2}a$. The trouble here is that $\mathbf{v} = 2\mathbf{u}$, and so any linear combination of **u** and $\mathbf{v}$ is just a multiple of $\mathbf{u}$, since $c\mathbf{u} + d\mathbf{v} = c\mathbf{u} + d(2\mathbf{u}) = (c + 2d)\mathbf{u}$. However the multiples of a single vector lie along a line and do not exhaust R^2.

Definition 1.7 *A pair of vectors* $\mathbf{u}$ *and* $\mathbf{v}$ *is a basis of* R^2 *if every vector* $\mathbf{w}$ *in* R^2 *can be expressed uniquely as a linear combination of* $\mathbf{u}$ *and* $\mathbf{v}$.

We have seen that some pairs of vectors in R^2 are a basis for R^2 and some are not.

Lemma 1.1 $\mathbf{u} = (a, b)$ *and* $\mathbf{v} = (c, d)$ *have the property that one is a scalar multiple of the other if and only if* $ad - bc = 0$.

Proof. If $\mathbf{u} = k\mathbf{v}$, we have $(a, b) = k(c, d)$, so that $a = kc$ and $b = kd$. Then $ad - bc = (kc)d - (kd)c = 0$. Conversely if $ad - bc = 0$, if $d \neq 0$, define k by $b = kd$. Then using $ad - bc = 0$, we have $ad = kdc$, and thus $a = kc$, so that $(a, b) = k(c, d)$. If $d = 0$ but $c \neq 0$, define k by $a = kc$, then $ad - bc = 0$ gives $kcd = bc$, or $kd = b$, and again we have $(a, b) = k(c, d)$. Finally if both c and d are 0 then $\mathbf{v} = \mathbf{O}$, and so $\mathbf{v} = 0\mathbf{u}$. ∎

Definition 1.8 *Two vectors* $\mathbf{u}$ *and* $\mathbf{v}$ *in* R^2 *are linearly independent if* $c\mathbf{u} + d\mathbf{v} = \mathbf{O}$ *only when* $c = 0$ *and* $d = 0$. *If* $\mathbf{u}$ *and* $\mathbf{v}$ *are not linearly independent, they are called linearly dependent.*

Clearly two vectors $\mathbf{u}$ and $\mathbf{v}$ in R^2 are linearly dependent if $c\mathbf{u}+d\mathbf{v} = \mathbf{O}$ for at least one choice of c and d not both zero.

Lemma 1.2 *The vectors* $\mathbf{u}$ *and* $\mathbf{v}$ *in* R^2 *are linearly dependent if and only if one is a scalar multiple of the other.*

Proof. Suppose $\mathbf{u} = k\mathbf{v}$, then $(-1)\mathbf{u} + k\mathbf{v} = \mathbf{O}$, so $\mathbf{u}$ and $\mathbf{v}$ are linearly dependent. Conversely, if $c\mathbf{u} + d\mathbf{v} = \mathbf{O}$, and either $c \neq 0$ or $d \neq 0$, then we can solve for one of the vectors. If for example $c \neq 0$, $\mathbf{u} = -\frac{d}{c}\mathbf{v}$. If $d \neq 0$, then $\mathbf{v} = -\frac{c}{d}\mathbf{u}$. ∎

Theorem 1.2 *If* $\mathbf{u} = (a, b)$ *and* $\mathbf{v} = (c, d)$ *and if* $ad - bc \neq 0$, *then* $\mathbf{u}$ *and* $\mathbf{v}$ *form a basis of* R^2.

Proof.   Let $\mathbf{w} = (e, f)$. Then $r(a, b) + s(c, d) = (e, f)$, or $(ra, rb) + (sc, sd) = (e, f)$. This leads to two equations

$$ra + sc = e \qquad (1.9)$$

$$rb + sd = f. \qquad (1.10)$$

In these equations a, b, c, d, e and f are regarded as given and the unknowns are r and s. Multiply the equation 1.9 by b and the equation 1.10 by a and subtract the second from the first. The r terms cancel and we get $s(bc - ad) = be - af$. Since $ad - bc \neq 0$, we can divide and get

$$s = \frac{be - af}{bc - ad}.$$

Similarly

$$r = \frac{de - cf}{ad - bc}.$$

These are the only possible solutions, and one can check by substituting them in 1.9 and 1.10 above that they are solutions. ∎

Combining the previous results and definitions we have:

Theorem 1.3 *If* $\mathbf{u} = (a, b)$ *and* $\mathbf{v} = (c, d)$, *the following are equivalent:*
1. $\mathbf{u}$ *and* $\mathbf{v}$ *are a basis of* R^2.
2. $\mathbf{u}$ *and* $\mathbf{v}$ *are linearly independent.*
3. $ad - bc \neq 0$.

Proof.   We shall show that $1 \Rightarrow 2 \Rightarrow 3 \Rightarrow 1$. If $\mathbf{u}$ and $\mathbf{v}$ are a basis for R^2, then every vector in R^2 can be expressed uniquely as a linear combination of $\mathbf{u}$ and $\mathbf{v}$. Thus the zero vector $\mathbf{O}$ can be expressed uniquely as $c\mathbf{u} + d\mathbf{v}$. But we know $\mathbf{O} = 0\mathbf{u} + 0\mathbf{v}$, thus if $\mathbf{O} = c\mathbf{u} + d\mathbf{v}$ we must have $c = 0$ and $d = 0$, hence $\mathbf{u}$ and $\mathbf{v}$ are linearly independent. Next assume that $\mathbf{u}$ and $\mathbf{v}$ are linearly independent. Then using Lemma 1.2, $\mathbf{u}$ is not a scalar multiple of $\mathbf{v}$. But now Lemma 1.1 implies that $ad - bc \neq 0$. That $3 \Rightarrow 1$ is Theorem 1.2. ∎

Exercises

1. If possible find scalars c and d such that:

   (a) $(2,5) = c(1,-1) + d(1,2)$.

   (b) $(2,5) = c(1,-1) + d(-1,1)$.

2. If possible express $(3,1)$ as a linear combination of $(1,1)$ and $(1,2)$.

3. Which of the following pairs of vectors is linearly independent?

   (a) $(1,1), (2,1)$,

   (b) $(1,-1), (-1,1)$,

   (c) $(2,4), (3,6)$.

4. Which of the pairs in problem 3 is a basis of R^2?

5. Use Theorem 1.3 to prove that if $\mathbf{u}$, $\mathbf{v}$ and $\mathbf{w}$ are any three vectors in R^2, then there exist scalars $c, d,$ and e not all 0 such that $c\mathbf{u} + d\mathbf{v} + e\mathbf{w} = \mathbf{O}$.

6. Show that the pair consisting of $\mathbf{O}$ and any other vector is always linearly dependent.

Section 1.3 Introduction to R^n

It is easy to see that what we have done in sections 1.1 and 1.2 for pairs of real numbers has its analog in triples. You are no doubt familiar with the idea of representing the points in 3 dimensional space by triples of real numbers, using the X, Y and Z axes. We define R^3 to be the set of triples of real numbers, the elements of R^3 will be called vectors and the vector (a,b,c) can be visualized as the directed line segment from the origin to the point (a,b,c). Vector addition and scalar multiplication can be defined componentwise as in R^2. But why stop there? Why not ordered 4-tuples and 5-tuples? While we are not able to visualize 4 or

5 mutually perpendicular axes to mimic what we did geometrically for pairs and triples, algebraically we encounter no problems at all when adding more components. Thus let us go directly to R^n, where n is a positive integer.

Definition 1.9 *The set of all ordered n-tuples of real numbers is called* R^n.

The elements of R^n will be called vectors. Let $\mathbf{u} = (a_1, a_2, \cdots, a_n)$ and $\mathbf{v} = (b_1, b_2, \cdots, b_n)$ be vectors in R^n and let c be a scalar, then we make the following definitions.

Definition 1.10

$$\mathbf{u} + \mathbf{v} = (a_1 + b_1, a_2 + b_2, \cdots, a_n + b_n).$$

Definition 1.11

$$c\mathbf{u} = (ca_1, ca_2, \cdots, ca_n).$$

Definition 1.12 *We write* $|\mathbf{u}|$ *for the length of* $\mathbf{u}$, *and define*

$$|\mathbf{u}| = \sqrt{a_1^2 + a_2^2 + \cdots + a_n^2}.$$

It is easy to verify that properties 1.1 through 1.8 remain true for these definitions of vector addition and scalar multiplication. Of course we must make the slight adjustment that the zero vector $\mathbf{O}$ in R^n is $(0, 0, \cdots, 0)$ and $-\mathbf{u}$ is $(-a_1, -a_2, \cdots, -a_n)$. We can also extend the definitions of linear combination, linear independence, and basis to R^n, and in a later chapter we will prove a result for R^n which will have Theorem 1.3 as a special case.

Definition 1.13 *If* $\mathbf{u}_1, \mathbf{u}_2, \cdots, \mathbf{u}_k$ *are k vectors in* R^n, *and* $c_1, c_2, \cdots, c_k$ *are k scalars, then the vector* $\mathbf{v} = c_1\mathbf{u}_1 + c_2\mathbf{u}_2 + \cdots + c_k\mathbf{u}_k$ *is called a linear combination of* $\mathbf{u}_1, \mathbf{u}_2, \cdots, \mathbf{u}_k$. *Note if* $k = 1$, *a linear combination of one vector is just a scalar multiple of that vector.*

Definition 1.14 *If* $c_1\mathbf{u}_1 + c_2\mathbf{u}_2 + \cdots + c_k\mathbf{u}_k = \mathbf{O}$ *only when* $c_1 = 0, c_2 = 0, \cdots, c_k = 0$, *then we say that* $\mathbf{u}_1, \mathbf{u}_2, \cdots, \mathbf{u}_k$ *are linearly independent.*

Definition 1.15 *If* $\mathbf{u}_1, \mathbf{u}_2, \cdots, \mathbf{u}_k$ *is a set of vectors in* R^n *such that every vector of* R^n *can be expressed uniquely as a linear combination of* $\mathbf{u}_1, \mathbf{u}_2, \cdots, \mathbf{u}_k$, *we say that the ordered set of vectors* $\{\mathbf{u}_1, \mathbf{u}_2, \cdots, \mathbf{u}_k\}$ *is a basis of* R^n.

Consider the n-tuples $\mathbf{e}_1 = (1,0,0,\cdots,0)$, $\mathbf{e}_2 = (0,1,0,\cdots,0)$, $\mathbf{e}_3 = (0,0,1,0,\cdots,0)$, $\cdots$, $\mathbf{e}_n = (0,0,0,\cdots,1)$, where $\mathbf{e}_k$ has a 1 as the kth component and all other components of $\mathbf{e}_k$ are 0. It is easy to see that $\{\mathbf{e}_1, \mathbf{e}_2, \cdots, \mathbf{e}_n\}$ is a basis of R^n. This basis is called the standard (or natural) basis of R^n. There are many other bases for R^n, but every basis of R^n has exactly n vectors. This is an important fact that we will prove in a later section.

Consider the triples of the form $(a,b,0)$, *i.e.* triples whose last component is 0. Note that the sum of two such vectors is again of the same form,

$$(a,b,0) + (c,d,0) = (a+c, b+d, 0),$$

and a scalar multiple

$$r(a,b,0) = (ra, rb, 0)$$

is also of the same form. This subset of R^3 satisfies properties 1.1 through 1.8 without any of the other triples being needed. Subsets of this type are called subspaces. Geometrically, this is the XY plane. It definitely resembles R^2. In fact just by dropping the last 0 we have a vector in R^2, and we could do all our calculations in R^2 and then put the 0 back as a third component, and this would give the same answer as computing with the triples. The correspondence $(a,b,0) \longleftrightarrow (a,b)$ is one to one and preserves addition and scalar multiplication. In this way R^2 can be identified with a subset of R^3. Note however that the triples of the form $(0,a,b)$ is also a subspace of R^3 which can be identified with R^2. Actually R^3 contains many copies of R^2. We explore these ideas further in the next section.

We note in conclusion that R^1 is just the real numbers, which in this case are being used as both the vectors and the scalars. Geometrically one can associate the real numbers with the points on a line by choosing an origin and a unit length. Then the vector a in R^1 can be viewed as the directed line segment from the origin to the point associated with the number a. Our definition of length then gives length of $a = \sqrt{a^2} = |a|$.

Exercises

1. Express $(2, 5, -3)$ as a linear combination of $(0, 0, 1)$, $(0, 1, 1)$, and $(1, 1, 1)$.

2. Express (a, b, c) as a linear combination of $(0, 0, 1)$, $(0, 1, 1)$, and $(1, 1, 1)$.

3. Is the set $\{(0, 0, 1), (0, 1, 1), (1, 1, 1)\}$ a basis of R^3? Explain.

4. Is the set in problem 3 a linearly independent set? Explain.

5. Let $\mathbf{u}$ be in R^n and c a scalar. Prove that if $c\mathbf{u} = \mathbf{O}$, we must have either $c = 0$ or $\mathbf{u} = \mathbf{O}$.

6. Prove that if $\mathbf{u}$ is in R^n and $\mathbf{u} \neq \mathbf{O}$, then $\{\mathbf{u}\}$ is a linearly independent set.

Section 1.4 Abstract Vector Space

The beauty and power of mathematics lies in its generality. We have discussed R^2 and seen how easily it generalized to R^n. Now we want to consider something even more general. We do this because some things are actually easier to see in the more general context. You may continue to visualize the vectors $\mathbf{u}, \mathbf{v}$, and $\mathbf{w}$ as ordered pairs or real numbers if that makes it easier for you, but we are now going to define an abstract vector space over the real numbers. In this case we don't specify what the vectors are, just that we have a set V of elements that we call vectors, together with a definition of vector addition that tells us how two of these vectors can be added, and a definition of scalar multiplication that tells us how to multiply one of these vectors by a scalar (real number).

Definition 1.16 *A vector space over the real numbers is a set V whose elements we will call vectors, together with a vector addition and a scalar multiplication such that for all $\mathbf{u}$, $\mathbf{v}$, and $\mathbf{w}$ in V, and a and b real numbers we have:*

A1
$$\mathbf{u} + \mathbf{v} \text{ is in } V.$$

A2
$$\mathbf{u} + \mathbf{v} = \mathbf{v} + \mathbf{u}.$$

A3
$$\mathbf{u} + (\mathbf{v} + \mathbf{w}) = (\mathbf{u} + \mathbf{v}) + \mathbf{w}$$

A4

There exists $\mathbf{O}$ in V such that $\mathbf{O} + \mathbf{u} = \mathbf{u}$ for all $\mathbf{u}$ in V.

A5

For every $\mathbf{u}$ in V there exists $\mathbf{z}$ in V such that $\mathbf{u} + \mathbf{z} = \mathbf{O}$.

A6
$$a\mathbf{u} \text{ is in } V$$

A7
$$a(\mathbf{u} + \mathbf{v}) = a\mathbf{u} + a\mathbf{v}.$$

A8
$$(a + b)\mathbf{u} = a\mathbf{u} + b\mathbf{u}.$$

A9
$$a(b\mathbf{u}) = (ab)\mathbf{u}.$$

A10
$$1\mathbf{u} = \mathbf{u}.$$

We use the same sign + for both vector addition and addition of real numbers, using the context to make clear which is meant. For example in A8 the + on the left hand side of the equation is between real numbers, while the + on the right hand side is between two elements of V and hence denotes vector addition. A1 says that the sum of two elements of V is again an element of V. Another way of saying this is to say V is "closed" under vector addition. Axiom A6 says that a scalar times a vector is a vector (because it's in V). We have not defined any multiplication between two vectors.

Example 1.1 *We have seen that R^n with the definitions of vector addition and scalar multiplication given in Section 1.3 satisfies A1 through A10.*

Example 1.2 *Let V be the set of all continuous functions from the real numbers to the real numbers. Let the vector addition be defined by $(f+g)(x) = f(x)+g(x)$, and scalar multiplication by $(af)(x) = a(f(x))$. In this case the zero vector $\mathbf{O}$ is the zero function which equals zero for all x, and $(-f)(x) = -(f(x))$. Since we know from calculus that the sum of two continuous functions is continuous, A1 is satisfied. It is left to the reader to check the remaining axioms.*

Remark   *In Example 1.2 we mean by a continuous function one that is continuous at r for every real number r. If we fix r, and consider the functions continuous at r, we get another example.*

Example 1.3 *Let V be the set of all polynomials. For vector addition, just add polynomials in the regular way. If*

$$p(x) = a_n x^n + a_{n-1} x^{n-1} + \cdots + a_1 x + a_0$$

and

$$q(x) = b_n x^n + b_{n-1} x^{n-1} + \cdots + b_1 x + b_0,$$

then

$$p(x) + q(x) =$$
$$(a_n + b_n)x^n + (a_{n-1} + b_{n-1})x^{n-1} + \cdots + (a_1 + b_1)x + (a_0 + b_0).$$

Define scalar multiplication by

$$cp(x) = ca_n x^n + ca_{n-1} x^{n-1} + \cdots + ca_1 x + ca_0.$$

Recall that the degree of a nonzero polynomial p is the largest positive integer n such that x^n has a nonzero coefficient in $p(x)$. If all the coefficients are zero, the degree is not defined. We can consider polynomials as a subset of the set of functions from the R to R, and so we use the same notion of equality, that two functions are equal if and only if they take the same value for every value of x. We know that a polynomial of degree n can have at most n distinct roots, thus the only polynomial that can equal the zero function is the polynomial with no degree, namely the one with all zero coefficients. If the polynomials aren't of the same degree, we can still add them by filling in powers of x with 0 coefficient in the polynomial of lower degree. Also when two polynomials of the same degree are added, the degree may go down, since we might have $b_n = -a_n$. However either $p + q$ is the zero polynomial or degree$(p + q) \leq$ max(degree p, degree q).

Example 1.4 *Let V be the set of all polynomials of degree 3 or less and the zero polynomial, and use the same definitions of vector addition and scalar multiplication as in Example 1.3. This is a vector space over R.*

Example 1.5 *Let V be the set of all solutions of the differential equation $f'' + f = 0$, where f'' indicates the second derivative of f, and addition and scalar multiplication are as in Example 1.2. It is left to the reader to check that A1 through A10 are satisfied.*

We now prove some useful properties that follow from A1-A10. Of course anything we can prove using just A1-A10 will be true in every vector space over the real numbers, and in particular in all the examples we have considered. Although many of these results are already known to you and quite trivial in the case of R^n, we are showing here that they follow just from A1-A10.

Proposition 1.1 *There is only one vector $\mathbf{O}$ in V which satisfies $\mathbf{O} + \mathbf{u} = \mathbf{u}$ for all $\mathbf{u}$ in V. (The zero vector is unique.)*

Proof.   Assume $\mathbf{O}_1$ and $\mathbf{O}_2$ are two such elements of V. Consider $\mathbf{O}_1 + \mathbf{O}_2$. Since $\mathbf{O}_1 + \mathbf{u} = \mathbf{u}$ for all $\mathbf{u}$, we must have $\mathbf{O}_1 + \mathbf{O}_2 = \mathbf{O}_2$. But $\mathbf{u} + \mathbf{O}_2 = \mathbf{O}_2 + \mathbf{u} = \mathbf{u}$ for all $\mathbf{u}$, implies that $\mathbf{O}_1 + \mathbf{O}_2 = \mathbf{O}_1$. Thus $\mathbf{O}_1 = \mathbf{O}_2$. ■

Proposition 1.2 *If $\mathbf{u}$ is any element of V there is one and only one element $\mathbf{z}$ of V such that $\mathbf{u} + \mathbf{z} = \mathbf{O}$. (The inverse of $\mathbf{u}$ is unique.)*

Proof.   Suppose z_1 and z_2 are two such vectors. Then note that because of commutativity (A2) we have $z_1 + u = O$. Because of the associative law (A3), we have $(z_1 + u) + z_2 = z_1 + (u + z_2)$. This gives $O + z_2 = z_1 + O$, or $z_2 = z_1$. ∎

Remark   *Since this element is unique we may give it a name. For a fixed u, the unique solution of $u + z = O$ is called the inverse of u and is written $-u$.*

Proposition 1.3 *For all u in V we have $0u = O$.*

Proof. $0u = (0 + 0)u = 0u + 0u$. Now adding the vector $-(0u)$ to both sides of this equation we get $O = 0u$. ∎

Proposition 1.4 *For all a in R we have $aO = O$.*

Proof. $aO = a(O + O) = aO + aO$. Adding $-(aO)$ to both sides gives $O = aO$. ∎

Proposition 1.5 *For every u in V and real number a, $(-a)u = -(au)$.*

Proof. $au + (-a)u = (a + -a)u = 0u = O$. Thus $(-a)u$ is a solution of $au + z = O$, but the unique solution to this equation is $-(au)$. Thus we have proved that $(-a)u = -(au)$. ∎

Proposition 1.6 *If c is a real number and u a vector in V, then $cu = O$ implies that either $c = 0$ or $u = O$.*

Proof.   If $c \neq 0$, then we can multiply both sides of $cu = O$ by $\frac{1}{c}$ to get $\frac{1}{c}(cu) = \frac{1}{c}O = O$. But by A9 and A10 we have $\frac{1}{c}(cu) = \frac{1}{c}cu = 1u = u$. Thus if $c \neq 0$ we have proved $u = O$. ∎

Exercises

1. Prove that the set V of Example 1.5 is closed under the vector addition and scalar multiplication given.

2. Let V be a vector space, a and b real numbers, and $\mathbf{u}$, $\mathbf{v}$ and $\mathbf{w}$ in V.

   (a) If $\mathbf{u} \neq \mathbf{O}$, prove that $a\mathbf{u} = b\mathbf{u}$ implies $a = b$.

   (b) If $a \neq 0$, prove that $a\mathbf{u} = a\mathbf{v}$ implies $\mathbf{u} = \mathbf{v}$.

   (c) Prove that $\mathbf{u} + \mathbf{v} = \mathbf{u} + \mathbf{w}$ implies $\mathbf{v} = \mathbf{w}$.

Section 1.5 Subspaces and Spanning Sets

If V and W are two sets such that every element of W is also an element of V, we say that W is a subset of V, or that W is contained in V, and write $W \subset V$. This allows for the possibility that they may be equal. If we wish to exclude that possibility we say that W is a proper subset of V and write $W \subset V, W \neq V$. If we let Φ denote the empty set, then $\Phi \subset V$ for every set V. We are interested in the case where V is a vector space over the real numbers and W is a nonempty subset of V. From now on we will use R to denote the set of real numbers.

Definition 1.17 *If V is a vector space over R and W is a subset of V such that W is also a vector space over R using the same definitions of vector addition and scalar multiplication, then W is a subspace of V.*

The idea of a subspace is a very simple one. It is just a subset that satisfies all the axioms A1-A10 on its own, without using any of the other vectors in V. At first it may seem that it will be a lot of work to determine whether a given subset of V is or is not a subspace, but in practice we do not need to check all the axioms A1-A10 to prove that W is a subspace of V. Some of them are automatic because W is contained in V. The following very useful theorem shows that we only have to check two things.

Theorem 1.4 *If W is a nonempty subset of V, then W is a subspace of V if and only if the following two conditions are satisfied:*
1. $\mathbf{u} + \mathbf{v}$ is in W for every $\mathbf{u}$ and $\mathbf{v}$ in W.
2. $c\mathbf{u}$ is in W for every $\mathbf{u}$ in W and every c in R.

Remark   *These two conditions can be expressed by saying that W is closed under vector addition and scalar multiplication.*

Proof.   Clearly if W is to be a vector space over R on its own, the above two conditions must hold because of A1 and A6. Conversely, if W satisfies conditions 1 and 2, then A1 and A6 are satisfied. The commutative law (A2) and the associative law (A3) for vector addition must hold for all elements of W, since it holds for all elements of V, and every element of W is an element of V. Now 0 is in R and since W is nonempty there is a vector $\mathbf{v}$ in W. But then by condition 2 we have $0\mathbf{v} = \mathbf{O}$ is in W. Since $\mathbf{O} + \mathbf{u} = \mathbf{u}$ for all $\mathbf{u}$ in V, we clearly have $\mathbf{O} + \mathbf{u} = \mathbf{u}$ for all $\mathbf{u}$ in W. Thus the zero vector $\mathbf{O}$ of V must be in W and A4 is satisfied. If $\mathbf{u}$ is in W then $(-1)\mathbf{u}$ is in W by condition 2, but we know (Proposition 1.5) that $(-1)\mathbf{u} = -(1\mathbf{u}) = -\mathbf{u}$, so that $-\mathbf{u}$ is in W. Thus we have A5. The remaining axioms A7-A10 are equations that hold for all vectors in V and so must hold for all vectors in W. ∎

Corollary 1   *If W is a subspace of V, W must contain the zero vector $\mathbf{O}$ of V.*

Remark   *This often allows us to see quickly that some subset of V is not a subspace by observing that it does not contain the zero vector of V.*

Example 1.6   *The set of all triples $(a, b, 0)$ such that the last component is zero is a subspace of R^3. The two conditions of Theorem 1.4 were verified in section 1.3.*

Example 1.7   *The set of all triples (x, y, z) which satisfy the equation $2x + y - (1/2)z = 0$ is a subspace of R^3. To see this we use Theorem 1.4. Let (a_1, a_2, a_3) and (b_1, b_2, b_3) be two triples that satisfy this equation. Then consider the sum $(a_1 + b_1, a_2 + b_2, a_3 + b_3)$. Substitute this into the equation. We get $2(a_1 + b_1) + a_2 + b_2 - \frac{1}{2}(a_3 + b_3) = [2a_1 + a_2 - \frac{1}{2}a_3] + [2b_1 + b_2 - \frac{1}{2}b_3] = 0 + 0 = 0$. So the sum of two triples satisfying this equation will also satisfy this equation. Now consider a scalar multiple $c(a_1, a_2, a_3) = (ca_1, ca_2, ca_3)$. Substituting this in the equation we get $2ca_1 + ca_2 - \frac{1}{2}ca_3 = c(2a_1 + a_2 - \frac{1}{2}a_3) = c0 = 0$. So the scalar multiple again satisfies the equation.*

Geometrically the above example is a plane through the origin. If the plane does not go through the origin, i.e. if the constant term is not zero the set of triples that satisfy it will not be a subspace since it will not contain the zero vector $(0, 0, 0)$.

Example 1.8 *The space of all polynomials (using the vector addition and scalar multiplication defined in Example 1.3) is a subspace of the space of all continuous functions from R to R, because polynomials are continuous functions.*

Example 1.9 *The space of all everywhere differentiable functions from R to R (using the vector addition and the scalar multiplication as defined in Example 1.2) is a subspace of the space of all continuous functions from R to R, because differentiablity implies continuity.*

Example 1.10 *The space of all solutions of the differential equation $f'' + f = 0$, again using the vector addition and scalar multiplication defined in Example 1.2) is a subspace of the space of all continuous functions from R to R.*

Because of the associative law A3, parentheses are not necessary in the expression $\mathbf{u} + \mathbf{v} + \mathbf{w}$. Using mathematical induction, one can prove that for any number of summands $\mathbf{u}_1 + \mathbf{u}_2 + \cdots + \mathbf{u}_k$ parentheses are also unnecessary. This is called the generalized associative law. We omit the proof.

Definition 1.18 *If $\mathbf{u}_1, \mathbf{u}_2, \cdots, \mathbf{u}_k$ are k vectors and $a_1, a_2, \cdots, a_k$ are k scalars, then the vector $\mathbf{v} = a_1\mathbf{u}_1 + a_2\mathbf{u}_2 + \cdots + a_k\mathbf{u}_k$ is called a linear combination of the vectors $\mathbf{u}_1, \mathbf{u}_2, \cdots, \mathbf{u}_k$.*

Remark *We introduce the use of the notation $\mathbf{u} - \mathbf{v}$ for $\mathbf{u} + (-\mathbf{v})$. And since $(-a)\mathbf{u} = -(a\mathbf{u})$, by Proposition 1.5, we can write $a_1\mathbf{u}_1 - a_2\mathbf{u}_2$ for $a_1\mathbf{u}_1 + (-a_2)\mathbf{u}_2$.*

Example 1.11 *The triple $(2, 0, -1)$ can be expressed as a linear combination of the triples $(1, 4, -3)$, $(2, 0, -1)$, and $(3, 4, -4)$ in many ways, the most obvious is $(2, 0, -1) = 0(1, 4, -3) + 1(2, 0, -1) + 0(3, 4, -4)$, but we also have $(2, 0, -1) = 2(1, 4, -3) + 3(2, 0, -1) - 2(3, 4, -4)$, and $(2, 0, -1) = -(1, 4, -3) + 0(2, 0, -1) + (3, 4, -4)$.*

Determining how to write one vector as a linear combination of others in R^n involves attempting to solve a system of n linear equations. In Chapter 2 we develop efficient methods of solving such systems.

Example 1.12 *Determine whether or not the triple* $(2, -1, 3)$ *is a linear combination of the triples* $(1, 0, 2)$ *and* $(1, 3, -1)$. *To solve this we must try to find real numbers* a *and* b *such that* $(2, -1, 3) = a(1, 0, 2) + b(1, 3, -1) = (a, 0, 2a) + (b, 3b, -b)$. *Equating components gives:*

$$\begin{aligned} 2 &= a + b \\ -1 &= 3b \\ 3 &= 2a - b. \end{aligned}$$

The first two equations imply that $b = -1/3$ *and* $a = 7/3$. *However now the third equation is not satisfied, since* $2(7/3) + 1/3 = 5 \neq 3$. *Thus there is no solution and so* $(2, -1, 3)$ *cannot be expressed as a linear combination of* $(1, 0, 2)$ *and* $(1, 3, -1)$.

Theorem 1.5 *If* $\mathbf{u}_1, \mathbf{u}_2, \cdots, \mathbf{u}_k$ *are any* k *elements of* V *and if we let* W *be the set of all linear combinations of the vectors* $\mathbf{u}_1, \mathbf{u}_2, \cdots, \mathbf{u}_k$, *then* W *is a subspace of* V.

Proof.   Let $\mathbf{w}$ and $\mathbf{x}$ be vectors in W. Then $\mathbf{w} = a_1\mathbf{u}_1 + a_2\mathbf{u}_2 + \cdots + a_k\mathbf{u}_k$, and $\mathbf{x} = b_1\mathbf{u}_1 + b_2\mathbf{u}_2 + \cdots + b_k\mathbf{u}_k$ for scalars $a_1, a_2, \cdots, a_k$ and $b_1, b_2, \cdots, b_k$. But then adding $\mathbf{w} + \mathbf{x}$ and using the axioms we have $\mathbf{w} + \mathbf{x} = (a_1 + b_1)\mathbf{u}_1 + (a_2 + b_2)\mathbf{u}_2 + \cdots + (a_k + b_k)\mathbf{u}_k$, so $\mathbf{w} + \mathbf{x}$ is in W. Similarly if c is any element of R we have $c\mathbf{w} = c(a_1\mathbf{u}_1 + a_2\mathbf{u}_2 + \cdots + a_k\mathbf{u}_k)$ and using A7 and A9, we get $c\mathbf{w} = (ca_1)\mathbf{u}_k + (ca_2)\mathbf{u}_k + \cdots + (ca_k)\mathbf{u}_k$, so $c\mathbf{w}$ is in W. Thus by Theorem 1.4, W is a subspace of V. ∎

Definition 1.19 *If* $\mathbf{u}_1, \mathbf{u}_2, \cdots, \mathbf{u}_k$ *are vectors in a vector space* V, *the set of all linear combinations of* $\mathbf{u}_1, \mathbf{u}_2, \cdots, \mathbf{u}_k$ *is called the span of* $\mathbf{u}_1, \mathbf{u}_2, \cdots, \mathbf{u}_k$, *denoted by* $sp\{\mathbf{u}_1, \mathbf{u}_2, \cdots, \mathbf{u}_k\}$.

We have just shown that $sp\{\mathbf{u}_1, \mathbf{u}_2, \cdots, \mathbf{u}_k\}$ is a subspace of V. We will now show that it is the smallest subspace of V that contains the vectors $\mathbf{u}_1, \mathbf{u}_2, \cdots, \mathbf{u}_k$.

Theorem 1.6 *If* S *is any subspace of* V *and if the vectors* $\mathbf{u}_1, \mathbf{u}_2, \cdots, \mathbf{u}_k$ *are all in* S, *then* $sp\{\mathbf{u}_1, \mathbf{u}_2, \cdots, \mathbf{u}_k\} \subset S$.

Proof.   To show that $a_1\mathbf{u}_1 + a_2\mathbf{u}_2 + \cdots + a_k\mathbf{u}_k$ is in S for all choices of scalars $a_1, a_2, \cdots, a_k$, note that $a_i\mathbf{u}_i$ is in S for all choices of a_i since by Theorem 1.4, S must be closed under scalar multiplication. Also $a_1\mathbf{u}_1 + a_2\mathbf{u}_2$ is in S because a subspace must be closed under vector addition. But then $(a_1\mathbf{u}_1 + a_2\mathbf{u}_2) + a_3\mathbf{u}_3$ is in S because this is a sum of two vectors that are in S. Continuing in this way we see that $a_1\mathbf{u}_1 + a_2\mathbf{u}_2 + \cdots + a_k\mathbf{u}_k$ is in S. ∎

Remark   *We introduce the symbol $\in$ to mean "is in " or "is an element of".*

Corollary 1 $Sp\{\mathbf{u}_1, \mathbf{u}_2, \cdots, \mathbf{u}_k\} = sp\{\mathbf{v}_1, \mathbf{v}_2, \cdots, \mathbf{v}_n\}$ *if and only if* $\mathbf{u}_i \in sp\{\mathbf{v}_1, \mathbf{v}_2, \cdots, \mathbf{v}_n\}$ *for all* $1 \le i \le k$, *and* $\mathbf{v}_j \in sp\{\mathbf{u}_1, \mathbf{u}_2, \cdots, \mathbf{u}_k\}$ *for all* $1 \le j \le n$.

Proof.   Let $W = sp\{\mathbf{u}_1, \mathbf{u}_2, \cdots, \mathbf{u}_k\}$, and $S = sp\{\mathbf{v}_1, \mathbf{v}_2, \cdots, \mathbf{v}_n\}$. If $W = S$, then $\mathbf{u}_i \in S$ because $\mathbf{u}_i \in W$. Similarly $\mathbf{v}_j \in W$. Conversely, if $\mathbf{u}_i \in S$ for all $1 \le i \le k$, then $W \subset S$ by the Theorem 1.6. Similarly if $\mathbf{v}_j \in W$ for all $1 \le j \le n$ then $S \subset W$. Thus $S = W$. ∎

The question of when two different sets of vectors span the same subspace is an interesting one. The above corollary gives one way of testing, but it is a laborious one. In Chapter 2 we will present a better method for answering this question.

Definition 1.20 *If $V = sp\{\mathbf{u}_1, \mathbf{u}_2, \cdots, \mathbf{u}_k\}$, we say that the set $\{\mathbf{u}_1, \mathbf{u}_2, \cdots, \mathbf{u}_k\}$ spans V.*

The set $\{\mathbf{u}_1, \mathbf{u}_2, \cdots, \mathbf{u}_k\}$ spans V if and only if every vector of V can be expressed as a linear combination of $\mathbf{u}_1, \mathbf{u}_2, \cdots, \mathbf{u}_k$, but the expression need not be unique. The words span and generate are synonyms. Of course by the very definition, $\{\mathbf{u}_1, \mathbf{u}_2, \cdots, \mathbf{u}_k\}$ spans $sp\{\mathbf{u}_1, \mathbf{u}_2, \cdots, \mathbf{u}_k\}$, so every set of vectors is a spanning set of some subspace, but we are usually interested in finding sets that span the whole vector space V.

It is possible to generalize the definition of the span of a set of vectors to infinite sets.

Definition 1.21 *If T is any set of vectors, we can define $sp\{T\}$ to be the set of all finite linear combinations of elements of T.*

One can still prove that $sp\{T\}$ is a subspace and that it is the smallest subspace containing T.

Example 1.13 *If V is any vector space, V itself is a spanning set of V, since $\mathbf{u} = 1\mathbf{u}$ is a linear combination of a finite number of vectors in V for all $\mathbf{u}$ in V.*

Definition 1.22 *If the vector space V has a finite spanning set, we say that V is finitely generated.*

Example 1.14 *R^n is finitely generated. The simplest spanning set for R^n is the set of natural basis vectors $\mathbf{e}_1, \mathbf{e}_2, \cdots, \mathbf{e}_n$ discussed in Section 1.3.*

If $\{\mathbf{u}_1, \mathbf{u}_2, \cdots, \mathbf{u}_k\}$ is a spanning set of V, and if T is any set of vectors in V, then $\{\mathbf{u}_1, \mathbf{u}_2, \cdots, \mathbf{u}_k\} \sqcup T$ is also a spanning set of V. In other words adding more vectors to a spanning set does not destroy the spanning property.

Example 1.15 *The triples $(1, 0, 2), (2, -1, 3)$ and $(1, 3, -1)$ are a spanning set of R^3. To prove this, let (a, b, c) be an arbitrary but fixed triple. We must show that the equation $(a, b, c) = x(1, 0, 2) + y(2, -1, 3) + z(1, 3, -1)$ has a solution for all choices of a, b, and c. Multiplying by the scalars gives $(a, b, c) = (x, 0, 2x) + (2y, -y, 3y) + (z, 3z, -z)$. Equating components we get:*

$$
\begin{array}{rcrcrcr}
a & = & x & + & 2y & + & z \\
b & = & & & -y & + & 3z \\
c & = & 2x & + & 3y & - & z
\end{array}
$$

From the second equation we have $y = 3z - b$. Substituting this in the first equation gives $a = x + 2(3z - b) + z$, or $a = x + 7z - 2b$, and solving for x gives $x = -7z + a + 2b$. Thus far we have solved for x and y in terms of a, b, and z. We now substitute this into the last of the original equations and solve for z. This gives $c = 2(-7z + a + 2b) + 3(3z - b) - z$, thus $c = -6z + 2a + b$, and solving for z gives $z = 1/6(2a + b - c)$. Substituting this into our previous equations for x and y and simplifying gives

$$
\begin{array}{rcrcrcr}
x & = & -\frac{4}{3}a & + & \frac{5}{6}b & + & \frac{7}{6}c \\
y & = & a & - & \frac{1}{2}b & - & \frac{1}{2}c \\
z & = & \frac{1}{3}a & + & \frac{1}{6}b & - & \frac{1}{6}c
\end{array}
$$

We have solved for x, y, and z in terms of a, b, and c. No restrictions have arisen on a, b, and c. This means that any triple can be expressed as a linear combination of $(1, 0, 2)$, $(2, -1, 3)$, and $(1, 3, -1)$. Thus these three triples form a spanning set of R^3. Our calculations actually tell us how to express any given triple as a linear combination of $(1, 0, 2)$, $(2, -1, 3)$, and $(1, 3, -1)$. For example consider the triple $(1, 2, -1)$. Substitute $a = 1$, $b = 2$, and $c = -1$ in the above equations. This gives $x = -\frac{5}{6}$, $y = \frac{1}{2}$, and $z = \frac{5}{6}$, and we have $(1, 2, -1) = -\frac{5}{6}(1, 0, 2) + \frac{1}{2}(2, -1, 3) + \frac{5}{6}(1, 3, -1)$, which you can check is correct. When we have proved some more theorems about vector spaces we will have easier ways of determining whether or not a set of n-tuples spans R^n.

Example 1.16 *We now look at a set of triples which does not span R^3, and we see what subspace of R^3 it does span. Consider the vectors $(1, 4, -3)$, $(2, 0, -1)$ and $(3, 4, -4)$. We try to solve the equation $(a, b, c) = x(1, 4, -3) + y(2, 0, -1) + z(3, 4, -4)$. Multiplying by the scalars and equating coefficients we get:*

$$
\begin{aligned}
a &= x + 2y + 3z \\
b &= 4x + 4z \\
c &= -3x - y - 4z
\end{aligned}
$$

From the second equation we get $x = \frac{1}{4}b - z$. Substituting this in the first equation gives $a = \frac{1}{4}b - z + 2y + 3z$, and solving for y we have $y = \frac{1}{2}a - \frac{1}{8}b - z$. Now substituting for x and y in the last of the centered equations above gives $c = -3(\frac{1}{4}b - z) - (\frac{1}{2}a - \frac{1}{8}b - z) - 4z$. When this is simplified, z cancels out, so that we cannot solve for it. Instead we are left with an equation involving only a, b, and c, namely

$$4a + 5b + 8c = 0.$$

This means we cannot write the triple (a, b, c) as a linear combination of the three given triples unless a, b, and c satisfy $4a + 5b + 8c = 0$. Since this is not true for all triples, we do not have a spanning set. The computations we have done are not wasted, however, because now we know exactly which triples can be expressed as linear combinations of $(1, 4, -3)$, $(2, 0, -1)$, and $(3, 4, -4)$. We have proved that $sp\{(1, 4, -3), (2, 0, -1), (3, 4, -4)\}$ equals the set of all triples (a, b, c) such that $4a + 5b + 8c = 0$. If we have a triple satisfying this equation, *such as* $(6, -5, \frac{1}{8})$, *the computing we have already done tells us how to express this triple as a linear combination of* $(1, 4, -3), (2, 0, -1)$, *and* $(3, 4, -4)$. *Since z canceled out, we can choose any number for z, but we must have* $x = \frac{1}{4}b - z$, *and* $y = \frac{1}{2}a - \frac{1}{8}b - z$. *Substituting* $a = 6$, $b = -5$, $c = \frac{1}{8}$ *and* $z = 0$, *we get* $x = -\frac{5}{4}$ *and* $y = \frac{29}{8}$, *and we see that indeed,* $(6, -5, \frac{1}{8}) = -\frac{5}{4}(1, 4, -3) + \frac{29}{8}(2, 0, -1)$. *Notice two things about this situation. First, since z can be chosen freely, there are many ways to express a vector in* $sp\{(1, 4, -3), (2, 0, -1), (3, 4, -4)\}$ *as a linear combination of these vectors. Second, since z can always be chosen to be zero if one wishes, the third vector isn't essential, in fact we have* $sp\{(1, 4, -3), (2, 0, -1)\} = sp\{(1, 4, -3), (2, 0, -1), (3, 4, -4)\}$.

Exercises

1. Prove that if W_1 and W_2 are subspaces of a vector space V, then $W_1 \sqcap W_2$ is a subspace of V. ($W_1 \sqcap W_2$ is the set of vectors common to both W_1 and W_2.)

2. Show that if $\{u_1, u_2, \cdots, u_k\}$ is a spanning set of V and if $u_1 \in sp\{u_2, u_3, \cdots, u_k\}$, then $\{u_2, u_3, \cdots, u_k\}$ is still a spanning set of V.

3. Show that if V contains a nonzero vector, then V must contain infinitely many different vectors.

4. Show that the set of triples of the form $(a, b, 1)$ do not form a subspace of R^3.

5. Show that the set of triples of the form $(a, 2a, a)$ form a subspace of R^3. Find a spanning set for this subspace.

6. Recall Definition 1.15 in Section 1.3 of a basis of R^n. Are the three vectors given in Example 1.15 a basis of R^3? Explain.

7. What is the difference between a basis of R^n and a spanning set of R^n? Is every basis of R^n a spanning set of R^n? Is every spanning set a basis?

Section 1.6 Linear Independence and Basis

Definition 1.23 *Let $\mathbf{u}_1, \mathbf{u}_2, \cdots, \mathbf{u}_k$ be k vectors in a vector space V. These vectors are linearly independent if $a_1\mathbf{u}_1 + a_2\mathbf{u}_2 + \cdots + a_k\mathbf{u}_k = \mathbf{O}$ only when all a_i are 0.*

If $\mathbf{u}_1, \mathbf{u}_2, \cdots, \mathbf{u}_k$ are not linearly independent, they are called linearly dependent. We can also state this as follows: $\mathbf{u}_1, \mathbf{u}_2, \cdots, \mathbf{u}_k$ are linearly dependent if there exist scalars $a_1, a_2, \cdots, a_k$, not all 0 such that $a_1\mathbf{u}_1 + a_2\mathbf{u}_2 + \cdots + a_k\mathbf{u}_k = \mathbf{O}$. This last equation is called a dependence relation. For any set of vectors $\mathbf{u}_1, \mathbf{u}_2, \cdots, \mathbf{u}_k$ we always have $0\mathbf{u}_1 + 0\mathbf{u}_2 + \cdots + 0\mathbf{u}_k = \mathbf{O}$. If this is the only way to express $\mathbf{O}$ as a linear combination of these vectors, then the vectors are linearly independent. If there is another way to express $\mathbf{O}$ as a linear combination of these vectors, then they are linearly dependent.

Example 1.17 *The set of vectors containing only the zero vector $\mathbf{O}$ is a linearly dependent set because $a\mathbf{O} = \mathbf{O}$ is a dependence relation for every choice of $a \neq 0$.*

Example 1.18 *Any set of vectors which contains the zero vector $\mathbf{O}$ is linearly dependent because $a\mathbf{O} + 0\mathbf{u}_1 + 0\mathbf{u}_2 + \cdots + 0\mathbf{u}_k = \mathbf{O}$ is a dependence relation for every $a \neq 0$.*

Example 1.19 *A set of vectors $\mathbf{u}_1, \mathbf{u}_2, \cdots, \mathbf{u}_k$ such that one is a scalar multiple of another, say $\mathbf{u}_1 = a\mathbf{u}_2$ is linearly dependent. (This includes the case where two or more of the vectors are equal.) To see this just note that $\mathbf{u}_1 - a\mathbf{u}_2 + 0\mathbf{u}_3 + \cdots + 0\mathbf{u}_k = \mathbf{O}$ is a linear dependence relation if $a \neq 0$, and if $a = 0$, then $\mathbf{u}_1 = \mathbf{O}$ and the set is linearly dependent as seen in the previous example.*

Example 1.20 *A set containing only one vector $\mathbf{u}$ with $\mathbf{u} \neq \mathbf{O}$ is a linearly independent set, since then $a\mathbf{u} = \mathbf{O}$ only if $a = 0$ by Proposition 1.6.*

If a set is linearly dependent and more vectors are added, the new set will still be linearly dependent, since we can use the same dependence relation, just putting zeros for the coefficients of the new vectors. Determining whether a set of vectors in R^n is linearly dependent is equivalent to determining if a certain system of equations has a nonzero solution.

Example 1.21 *The triples* $(1, 1, 1)$, $(2, 1, 0)$ *and* $(3, 1, 1)$ *are linearly independent. To prove this we must show that* $a(1, 1, 1) + b(2, 1, 0) + c(3, 1, 1) = (0, 0, 0)$ *only if* a, b, *and* c *are all equal to zero. Performing the scalar multiplications gives* $(a, a, a) + (2b, b, 0) + (3c, c, c) = (0, 0, 0)$. *Equating components gives the following system of equations*

$$\begin{aligned} a + 2b + 3c &= 0 \\ a + b + c &= 0 \\ a \quad\quad + c &= 0 \end{aligned}$$

The last equation gives $c = -a$. *Substituting this in the second equation we get* $b = 0$. *Using all this in the first equation gives* $a + 0 - 3a = 0$, *or* $-2a = 0$. *Thus we must have* $a = 0$. *But since* $c = -a$, *this means* $c = 0$. *We have shown that* a, b, *and* c *must all be zero, so the vectors are linearly independent.*

Example 1.22 *The set of triples* $(1, 2, 1)$, $(2, 4, 1)$, *and* $(3, 6, 0)$ *is a linearly dependent set. This is not obvious at first glance, since no one of them is a scalar multiple of another, nor is one of them the sum of the other two. To see that they are linearly dependent we seek scalars* a, b *and* c, *not all zero such that* $a(1, 2, 1) + b(2, 4, 1) + c(3, 6, 0) = (0, 0, 0)$. *Multiplying by the scalars and equating components leads to the equations*

$$\begin{aligned} a + 2b + 3c &= 0 \\ 2a + 4b + 6c &= 0 \\ a + b \quad\quad &= 0 \end{aligned}$$

The last equation implies that $b = -a$. *Substituting this in the first equation we see that* $c = (1/3)a$. *Since the second equation is just twice the first, it gives no further restriction on* a, b *and* c. *Thus we may choose* a *freely and use the fact that* $b = -a$ *and* $c = (1/3)a$ *to determine* b *and* c. *Let us try* $a = 3$, $b = -3$, *and* $c = 1$. *We check to see if this is a solution. Substituting in the vector equation gives* $3(1,2,1) - 3(2,4,1) + 1(3,6,0) = \mathbf{O}$, *which is a dependence relation of our triples.*

Example 1.23 *The standard basis of* R^n, $\mathbf{e}_1 = (1,0,0,\cdots 0)$, $\mathbf{e}_2 = (0,1,0,\cdots 0)$, $\cdots$, $\mathbf{e}_n = (0,0,\cdots 0,1)$ *discussed in Section 1.3 is a linearly independent set. Since* $a_1\mathbf{e}_1 + a_2\mathbf{e}_2 + \cdots + a_n\mathbf{e}_n = (a_1, a_2, \cdots, a_n)$, *it is clear that this can only be the zero vector if all* a_i *are 0.*

Our definitions of linear independence and dependence have been given only for finite sets, but can be extended to infinite sets as follows: A set T of vectors is a linearly independent set if every finite subset of T is linearly independent.

Lemma 1.3 *The set* $\{\mathbf{u}_1, \mathbf{u}_2, \cdots, \mathbf{u}_k\}$ *is linearly dependent if and only if one of the vectors in the set can be expressed as a linear combination of the others.*

Proof.   If the set $\{\mathbf{u}_1, \mathbf{u}_2, \cdots, \mathbf{u}_k\}$ is linearly dependent, we have scalars $a_1, a_2, \cdots, a_k$, not all zero such that

$$a_1\mathbf{u}_1 + a_2\mathbf{u}_2 + \cdots + a_k\mathbf{u}_k = \mathbf{O}.$$

Without loss of generality we may assume a_1 is not zero, since we could renumber the vectors if necessary. But then we can solve the dependence relation for $\mathbf{u}_1$, obtaining

$$\mathbf{u}_1 = -\frac{a_2}{a_1}\mathbf{u}_2 - \frac{a_3}{a_1}\mathbf{u}_3 - \cdots - \frac{a_k}{a_1}\mathbf{u}_k,$$

so that $\mathbf{u}_1$ is expressed as a linear combination of the others. Conversely, suppose that one of the vectors equals a linear combination of the others. Again we can assume it is $\mathbf{u}_1$. Then there exist scalars $b_1, b_2, \cdots, b_{k-1}$ such that

$$\mathbf{u}_1 = b_1\mathbf{u}_2 + b_2\mathbf{u}_3 + \cdots + b_{k-1}\mathbf{u}_k.$$

But then

$$\mathbf{u}_1 - b_1\mathbf{u}_2 - b_2\mathbf{u}_3 - \cdots - b_{k-1}\mathbf{u}_k = \mathbf{O}.$$

Not all the coefficients in this last equation are 0, since the coefficient of $\mathbf{u}_1$ is -1. Thus the set $\{\mathbf{u}_1, \mathbf{u}_2, \cdots, \mathbf{u}_k\}$ is linearly dependent. ∎

Lemma 1.4 *A set of vectors* $\{u_1, u_2, \cdots, u_k\}$ *with* $u_1 \neq O$ *is linearly dependent if and only if one of the vectors is a linear combination of vectors which precede it in the list.*

Proof.   If one of the vectors is a linear combination of those which precede it, we use Lemma 1.3 to conclude that the set is linearly dependent. Conversely, assume the set is linearly dependent. Then we have a dependence relation

$$a_1 u_1 + a_2 u_2 + \cdots + a_k u_k = O.$$

We know $a_i \neq 0$ for at least one value if i between 1 and k. Now let j be the largest integer such that $a_j \neq 0$. Since $u_1 \neq O$, we may assume that $j \geq 2$. Then the dependence relation is

$$a_1 u_1 + a_2 u_2 + \cdots + a_j u_j + 0 u_{j+1} + 0 u_{j+2} + \cdots + 0 u_k = O.$$

We can solve for u_j and not use $u_{j+1}, u_{j+2}, \cdots, u_k$, since they have zero coefficients. We get

$$u_j = -\frac{a_1}{a_j} u_1 - \frac{a_2}{a_j} u_2 - \cdots - \frac{a_{j-1}}{a_j} u_{j-1},$$

and u_j equals a linear combination of vectors that precede it in the list. ∎

Remark   *Lemma 1.4 does not say that if the set* $\{u_1, u_2, \cdots, u_k\}$ *is linearly dependent you can choose any* u_i *and it will be a linear combination of preceding vectors. This is not true. All you can conclude is that some one of the vectors is a linear combination of vectors that precede it, but you don't know which vector has this property. Of course more than one vector could have this property, but the lemma only asserts the existence of at least one.*

Example 1.24 *We can use Lemma 1.4 to prove that the triples* $(1, 1, 1)$, $(2, 1, 0)$, *and* $(3, 1, 1)$ *are linearly independent. Note that for no real number* r *can we have* $(2, 1, 0) = r(1, 1, 1) = (r, r, r)$, *since the second component gives* $r = 1$, *while the third component gives* $r = 0$. *If* $(3, 1, 1) = a(1, 1, 1) + b(2, 1, 0)$ *we would have* $3 = a + 2b$, $1 = a + b$, *and* $1 = a$. *The second and third imply* $a = 1$ *and* $b = 0$, *but then the first equation is not satisfied.*

Example 1.25 *We proved in a previous example that the set of triples* $(1, 2, 1)$, $(2, 4, 1)$, *and* $(3, 6, 0)$ *is a linearly dependent set. Lemma 1.4 says that one of these must be a linear combination of the ones preceding it. Notice that* $(2, 4, 1) = a(1, 2, 1)$ *is not possible for any real number* a. *Thus it must be the case that* $(3, 6, 0) = a(1, 2, 1) + b(2, 4, 1)$, *and indeed one can see that* $a = -3$, $b = 3$ *is a solution.*

We have already defined what is meant by a basis of R^n. We extend the idea to the more general setting.

Definition 1.24 *A set of vectors* $\{u_1, u_2, \cdots, u_k\}$ *is a basis of* V *if every vector in* V *can be expressed uniquely as a linear combination of* $u_1, u_2, \cdots, u_k$.

Theorem 1.7 *The set of vectors* $\{u_1, u_2, \cdots, u_k\}$ *is a basis for* V *if and only if* $\{u_1, u_2, \cdots, u_k\}$ *is a linearly independent set which spans* V.

Proof. Assume that $u_1, u_2, \cdots, u_k$ is a basis of V. Then every vector can be expressed uniquely as a linear combination of $u_1, u_2, \cdots, u_k$ so that $V = sp\{u_1, u_2, \cdots, u_k\}$, whence $\{u_1, u_2, \cdots, u_k\}$ is a spanning set of V. To see that $\{u_1, u_2, \cdots, u_k\}$ is a linearly independent set we use the uniqueness of the expression of the zero vector, O. Since $O = 0u_1 + 0u_2 + \cdots + 0u_k$, and since every vector in V can be expressed in only one way as a linear combination of $u_1, u_2, \cdots, u_k$, we have $a_1u_1 + a_2u_2 + \cdots + a_ku_k = O$ only if $a_i = 0$ for all i, $1 \leq i \leq k$. Conversely if $\{u_1, u_2, \cdots, u_k\}$ is a linearly independent set that spans V, then every vector of V can be expressed as a linear combination of $u_1, u_2, \cdots, u_k$, and all we need to show is the uniqueness of the expression. Suppose that for some v in V we have

$$v = a_1u_1 + a_2u_2 + \cdots + a_ku_k,$$

and also

$$v = b_1u_1 + b_2u_2 + \cdots + b_ku_k.$$

Then subtracting the second equation from the first gives

$$O = (a_1 - b_1)u_1 + (a_2 - b_2)u_2 + \cdots + (a_k - b_k)u_k.$$

But since $\{u_1, u_2, \cdots, u_k\}$ is a linearly independent set, this can only happen if all the coefficients are zero, thus $a_i - b_i = 0$ for all i, $1 \le i \le k$, which means $a_i = b_i$, for all i, $1 \le i \le k$. This proves the expression is unique. ∎

Exercises

1. Is the set $\{(2, 1, 6), (1, 1, 1), (1, 2, 3)\}$ a linearly dependent set?

2. Find conditions on a, b, and c so that $(1, 0, 1)$, $(0, 1, 2)$ and (a, b, c) are linearly independent.

3. What is the maximum number of linearly independent vectors in R^2? Prove your answer. (Hint: See Theorem 1.3).

4. Prove that $\{1, x, x^2, x^3\}$ is a basis for the space consisting of all polynomials of degree 3 or less and the zero polynomial. (See the remarks following Example 1.3, Section 1.4.)

Section 1.7 Dimension

In this section we will prove that every finitely generated nonzero vector space V has a basis, and that any two bases of such a space must have the same number of vectors. The number of vectors in a basis will be the dimension of the vector space. If the vector space is not finitely generated, we will say that its dimension is infinite. The vector space consisting of a single vector, which is then necessarily the zero vector, has no basis, since it has no nonempty linearly independent subset. We need the following result:

Lemma 1.5 *If the set of vectors $\{u_1, u_2, \cdots, u_k\}$ spans V and if for some j we have $u_j \in sp\{u_1, u_2, \cdots, u_{j-1}\}$, then $\{u_1, u_2, \cdots, u_{j-1}, u_{j+1}, \cdots, u_k\}$ also spans V.*

Proof. This follows from the corollary to Theorem 1.6. ∎

Theorem 1.8 *If $S = \{u_1, u_2, \cdots, u_k\}$ is a spanning set of V such that $u_1 \neq O$, then a subset T which is a basis of V can be selected from S by omitting exactly those u_i which can be expressed as linear combinations of preceding vectors in the given ordering of S.*

Proof. If no vector in S is a linear combination of preceding vectors, then S is a linearly independent set, by Lemma 1.4, and hence S is a basis. If there is a vector, say u_j that is a linear combination of preceding vectors, we can remove it and $S' = S - \{u_j\}$ is still spans V by Lemma 1.5. If no vector of S' is a linear combination of preceding vectors, then S' is linearly independent and hence a basis. Continue in this way. Either at some stage you arrive at a basis, or you keep going until only u_1 is left. Since $u_1 \neq O$, the set $T = \{u_1\}$ is linearly independent, and since at each stage we preserve the spanning property, T will be a basis of V. ∎

The basis resulting from the procedure outlined in the proof of Theorem 1.8 will depend on the ordering of the vectors $u_1, u_2, \cdots, u_k$, but it will not depend on the order in which the eliminations are performed.

Corollary 1 *If $V \neq \{O\}$, every finite spanning set of V contains a basis of V.*

Corollary 2 *Every nonzero finitely generated vector space has a basis.*

Remark *Since the set $\{O\}$ is linearly dependent, the zero vector space $V = \{O\}$ has no basis.*

Theorem 1.9 *If V is a vector space with a basis $B = \{u_1, u_2, \cdots, u_n\}$, then every vector in V can be represented uniquely by an n-tuple of real numbers, moreover any computation in V involving only the vector addition and scalar multiplication can be done with the n-tuple representatives of the vectors.*

Proof. From the definition of basis we know that every vector $\mathbf{v}$ of V can be expressed uniquely as a linear combination of $\mathbf{u}_1, \mathbf{u}_2, \cdots, \mathbf{u}_n$. If $\mathbf{v} = a_1\mathbf{u}_1 + a_2\mathbf{u}_2 + \cdots + a_n\mathbf{u}_n$, we let the n-tuple $(a_1, a_2, \cdots, a_n)$ represent $\mathbf{v}$. This n-tuple is called the coordinate vector of $\mathbf{v}$ with respect to the basis B. If $\mathbf{w}$ is another vector in V and if the coordinate vector of $\mathbf{w}$ is $(b_1, b_2, \cdots, b_n)$, it is easy to see that the coordinate vector of $\mathbf{v} + \mathbf{w}$ is $(a_1, a_2, \cdots, a_n) + (b_1, b_2, \cdots, b_n)$. It is also easy to see that the coordinate vector of $r\mathbf{v}$ is $r(a_1, a_2, \cdots, a_n)$. Thus if we want to add two vectors in V, we can add their coordinate vectors and then find the vector of V represented by that n-tuple. A vector $\mathbf{w}$ is a linear combination of vectors $\mathbf{v}_i$ if and only if the coordinate vector of $\mathbf{w}$ is a linear combination of the coordinate vectors of the $\mathbf{v}_i$. If we use the notation $[\mathbf{v}]_B$ to denote the coordinate vector of $\mathbf{v}$ with respect to the basis B, then what we have said can be restated as: $[\mathbf{v}+\mathbf{w}]_B = [\mathbf{v}]_B + [\mathbf{w}]_B$, and $[r\mathbf{v}]_B = r[\mathbf{v}]_B$, and $[a_1\mathbf{v}_1 + a_2\mathbf{v}_2 + \cdots + a_k\mathbf{v}_k] = a_1[\mathbf{v}_1] + a_2[\mathbf{v}_2] + \cdots + a_k[\mathbf{v}_k]$, (When there is no possibility of confusion, we can omit the subscript B.) Since a dependence relation among the vectors $\mathbf{v}_i$ translates into a dependence relation between $[\mathbf{v}_i]$, and vice-versa, we see that vectors in V are linearly dependent if and only if their coordinate vectors are linearly dependent in R^n. ∎

Because of Theorem 1.9, any vector space V over R which has a basis containing n vectors will behave exactly like R^n. Thus we see that our more general definition of a vector space over R is not more general after all, except in the case where V is not finitely generated. Any result that we can prove for R^n will be true for any finitely generated vector space, and problems in such a vector space can always be translated into problems involving n-tuples. This makes the techniques that we will develop in Chapter 2 very important, but there will still be times when the abstract viewpoint simplifies, clarifies, or gives more insight. We feel the proof of the following result is such a time.

Theorem 1.10 (Replacement Theorem) *If $S = \{\mathbf{w}_1, \mathbf{w}_2, \cdots, \mathbf{w}_s\}$ is a spanning set for the nonzero space V, and if $\{\mathbf{v}_1, \mathbf{v}_2, \cdots, \mathbf{v}_r\}$ is any linearly independent set of vectors in V, then $s \geq r$, and there exists a subset T (possibly empty) of S such that $\{\mathbf{v}_1, \mathbf{v}_2, \cdots, \mathbf{v}_r\} \sqcup T$ is a basis of V.*

Proof. The proof will proceed by repeated application of the following process:

Step 1 Adjoin $\mathbf{v}_1$ to S. The set $\{\mathbf{v}_1, \mathbf{w}_1, \mathbf{w}_2, \cdots, \mathbf{w}_s\}$ is still a spanning set, but it is linearly dependent by Lemma 1.3, since $\mathbf{v}_1 \in sp\{S\}$.
Step 2 Eliminate one of the $\mathbf{w}_i$. Since the set $\{\mathbf{v}_1, \mathbf{w}_1, \mathbf{w}_2, \cdots, \mathbf{w}_s\}$ is linearly dependent, one of these vectors is a linear combination of preceding vectors by Lemma 1.4. It can't be $\mathbf{v}_1$, since no vectors precede $\mathbf{v}_1$, thus it must be $\mathbf{w}_k$ for some k between 1 and s. Now by Lemma 1.5, the new set $S' = \{\mathbf{v}_1, \mathbf{w}_1, \mathbf{w}_2, \cdots, \mathbf{w}_{k-1}, \mathbf{w}_{k+1}, \cdots, \mathbf{w}_s\}$ is still a spanning set.
Step 3 If necessary eliminate more of the $\mathbf{w}_i$. The new set S' may or may not be linearly independent, but by Theorem 1.8, we can find a subset S'' of S' such that S'' is a basis of V. S'' will contain $\mathbf{v}_1$ since $\mathbf{v}_1$ is not a linear combination of preceding vectors.

We have completed the description of the process. The effect of one application of this process is to add to the set S one $\mathbf{v}$ and to eliminate at least one $\mathbf{w}$, and to do this in such a way as to insure that the resulting set will be a basis of V. Repeat this process until all the vectors $\mathbf{v}_1, \mathbf{v}_2, \cdots, \mathbf{v}_r$ have been adjoined to S. Let us look at the stage where we adjoin $\mathbf{v}_j$ for some j satisfying $1 \leq j \leq r$. As we start Step 1, we have a basis consisting of

$$\{\mathbf{v}_{j-1}, \cdots, \mathbf{v}_2, \mathbf{v}_1, \mathbf{w}_{i_1}, \mathbf{w}_{i_2}, \cdots, \mathbf{w}_{i_m}\}.$$

We adjoin $\mathbf{v}_j$ at the front of this list and obtain the new set

$$\{\mathbf{v}_j, \mathbf{v}_{j-1}, \cdots, \mathbf{v}_2, \mathbf{v}_1, \mathbf{w}_{i_1}, \mathbf{w}_{i_2}, \cdots, \mathbf{w}_{i_m}\}.$$

This new set is linearly dependent and a spanning set. We perform steps 2 and 3 to reduce it to a basis of V. We must be sure that no $\mathbf{v}$ is eliminated in this process. The reason we can be sure of this is that $\{\mathbf{v}_1, \mathbf{v}_2, \cdots, \mathbf{v}_r\}$ is a linearly independent set, so no $\mathbf{v}$ is a linear combination of the others. Since in Step 1 we always adjoin the new $\mathbf{v}$ at the front of the list, each $\mathbf{v}$ in the list is preceded only by other $\mathbf{v}_h$, thus no $\mathbf{v}$ is eliminated in steps 2 and 3. Notice that each time a new $\mathbf{v}$ is added, the set becomes linearly dependent, which means that there must then be a $\mathbf{w}$ to eliminate, or we would contradict the fact that $\{\mathbf{v}_1, \mathbf{v}_2, \cdots, \mathbf{v}_r\}$ are linearly independent. We repeat the process

r times, so that $\mathbf{v}_1, \mathbf{v}_2, \cdots, \mathbf{v}_r$ have all been adjoined. Since in each repetition of the process we eliminate at least one $\mathbf{w}$ and the process is done r times, we must have $s \geq r$. Also at the end we have a basis consisting of $\mathbf{v}_1, \mathbf{v}_2, \cdots, \mathbf{v}_r$ and a subset of S. ■

If V is a nonzero finitely generated vector space we have the following corollaries:

Corollary 1 *Any linearly independent set of vectors in V can be extended to a basis of V.*

Corollary 2 *If $\mathbf{w}_1, \mathbf{w}_2, \cdots, \mathbf{w}_n$ is a basis of V and $\mathbf{v}_1, \mathbf{v}_2, \cdots, \mathbf{v}_m$ is another basis of V, then $n = m$.*

Proof.   Apply the Replacement Theorem to get $n \geq m$. Now reverse the roles of the two sets of vectors to get $m \geq n$. This is possible since both sets are spanning sets, and both sets are linearly independent. ■

Definition 1.25 *If V is a nonzero finitely generated vector space, the dimension of V is the number of vectors in any basis of V. The dimension of the trivial vector space $\{\mathbf{O}\}$ consisting of the zero vector alone is defined to be 0.*

Corollary 3 *If V has dimension n then any set of $n+1$ or more vectors in V is a linearly dependent set.*

Proof.   A basis is a spanning set, so the assumption says we have a spanning set with n vectors. Thus a set of $n + 1$ or more linearly independent vectors would contradict the Replacement Theorem. ■

Corollary 4 *If V has dimension n then any set of $n - 1$ or fewer vectors cannot be a spanning set of V.*

Proof.   A basis is a linearly independent set, and by the Replacement Theorem the number of elements in any spanning set must be greater than or equal to the number of elements in a basis which is n. ■

Corollary 5 *If V has dimension n any linearly independent set of n vectors in V is a basis of V. Thus in R^n any n linearly independent vectors is a basis. (In Theorem 1.3 we proved this for $n = 2$.)*

Proof.   If the linearly independent set is not a basis, it can be extended to a basis. But if any vectors are actually added, we would have a basis with more than n vectors, which is not possible in a space of dimension n because of Corollary 2 ∎.

Corollary 6 *If V is a vector space of dimension n then any spanning set with n vectors in V is a basis of V.*

Proof.   If the spanning set is not a basis of V, it contains a subset which is a basis of V. If this subset has fewer than n vectors, we contradict Corollary 2. ∎

Corollary 7 *If W is a subspace of V and $\dim V = n$, then W is finitely generated and $\dim W \leq \dim V$; moreover $\dim W = \dim V$ if and only if $W = V$.*

Proof.   If $W = \{O\}$ the result is obvious. If $W \neq \{O\}$, let $\{v_1, v_2, \cdots, v_s\}$ be a maximal linearly independent set in W. We know this exists and that $s \leq n$, since there can't be more than n linearly independent vectors in V. Now the set $\{v_1, v_2, \cdots, v_s\}$ must span W, since for any vector $w \in W$ the set $\{v_1, v_2, \cdots, v_s, w\}$ is linearly dependent which implies by Lemma 1.4 that $w \in sp\{v_1, v_2, \cdots, v_s\}$. This proves the first part of the theorem. If $\dim W = \dim V = n$, then a basis of W (being n linearly independent vectors) must be a basis of V and hence $V \subset W$, so $W = V$. ∎

Definition 1.26 *In a vector space V, the zero vector alone $\{O\}$, and the whole space V are always subspaces of V. Any other subspaces of V are called proper subspaces.*

What are the proper subspaces of R^2? If W is a proper subspace of R^2, W must have a basis of 1 or 2 vectors. If W has a basis of 2 vectors, then $W = R^2$. If W has a basis of one vector $u = (a, b)$, then W is just the set of all scalar multiples of u, geometrically this is the line through the origin and the point (a, b). Similarly, a proper subspace of R^3 must have a basis of one vector (line through the origin), or a basis of two vectors. If $W = sp\{u, v\}$, then W is the plane containing the directed line segments representing u and v. If $u = (a, b, c)$ and $v = (d, e, f)$, this will be the plane containing the origin and the points (a, b, c) and (d, e, f). Thus the proper subspaces of R^3 are lines through the origin and planes through the origin.

Corollary 8 *The space of continuous functions from R to R is not finitely generated.*

Proof.   If this space were finitely generated, the subspace of all polynomials would be finitely generated by Cor. 7. The polynomials cannot be finitely generated because if $p \in sp\{p_1, p_2, \cdots p_k\}$, then degree $p \leq \max\{\deg p_1, \deg p_2, \cdots \deg p_k\}$. ■

Example 1.26 *In Ex.1.21 we showed that $\{(1, 1, 1), (2, 1, 0), (3, 1, 1)\}$ is a linearly independent set. Using Corollary 5, we can now conclude that this set is a basis of R^3.*

Example 1.27 *In Ex.1.14 we showed that $\{(1, 0, 2), (2, -1, 3), (1, 3, -1)\}$ is a spanning set of R^3. Using Corollary 6, we can now conclude that this set is a basis of R^3.*

Remark   *From now on we will use the symbol P_n to refer to the set consisting of all polynomials of degree $\leq n$ and the zero polynomial.*

Example 1.28 *Using the basis $\{1, x, x^2\}$ of the space P_2, we can represent the polynomials $p(x) = 1 + 2x^2$, $q(x) = 2 - x + 3x^2$, and $s(x) = 1 + 3x - x^2$ by the triples (1,0,2),(2,-1,3), and (1,3,-1). Since the triples are a basis of R^3 we can conclude that the polynomials $p(x)$, $q(x)$, and $s(x)$ form a basis of P_2.*

Procedure 1.1 To check if a set of vectors is a basis of R^n, first count the vectors. If there are more or fewer than n, the set cannot be a basis. If there are exactly n vectors, you need only check one of the two properties, linear independence or spanning. Usually it is easier to check for linear independence.

Procedure 1.2 To extend a linearly independent set $\{u_1, u_2, \cdots, u_k\}$ in R^n to a basis of R^n, we use the natural basis vectors $e_1, e_2, \cdots, e_n$. Then $\{u_1, u_2, \cdots, u_k, e_1, e_2, \cdots, e_n\}$ is a spanning set of R^n. Remove each e_i which can be expressed as a linear combination of preceding vectors in this list, and the remaining vectors form a basis of R^n.

Exercises

1. Prove that the triples $(2,0,1)$, $(0,2,1)$, $(1,1,1)$ and $(1,1,3)$ are linearly dependent. Are these vectors a spanning set for R^3? If possible find a subset which is a basis of R^3.

2. Prove that the 4-tuples $(1,3,-2,,0)$, $(1,-1,0,5)$ and $(3,6,2,1)$ cannot be a spanning set of R^4. Are these vectors linearly independent? If possible extend this set to a basis of R^4.

3. Find a basis for R^4 containing the vectors $(1,1,-1,1)$ and $(1,0,1,1)$.

4. Find a basis for P_3 containing the vectors $1 + x - x^2 + x^3$ and $1 + x^2 + x^3$. (Hint: Compare with problem 3.)

Section 1.8 Sums and Intersections of Subspaces

We introduce the notation $S = \{x|$ conditions on $x\}$, to mean the set of all x such that the given conditions on x are satisfied.

Definition 1.27 *If S and T are subspaces of a vector space V, then $S + T = \{u + v | u \in S$ and $v \in T\}$.*

Definition 1.28 *If S and T are any two sets, $S \sqcap T = \{x | x \in S$ and $x \in T\}$.*

Definition 1.29 *If S and T are any two sets, $S \sqcup T = \{x | x \in S$ or $x \in T$ (or both)$\}$*

Theorem 1.11 *If V is a vector space with subspaces S and T, then $S \sqcap T$ and $S + T$ are also subspaces of V.*

Proof. That $S \sqcap T$ is a subspace was Exercise 1, Section 1.5. To show $S + T$ is a subspace we need only show it is closed under vector addition and scalar multiplication. Let $\mathbf{x}$ and $\mathbf{y}$ be elements of $S + T$. Then $\mathbf{x} = \mathbf{u} + \mathbf{v}$, and $\mathbf{y} = \mathbf{u}' + \mathbf{v}'$, with $\mathbf{u}$ and $\mathbf{u}'$ in S and $\mathbf{v}$ and $\mathbf{v}'$ in T. But then $\mathbf{x} + \mathbf{y} = \mathbf{u} + \mathbf{v} + \mathbf{u}' + \mathbf{v}' = \mathbf{u} + \mathbf{u}' + \mathbf{v} + \mathbf{v}'$. Now $\mathbf{u} + \mathbf{u}' \in S$, and $\mathbf{v} + \mathbf{v}' \in T$, because S and T are subspaces and so closed under vector addition. Thus we have $\mathbf{x} + \mathbf{y} \in S + T$. Also $r\mathbf{x} = r\mathbf{u} + r\mathbf{v}$, and since $r\mathbf{u} \in S$ and $r\mathbf{v} \in T$, $r\mathbf{x} \in S + T$. ∎

The union, $S \sqcup T$ need not be a subspace, as is shown by the following:

Example 1.29 *Let S and T be two different lines through the origin in R^2. Then $S \sqcup T$ is just the two lines, but the sum $\mathbf{u} + \mathbf{v}$ of a vector $\mathbf{u}$ on one line and a vector $\mathbf{v}$ on the other line does not lie on either line (Figure 1.3), and so is not in $S \sqcup T$. In this case $\mathbf{u}$ and $\mathbf{v}$ generate the whole plane, so we have $S + T = R^2$ and $S \sqcap T = \{\mathbf{O}\}$.*

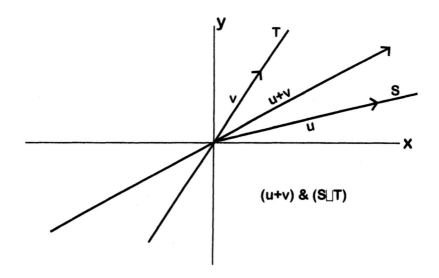

Figure 1.3

Theorem 1.12 *If S and T are finite dimensional subspaces of a vector space V, then $\dim(S + T) = \dim(S) + \dim(T) - \dim(S \sqcap T)$.*

Proof. Let $\dim(S) = k$, and $\dim(T) = j$. First we assume $S \sqcap T = \{O\}$ Then $\dim(S \sqcap T) = 0$, so we must prove that $\dim(S + T) = \dim(S) + \dim(T)$. Let $\{s_1, s_2, \cdots, s_k\}$ be a basis of S, and let $\{t_1, t_2, \cdots, t_j\}$ be a basis for T. Now clearly $\{s_1, s_2, \cdots, s_k, t_1, t_2, \cdots, t_j\}$ is a spanning set for $S + T$. We will show it is also a linearly independent set. Suppose that

$$a_1 s_1 + a_2 s_2 + \cdots + a_k s_k + b_1 t_1 + b_2 t_2 + \cdots + b_j t_j = O. \qquad (1.11)$$

Then

$$a_1 s_1 + a_2 s_2 + \cdots + a_k s_k = -(b_1 t_1 + b_2 t_2 + \cdots + b_j t_j).$$

Now the left hand side of this equation is a vector $\mathbf{u}$ in S, but it is equal to a linear combination of vectors in T, thus $\mathbf{u}$ is in both S and T. Since we are assuming $S \sqcap T = \{O\}$, this means

$$\mathbf{u} = a_1 s_1 + a_2 s_2 + \cdots + a_k s_k = O.$$

But $\{s_1, s_2, \cdots, s_k\}$ is a basis for S, and so is a linearly independent set, thus $O = a_1 s_1 + a_2 s_2 + \cdots + a_k s_k$ implies $a_i = 0$ for $i = 1$ to k. Then eq.1.11 becomes

$$b_1 t_1 + b_2 t_2 + \cdots + b_j t_j = O,$$

which implies $b_i = 0$ for $i = 1$ to j. Thus all the coefficients in eq. 1.11 are 0, and we have proved $\{s_1, s_2, \cdots, s_k, t_1, t_2, \cdots, t_j\}$ is a linearly independent set and hence a basis of $S + T$. Since it has $k + j$ vectors, we have $\dim(S + T) = k + j = \dim(S) + \dim(T)$. We now consider the case where $S \sqcap T \neq \{O\}$. In this case let $\{x_1, x_2, \cdots, x_r\}$ be a basis for $S \sqcap T$. Since this is a linearly independent set of vectors in S, we can extend it to a basis of S. Let $\{x_1, x_2, \cdots, x_r, q_1, q_2, \cdots, q_{k-r}\}$, be such a basis of S. We also observe that $x_1, x_2, \cdots, x_r$ are in T, so we can extend to a basis of T. We get $\{x_1, x_2, \cdots, x_r, p_1, p_2, \cdots, p_{j-r}\}$, a basis of T. As before, {basis of S} $\sqcup$ {basis of T} will span $S + T$, so that to show that the set $B = \{x_1, x_2, \cdots, x_r, q_1, q_2, \cdots, q_{k-r}, p_1, p_2, \cdots, p_{j-r}\}$ is a basis of $S + T$ it is only necessary to show that these vectors are linearly independent. To see this suppose

$$a_1\mathbf{x}_1 + \cdots + a_r\mathbf{x}_r + b_1\mathbf{q}_1 + \cdots + b_{k-r}\mathbf{q}_{k-r} + c_1\mathbf{p}_1 + \cdots + c_{j-r}\mathbf{p}_{j-r} = \mathbf{O}.$$
$$(1.12)$$

This gives

$$a_1\mathbf{x}_1 + \cdots + a_r\mathbf{x}_r + b_1\mathbf{q}_1 + \cdots + b_{k-r}\mathbf{q}_{k-r} = -(c_1\mathbf{p}_1 + \cdots + c_{j-r}\mathbf{p}_{j-r}).$$

But now the left hand side of this equation is in S, and the right hand side is in T, thus the vector

$$\mathbf{v} = a_1\mathbf{x}_1 + \cdots + a_r\mathbf{x}_r + b_1\mathbf{q}_1 + \cdots + b_{k-r}\mathbf{q}_{k-r}$$

is in $S \sqcap T$. But $\{\mathbf{x}_1, \mathbf{x}_2, \cdots, \mathbf{x}_r\}$ is a basis of $S \sqcap T$, so $\mathbf{v}$ can be expressed using only the $\mathbf{x}_i$. Moreover $\mathbf{v}$ is in T and the expression for $\mathbf{v}$ in terms of our basis of T is unique, therefore we must have $\mathbf{q}_i = 0$ for $i = 1$ to $k - r$. Substituting this in 1.12 gives

$$a_1\mathbf{x}_1 + \cdots + a_r\mathbf{x}_r + c_1\mathbf{p}_1 + \cdots + c_{j-r}\mathbf{p}_{j-r} = \mathbf{O}. \qquad (1.13)$$

But the set $\{\mathbf{x}_1, \mathbf{x}_2, \cdots, \mathbf{x}_r, \mathbf{p}_1, \mathbf{p}_2, \cdots, \mathbf{p}_{j-r}\}$ is linearly independent, so $a_i = 0$ for $i = 1, \cdots, r$ and $c_i = 0$ for $i = 1, \cdots, j - r$. Thus the $r + (k - r) + (j - r)$ vectors in B are a basis of $S + T$. So $\dim(S + T) = k + j - r = \dim(S) + \dim(T) - \dim(S \sqcap T)$. ∎

Definition 1.30 *If S and T are proper subspaces of a vector space V such that $S + T = V$ and $S \sqcap T = \{\mathbf{O}\}$, we say that V is the direct sum of S and T and we write $V = S \oplus T$.*

Theorem 1.13 *If $V = S \oplus T$, then every vector in V can be expressed uniquely as a sum of a vector in S and a vector in T.*

Proof.   Since $V = S + T$, every vector of V can be expressed as a sum of a vector from S and a vector from T. To show that this expression is unique, suppose that $\mathbf{v} = \mathbf{s} + \mathbf{t}$ and $\mathbf{v} = \mathbf{s}' + \mathbf{t}'$. Then $\mathbf{s} + \mathbf{t} = \mathbf{s}' + \mathbf{t}'$, which gives $\mathbf{s} - \mathbf{s}' = \mathbf{t} - \mathbf{t}'$. Let $\mathbf{u} = \mathbf{s} - \mathbf{s}'$. Then $\mathbf{u} = \mathbf{t} - \mathbf{t}'$ also, but $\mathbf{s} - \mathbf{s}' \in S$, and $\mathbf{t} - \mathbf{t}' \in T$. It follows that $\mathbf{u} \in S \sqcap T$. By assumption $S \sqcap T = \{\mathbf{O}\}$, thus $\mathbf{u} = \mathbf{O}$, so $\mathbf{s} = \mathbf{s}'$ and $\mathbf{t} = \mathbf{t}'$. ∎

Exercises

1. If $S = sp\{(1,1,2)\}$ and $T = sp\{(1,3,1),(1,1,1),(2,4,2)\}$, find $\dim(S)$, $\dim(T)$, $\dim(S+T)$ and $\dim(S \sqcap T)$.

2. If $S = sp\{(1,0,1),(1,2,1)\}$ and $T = sp\{(1,-2,1),(2,2,0)\}$, find $\dim(S)$, $\dim(T)$, $\dim(S+T)$ and $\dim(S \sqcap T)$. Find all vectors in $S \sqcap T$.

Chapter 2

Matrices and Systems of Linear Equations

Section 2.1 Matrix Operations

Definition 2.1 *A matrix is a rectangular array of real numbers.*

Example 2.1 *The matrix* $\begin{bmatrix} 1 & 2 \\ -1 & 3 \end{bmatrix}$ *is a 2 × 2 matrix.*

Example 2.2 *The matrix* $\begin{bmatrix} 2 & 1 & 1 \\ 0 & -1 & 0 \end{bmatrix}$ *is a 2 × 3 matrix.*

Example 2.3 *The generic m × n matrix is*

$$A = \begin{bmatrix} a_{11} & a_{12} & a_{13} & \cdots & a_{1n} \\ a_{21} & a_{22} & a_{23} & \cdots & a_{2n} \\ \vdots & \vdots & \vdots & & \vdots \\ a_{m1} & a_{m2} & a_{m3} & \cdots & a_{mn} \end{bmatrix}.$$

We sometimes abbreviate this to $A = [a_{ij}]$.

A matrix has rows and columns. We use double subscripts to denote the entries in the matrix. The first subscript designates the row and the second subscript designates the column. An mxn matrix A has m

rows and n columns, and a_{ij} is the entry in the ith row, j^{th} column. We also use the notation $[A]_{ij}$ to designate the entry in the ith row, j^{th} column of A.

The rows of the $m \times n$ matrix A are n-tuples, and the columns of A are m-tuples. A matrix can be thought of as an ordered set of m n-tuples (the rows), or as an ordered set of n m-tuples (the columns). We sometimes write

$$A = \begin{bmatrix} \mathbf{u}_1 \\ \mathbf{u}_2 \\ \vdots \\ \mathbf{u}_m \end{bmatrix},$$

where $\mathbf{u}_i$ is the ith row of A,

$$\mathbf{u}_1 = (a_{11}, a_{12}, \cdots, a_{1n}),$$

$$\mathbf{u}_2 = (a_{21}, a_{22}, \cdots, a_{2n}), etc.$$

Thinking of the matrix A as an ordered set of columns, we sometimes write

$$A = \begin{bmatrix} \mathbf{v}_1 & \mathbf{v}_2 & \cdots & \mathbf{v}_n \end{bmatrix},$$

where $\mathbf{v}_i$ is the ith column of A,

$$\mathbf{v}_1 = \begin{bmatrix} a_{11} \\ a_{21} \\ \vdots \\ a_{m1} \end{bmatrix}, \mathbf{v}_2 = \begin{bmatrix} a_{12} \\ a_{22} \\ \vdots \\ a_{m2} \end{bmatrix}, etc.$$

An n-tuple written as a row is sometimes called a row vector. It is also a $1 \times n$ matrix. An m-tuple written as a column is called a column vector. It is also an $m \times 1$ matrix. When we are working with n-tuples in the vector space R^n we may write them either as rows or columns, but when we are working with matrices, a $1 \times n$ matrix is different from an $n \times 1$ matrix. We have the following definition of equality for matrices.

Definition 2.2 *Two matrices A and B are equal if they are of the same size (same number of rows and same number of columns) and if $a_{ij} = b_{ij}$ for all i and j.*

Definition 2.3 *We add two matrices of the same size by adding corresponding entries. If $C = A + B$, then $c_{ij} = a_{ij} + b_{ij}$.*

If the matrices are not of the same size, addition is not defined.

Definition 2.4 *We can multiply a matrix A (on either side) by a real number r by multiplying each entry of A by the number r. Thus $rA = Ar$, and $[rA]_{ij} = ra_{ij}$.*

Example 2.4

$$\begin{bmatrix} 2 & 1 & 0 \\ 3 & 1 & 1 \end{bmatrix} + \begin{bmatrix} 1 & 0 & 1 \\ -1 & 2 & 1 \end{bmatrix} = \begin{bmatrix} 3 & 1 & 1 \\ 2 & 3 & 2 \end{bmatrix}.$$

Example 2.5

$$2\begin{bmatrix} 1 & 3 & 0 \\ 0 & 1 & 2 \end{bmatrix} = \begin{bmatrix} 2 & 6 & 0 \\ 0 & 2 & 4 \end{bmatrix}.$$

Consider for a moment the set of all 2×3 matrices. We have a set with an addition and a scalar multiplication. One can show that all the axioms of a vector space are satisfied. If we let

$$\mathbf{O} = \begin{bmatrix} 0 & 0 & 0 \\ 0 & 0 & 0 \end{bmatrix},$$

we have

$$A + \mathbf{O} = \mathbf{O} + A$$

for every 2×3 matrix A. Also note that

$$\begin{bmatrix} -a_{11} & -a_{12} & -a_{13} \\ -a_{21} & -a_{22} & -a_{23} \end{bmatrix} + A = \mathbf{O}.$$

Thus the matrices of a fixed size form a vector space over R. The correspondence

$$\begin{bmatrix} a & b & c \\ d & e & f \end{bmatrix} \longleftrightarrow (a, b, c, d, e, f)$$

is 1-1 and onto and preserves addition and scalar multiplication, so that when viewed as a vector space, the set of 2×3 matrices is just like R^6. In this case there is no advantage in writing the 6-tuple as two rows of triples. Things become more interesting when we define matrix multiplication.

Definition 2.5 *If A is mxn matrix and B an $n \times k$ matrix, then the product AB is an $m \times k$ matrix defined by*

$$[AB]_{ij} = a_{i1}b_{1j} + a_{i2}b_{2j} + a_{i3}b_{3j} + \cdots + a_{in}b_{nj}. \qquad (2.1)$$

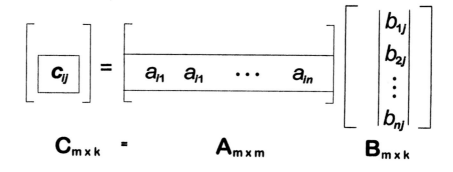

Figure 2.1

To get the entry in the ith row, j^{th} column of AB, we use the ith row of A and the jth column of B, multiply corresponding entries and add, moving across the ith row of A and down the jth column of B as in Figure 2.1. If the number of columns of A is not the same as the number of rows of B, the product AB is not defined.

An $m \times n$ matrix A times an $n \times 1$ column vector is an $m \times 1$ column vector. It is clear from the definition of matrix multiplication that the first column of AB equals A times the first column of B. Since a row vector is an $1 \times m$ matrix, a row vector must come in front of the mxn matrix A otherwise the product will not be defined (unless $m = 1$). A $1 \times m$ row vector times an $m \times n$ matrix A is a $1 \times n$ row vector. It is clear from the definition of matrix multiplication that the first row of

AB equals the first row of A times the matrix B. We have the following lemma.

Lemma 2.1 *If A is an $m \times n$ matrix with rows $\mathbf{u}_i$ and B an $n \times k$ matrix with columns $\mathbf{v}_i$, then*

$$AB = \left[\begin{array}{cccc} A\mathbf{v}_1 & A\mathbf{v}_2 & \cdots & A\mathbf{v}_k \end{array} \right]$$

and

$$AB = \left[\begin{array}{c} \mathbf{u}_1 B \\ \mathbf{u}_2 B \\ \vdots \\ \mathbf{u}_m B \end{array} \right].$$

Proof. When calculating the ith column of AB, one uses only the entries of A and the ith column of B. The calculations are exactly the same as one would do to calculate $A\mathbf{v}_i$. Similarly when calculating the ith row of AB, one uses only the entries of B and the ith row of A, and the calculations are exactly the same as would be done to calculate $\mathbf{u}_i B$. ∎

Definition 2.6 *If $\mathbf{u}$ and $\mathbf{v}$ vectors in R^n, with $\mathbf{u} = (a_1, a_2, \cdots, a_n)$ and $\mathbf{v} = (b_1, b_2, \cdots b_n)$, we define the dot product,*

$$\mathbf{u} \cdot \mathbf{v} = a_1 b_1 + a_2 b_2 + \cdots + a_n b_n.$$

The dot product of two vectors in R^n is not another vector in R^n, rather it is a real number. Restating Definition 1.12 in terms of the dot product, we see that

$$|\mathbf{u}| = \sqrt{\mathbf{u} \cdot \mathbf{u}}.$$

It is easy to see that

$$\mathbf{u} \cdot \mathbf{v} = \mathbf{v} \cdot \mathbf{u},$$

since $a_i b_i = b_i a_i$. It is also not hard to show that

$$\mathbf{u} \cdot (\mathbf{v} + \mathbf{w}) = \mathbf{u} \cdot \mathbf{v} + \mathbf{u} \cdot \mathbf{w}$$

and

$$(r\mathbf{u}) \cdot \mathbf{w} = \mathbf{u} \cdot (r\mathbf{w}) = r(\mathbf{u} \cdot \mathbf{w}).$$

Note that

$$\mathbf{u} \cdot \mathbf{u} \geq 0 \text{ for all } \mathbf{u} \text{ and } \mathbf{u} \cdot \mathbf{u} = 0 \text{ only if } \mathbf{u} = \mathbf{O}.$$

We leave the proof of these facts to the reader.

The dot product gives us another way to think of matrix multiplication. If A is $m \times n$, B is $n \times k$ and if

$$A = \begin{bmatrix} \mathbf{u}_1 \\ \mathbf{u}_2 \\ \vdots \\ \mathbf{u}_m \end{bmatrix}$$

and

$$B = \begin{bmatrix} \mathbf{v}_1 & \mathbf{v}_2 & \cdots & \mathbf{v}_n \end{bmatrix},$$

we see that

$$[AB]_{ij} = \mathbf{u}_i \cdot \mathbf{v}_j,$$

the dot product of the ith row of A with the jth column of B.

Example 2.6 Let $A = \begin{bmatrix} 0 & 0 & 0 \\ 2 & 1 & 1 \\ 1 & 3 & 2 \end{bmatrix}$, $B = \begin{bmatrix} 1 & 2 & 1 \\ 3 & 0 & 1 \\ 1 & 1 & 2 \end{bmatrix}$, and

$C = \begin{bmatrix} 2 & 2 & 0 \\ 3 & -1 & 0 \\ 4 & 5 & 0 \end{bmatrix}$, then

$$AB = \begin{bmatrix} 0 & 0 & 0 \\ 2 & 1 & 1 \\ 1 & 3 & 2 \end{bmatrix} \begin{bmatrix} 1 & 2 & 1 \\ 3 & 0 & 1 \\ 1 & 1 & 2 \end{bmatrix} = \begin{bmatrix} 0 & 0 & 0 \\ 6 & 5 & 5 \\ 12 & 4 & 8 \end{bmatrix}.$$

Notice how the zero row of A is reflected in the zero row of the product. It is easy to see from the definition of matrix multiplication that if the ith row of A is $\mathbf{O}$, then the ith row of AB will also be $\mathbf{O}$ for any B where the multiplication is defined. Now we multiply the same two matrices in reverse order.

$$BA = \begin{bmatrix} 1 & 2 & 1 \\ 3 & 0 & 1 \\ 1 & 1 & 2 \end{bmatrix} \begin{bmatrix} 0 & 0 & 0 \\ 2 & 1 & 1 \\ 1 & 3 & 2 \end{bmatrix} = \begin{bmatrix} 5 & 5 & 4 \\ 1 & 3 & 2 \\ 4 & 7 & 5 \end{bmatrix}.$$

We see that matrix multiplication is not commutative. The zero row in the second matrix is not reflected in the product. One can easily see however that a zero column in the matrix on the right will produce a zero column in the product in the same position, for example

$$BC = \begin{bmatrix} 1 & 2 & 1 \\ 3 & 0 & 1 \\ 1 & 1 & 2 \end{bmatrix} \begin{bmatrix} 2 & 2 & 0 \\ 3 & -1 & 0 \\ 4 & 5 & 0 \end{bmatrix} = \begin{bmatrix} 12 & 5 & 0 \\ 10 & 11 & 0 \\ 13 & 11 & 0 \end{bmatrix}.$$

Example 2.7 *A special case of interest is the product of a matrix A and a column vector* $\mathbf{x}$*. Let A be an* $m \times n$ *matrix and let* $\mathbf{x}$ *be the* $n \times 1$ *column vector with entries* x_j*. The product* $A\mathbf{x}$ *will be an* $m \times 1$ *column vector, and we have*

$$A\mathbf{x} = \begin{bmatrix} a_{11} & a_{12} & \cdots & a_{1n} \\ a_{21} & a_{22} & \cdots & a_{2n} \\ \vdots & \vdots & & \vdots \\ a_{m1} & a_{m2} & \cdots & a_{mn} \end{bmatrix} \begin{bmatrix} x_1 \\ x_2 \\ \vdots \\ x_n \end{bmatrix} \tag{2.2}$$

$$= \begin{bmatrix} a_{11}x_1 + a_{12}x_2 + \cdots a_{1n}x_n \\ a_{21}x_1 + a_{22}x_2 + \cdots a_{2n}x_n \\ \vdots \\ a_{m1}x_1 + a_{m2}x_2 + \cdots a_{mn}x_n \end{bmatrix}$$

$$A\mathbf{x} = x_1 \begin{bmatrix} a_{11} \\ a_{21} \\ \vdots \\ a_{m1} \end{bmatrix} + \cdots + x_n \begin{bmatrix} a_{1n} \\ a_{2n} \\ \vdots \\ a_{mn} \end{bmatrix} \tag{2.3}$$

Thus we see that the column $A\mathbf{x}$ is a linear combination of the columns of A with coefficients x_i.

In the following two lemmas, assume that A is $m \times n$ and B is $n \times k$.

Lemma 2.2 *The* i^{th} *column of AB is a linear combination of the columns of A where the coefficients are the entries of the* i^{th} *column of B. More exactly if* $\mathbf{v}_j$ *are the columns of A, then*

$$i^{th} \text{ column of } AB = b_{1i}\mathbf{v}_1 + b_{2i}\mathbf{v}_2 + \cdots + b_{ni}\mathbf{v}_n.$$

Proof.   This is immediate from Lemma 2.1 and Example 2.7. ∎

Lemma 2.3 *The ith row of AB is a linear combination of the rows of B where the coefficients are the entries of the ith row of A. More exactly, let $\mathbf{w}_j$ be the rows of B, then*

$$i^{th} row\ of\ AB = a_{i1}\mathbf{w}_1 + a_{i2}\mathbf{w}_2 + \cdots + a_{in}\mathbf{w}_n.$$

Proof.   Write out the product $\begin{bmatrix} c_1 & c_2 & \cdots & c_n \end{bmatrix} B$, and observe that this row vector is a linear combination of the rows of B with coefficients c_i. Then combine this result with Lemma 2.1. (This can also be proved from Lemma 2.2 using Theorem 2.1 below. See Exercise 5.) ∎

Definition 2.7 *If A is an $m \times n$ matrix, the transpose of A, written A^T, is the $n \times m$ matrix whose rows are the columns of A. In other words $(A^T)_{ij} = a_{ji}$.*

The columns of A become the rows of A^T and the rows of A become the columns of A^T. The entry in the ij position of A is found in the ji position of A^T.

Example 2.8 *If $A = \begin{bmatrix} 2 & 1 & 0 \\ 3 & 1 & 1 \end{bmatrix}$, then $A^T = \begin{bmatrix} 2 & 3 \\ 1 & 1 \\ 0 & 1 \end{bmatrix}$.*

Example 2.9 *For a square matrix, we can get A^T by flipping A around its main (left-right) diagonal. If*

$$A = \begin{bmatrix} 0 & 0 & 0 \\ 2 & 1 & 1 \\ 1 & 3 & 2 \end{bmatrix},$$

then

$$A^T = \begin{bmatrix} 0 & 2 & 1 \\ 0 & 1 & 3 \\ 0 & 1 & 2 \end{bmatrix}.$$

The entries on the main diagonal stay the same.

Definition 2.8 *A square matrix is symmetric if $A = A^T$.*

Theorem 2.1 $(AB)^T = B^T A^T$.

Proof. Let the rows of A be $\mathbf{u}_i$ and let the columns of B be $\mathbf{v}_j$. To show two matrices are equal it is sufficient to show the ij^{th} entries are the same for all i and j. Let us investigate the ij^{th} entry of $(AB)^T$, it is equal to the ji^{th} entry of AB, which equals $\mathbf{u}_j \cdot \mathbf{v}_i$. Now the ij^{th} element of $B^T A^T$ is $\mathbf{v}_i \cdot \mathbf{u}_j$. But the dot product is commutative, so these are equal. ∎

Definition 2.9 *The $n \times n$ (square) matrix A is called diagonal if $a_{ij} = 0$ for $i \neq j$.*

Definition 2.10 *The $n \times n$ (square) matrix A is called upper triangular if all entries below the main diagonal are zero, i.e. if $a_{ij} = 0$ for $i > j$. A square matrix is called lower triangular if its transpose is upper triangular.*

Exercises

1. Perform the following matrix operations where possible:

(a) $\begin{bmatrix} -1 & 3 & 0 \\ 0 & 0 & 0 \\ 2 & -1 & 2 \end{bmatrix} \begin{bmatrix} 1 & 0 & 4 \\ -2 & 0 & 1 \\ 3 & 1 & 2 \end{bmatrix}$

(b) $\begin{bmatrix} 1 & 3 & 5 \end{bmatrix} \begin{bmatrix} 2 & 1 & -1 \\ 1 & 2 & 4 \\ 0 & 0 & 0 \end{bmatrix}$

(c) $\begin{bmatrix} 2 & 1 & -1 \\ 1 & 2 & 4 \\ 0 & 0 & 0 \end{bmatrix} \begin{bmatrix} 1 \\ 3 \\ 5 \end{bmatrix}$

(d) $\begin{bmatrix} 2 & 1 & -1 \\ 1 & 2 & 4 \\ 0 & 0 & 0 \end{bmatrix} \begin{bmatrix} 1 & 3 & 5 \end{bmatrix}$

(e) $\begin{bmatrix} 2 & 4 & 1 \end{bmatrix} \begin{bmatrix} 1 \\ 3 \\ 5 \end{bmatrix}$

(f) $\begin{bmatrix} 1 \\ 3 \\ 5 \end{bmatrix} \begin{bmatrix} 2 & 4 & 1 \end{bmatrix}$

2. If $C = \begin{bmatrix} 3 & 2 & 1 & 1 \\ 1 & 0 & 0 & -1 \end{bmatrix} \begin{bmatrix} \mathbf{u}_1 \\ \mathbf{u}_2 \\ \mathbf{u}_3 \\ \mathbf{u}_4 \end{bmatrix}$, where $\mathbf{u}_i$ are the rows of a matrix

A.

(a) How many rows does C have?

(b) Find row 2 of C in terms of the rows of A.

3. Let A and B be diagonal matrices and let $C = AB$. Prove that C is diagonal with $c_{ii} = a_{ii}b_{ii}$.(One way to do this is to use Lemma 2.2).

4. Let A and B be upper triangular matrices. Prove that AB is upper triangular and that if $C = AB$ then $c_{ii} = a_{ii}b_{ii}$. (Lemma 2.2 can also be used here.).

5. Prove Lemma 2.3 using Lemma 2.2 and Theorem 2.1

Section 2.2 More on Matrix Operations

Some matrices are easy to multiply, especially using Lemmas 2.2 and 2.3.

Definition 2.11 *If C is an $n \times n$ matrix with c on the diagonal and zeros elserwhere,*

$$C = \begin{bmatrix} c & 0 & \cdots & 0 \\ 0 & c & & \vdots \\ \vdots & & \ddots & 0 \\ 0 & 0 & \cdots & c \end{bmatrix},$$

then C is called a scalar matrix. The special case of an $n \times n$ scalar matrix with $c = 1$ is denoted by I_n, and called the $n \times n$ identity matrix. Usually the size is made clear by the context so we omit the n and just write I.

Lemma 2.4 *Let C be the scalar matrix with c on the main diagonal, then C commutes with all square matrices A of the same size and $CA = cA = AC$.*

Proof. Using Lemma 2.3, the ith row of CA is c times the ith row of A, which is the same as the ith row of cA. Thus $CA = cA$. Similarly using Lemma 2.2, the ith column of AC is c times the ith column of A, which is the same as the ith column of cA. ∎

Corollary 1 *If I is the $n \times n$ identity matrix then $IA = A = AI$ for every $n \times n$ matrix A.*

Remark Note that even for matrices A and B that are not square, $CA = cA$ and $BC = cB$ when sizes are such that the multiplication are defined. Thus if I is the $n \times n$ identity, $IA = A$ for all matrices A with n rows and $BI = B$ for all matrices B with n columns.

When the size is made clear by the context we use $[0]$ to denote the zero matrix of any size. Clearly $[0]A = [0]$ and $B[0] = [0]$ whenever sizes are such that the product is defined.

Definition 2.12 *The matrix $E_{ij}(n)$ will denote the $n \times n$ matrix that has a 1 in row i, column j, and 0 in every other position. When n remains constant throughout a discussion, we surpress the n and just write E_{ij}.*

Lemma 2.5 *Let A be a matrix with rows $\mathbf{u}_1, \mathbf{u}_2, \cdots \mathbf{u}_m$. Then $E_{ij}A$ is the matrix with all rows zero except possibly row i, in which we find $\mathbf{u}_j$.*

Proof. If $k \neq i$, E_{ij} the kth row of E_{ij} is zero, so the kth row of $E_{ij}A$ will be zero for $k \neq i$. If $k = i$, we have by Lemma 2.3 that the ith row of $E_{ij}A$ is the linear combination $c_1\mathbf{u}_1 + c_2\mathbf{u}_2 + \cdots + c_n\mathbf{u}_n$, where $(c_1, c_2, \cdots, c_n)$ is the ith row of E_{ij}. By definition E_{ij} is zero everywhere except in the ij^{th} position, where there is a 1, so $c_k = 0$ for $k \neq j$ and $c_j = 1$. Thus the ith row of the product $E_{ij}A$ is $\mathbf{u}_j$. ∎

Lemma 2.6 $E_{ij}E_{km} = \begin{cases} E_{im} & \text{if } j = k \\ [0] & \text{if } j \neq k. \end{cases}$

Proof.   Using Lemma 2.5 we see that $E_{ij}A$ gives a matrix with every row zero except possibly row i, in which we find the jth row of A. Applying this to the product $E_{ij}E_{km}$, we have the ith row of the product is the jth row of E_{km}, which is zero if $j \neq k$. Thus if $j \neq k$ we have $E_{ij}E_{km} = [0]$. If $j = k$, then the ith row of the product (being the $j^{th} = k^{th}$ row of E_{km}) has a 1 in the mth column, and hence equals E_{im}. ∎

We use Lemma 2.6 to get the following example.

Example 2.10 *Consider the 2×2 matrices*

$$E_{12} = \begin{bmatrix} 0 & 1 \\ 0 & 0 \end{bmatrix} \text{ and } E_{11} = \begin{bmatrix} 1 & 0 \\ 0 & 0 \end{bmatrix}.$$

We see that
$$E_{12}E_{11} = [0],$$
but
$$E_{11}E_{12} = E_{12} \neq [0].$$

Thus not only do the matrices fail to commute, but the product in one way is the zero matrix, while the other is not. This example also shows that the product of two 2×2 matrices can be $[0]$ even if neither matrix is $[0]$. This cannot happen for real numbers, where we have $ab = 0$ only if $a = 0$ or $b = 0$. This example also shows that we do not have the cancellation law for matrices. For real numbers we know that if $ab = ac$ and $a \neq 0$, then $b = c$. Observe however that

$$E_{12}E_{11} = E_{12}[0],$$

$E_{12} \neq [0]$, but
$$E_{11} \neq [0].$$

We can't cancel the E_{12}.

Example 2.11 *The 2×2 matrix E_{11} above satisfies $E_{11}E_{11} = E_{11}$. An element that satisfies $x^2 = x$ is called an idempotent. In the real numbers, the only idempotents are 0 and 1. We see that in the $n \times n$ matrices there are lots of idempotents, in particular $E_{ii}E_{ii} = E_{ii}$ for $i = 1, \cdots, n$.*

Example 2.12 *The* 2×2 *matrix* E_{12} *above satisfies* $E_{12}E_{12} = [0]$. *An element that satisfies* $x^k = 0$ *for some positive integer* k *is called nilpotent. The real numbers have no nonzero nilpotent elements.*

Although we see that matrix arithmetic does not have all the properties of ordinary arithmetic (of real numbers), it does have many of them. Matrix addition is commutative, associative, has $[0]$ (additive identity), and for each A we have $-A$ (additive inverse). Although matrix multiplication is not commutative, it is associative, $A(BC) = (AB)C$ whenever the multiplications are defined, and we have the distributive laws $A(B + C) = AB + AC$, and $(B + C)A = BA + CA$ whenever the multiplications and additions involved are defined.

These properties are not hard to prove, and we include proofs here for completeness.

Theorem 2.2 (Associative Law) *Let* A *be* $m \times n$, *let* B *be* $n \times l$ *and let* C *be* $l \times s$, *then* $A(BC) = (AB)C$.

Proof. BC is $n \times s$, so $A(BC)$ is defined and is $m \times s$. AB is $m \times l$, so $(AB)C$ is defined and is also $m \times s$. To show that two matrices are equal we must show their ij^{th} entries are equal for all i and j. Consider

$$
\begin{aligned}
[(AB)C]_{ij} &= \sum_{k=1}^{l} [AB]_{ik} c_{kj} \\
&= \sum_{k=1}^{l} \left(\sum_{t=1}^{n} a_{it} b_{tk} \right) c_{kj} \\
&= \sum_{k=1}^{l} \sum_{t=1}^{n} (a_{it} b_{tk}) c_{kj}.
\end{aligned}
$$

We have used the definition of matrix multiplication, the distributive law for real numbers. Now we exchange the order of the summations. This just amounts to a reordering of the terms of the sum, which is OK since addition of real numbers is commutative. So we get

$$
\sum_{k=1}^{l} \sum_{t=1}^{n} (a_{it} b_{tk}) c_{kj}
$$

$$= \sum_{t=1}^{n} \sum_{k=1}^{l} (a_{it} b_{tk}) c_{kj}$$

$$= \sum_{t=1}^{n} \sum_{k=1}^{l} a_{it} (b_{tk} c_{kj}).$$

The last step uses the associative law for multiplication of real numbers. But

$$\sum_{t=1}^{n} \sum_{k=1}^{l} a_{it} (b_{tk} c_{kj})$$

$$= \sum_{t=1}^{n} a_{it} \left(\sum_{k=1}^{l} b_{tk} c_{kj} \right)$$

$$= \sum_{t=1}^{n} a_{it} [BC]_{tj}$$

$$= [A(BC)]_{ij}.$$

We have used the distributive law to factor out the a_{it} from the summation on k, and then twice used the definition of matrix multiplication. Putting it all together we see that $[(AB)C]_{ij} = [A(BC)]_{ij}$ for all i and j, hence the matrices are equal. ∎

Remark   *Theorem 2.2 shows that we don't need parentheses in products of three matrices. Parentheses are also unnecessary in products of k matrices for any positive integer k. This is called the generalized associative law.*

Theorem 2.3 (Left Distributive Law) *If A is $m \times n$, B is $n \times l$, and C is $n \times l$ then $A(B + C) = AB + AC$.*

Proof.   It is clear that all operations are defined and that the result is an $m \times l$ matrix on both sides of the equation. Let us consider the ij^{th} entry of $A(B + C)$.

$$[A(B + C)]_{ij} = \sum_{k=1}^{n} a_{ik} [B + C]_{kj}$$

$$= \sum_{k=1}^{n} a_{ik} (b_{kj} + c_{kj})$$

$$= \sum_{k=1}^{n}(a_{ik}b_{kj} + a_{ik}c_{kj}).$$

We have used the definition of matrix multiplication, the definition of matrix addition, and then the distributive law for real numbers. We now want to sum first all the ab products and then all the ac products. This just involves a reordering of the summands, so we can do it using the commutative law of addition of real numbers. We get

$$\sum_{k=1}^{n}(a_{ik}b_{kj} + a_{ik}c_{kj}) = \sum_{k=1}^{n} a_{ik}b_{kj} + \sum_{k=1}^{n} a_{ik}c_{kj}$$
$$= [AB + AC]_{ij}.$$

Putting it all together gives $[A(B + C)]_{ij} = [AB + AC]_{ij}$. ∎

The right distributive law $(B+C)A = BA+CA$ is proved similarly, but it does not follow directly from the left distributive law because matrix multiplication is not commutative.

Definition 2.13 *If A and B are $n \times n$ matrices and if $BA = I$, we say that B is a left inverse for A and A is a right inverse for B. If $AB = BA = I$, we say that B is an inverse (two-sided) of A. If A has an inverse we say A is invertible.*

Theorem 2.4 *Let A be a square matrix. If B is a left inverse for A and C a right inverse for A, then $B = C$ and B is an inverse (two-sided) of A.*

Proof.   We are given that $AB = I$ and $CA = I$. Consider the product CAB. We use the associative law of matrix multiplication. $(CA)B = IB = B$, while $C(AB) = CI = C$. Since $(CA)B = C(AB)$, we conclude that $C = B$. ∎

Corollary 1 *If A has an inverse (two-sided), it is unique, and we call it A^{-1}.*

Proof.   If B and C are both inverses of A that work on both sides, we can use the above result to show that $B = C$. ∎

Thus we see that if A has both a left and a right inverse, they must be the same and hence A is invertible. Is it possible that A might have a left inverse but no right inverse? Soon we will see that this is not possible, that any left (or right) inverse of A be an (two-sided) inverse of A. When we refer to the inverse of a matrix A, we mean the multiplicative inverse A^{-1}, not the additive inverse, $-A$.

Definition 2.14 *If k is a positive integer and A a square matrix, we define A^k as A multiplied by itself k times. If A is invertible we define A^{-k} to be A^{-1} multiplied times itself k times. We define $A^0 = I$.*

One can show that the usual laws of exponents hold. $A^k A^m = A^{k+m}$, $(A^k)^m = A^{km}$.

Theorem 2.5 *If A and B are $n \times n$ invertible matrices, the AB is invertible and $(AB)^{-1} = B^{-1}A^{-1}$.*

Proof.   Because of the uniqueness of the inverse it is only necessary to show that $(AB)(B^{-1}A^{-1}) = I$. But this follows quickly from the associative law. ∎

Corollary 1 *If $A_1, A_2, \cdots, A_k$ are invertible matrices, the the product $A_1 A_2 \cdots A_k$ is invertible and $(A_1 A_2 \cdots A_k)^{-1} = A_k^{-1} \cdots A_2^{-1} A_1^{-1}$.*

Proof.   This can be proved formally from Theorem 2.5 using mathematical induction. However it is also easy to see that $(A_1 A_2 \cdots A_k)(A_k^{-1} \cdots A_2^{-1} A_1^{-1}) = I$. ∎

If a matrix is viewed as being made up of a number of smaller matrices, called blocks, it is sometimes possible to shorten calculations of products. If we divide a matrix A into blocks by drawing horizontal and vertical lines, this is called partitioning the matrix. If two matrices A and B are partitioned and if we label the blocks by row and column, for example if

$$A = \begin{bmatrix} A_{11} & A_{12} & A_{13} \\ A_{21} & A_{22} & A_{23} \\ A_{31} & A_{32} & A_{33} \end{bmatrix},$$

where each of the A_{ij} is itself a matrix, and if

$$B = \begin{bmatrix} B_{11} & B_{12} \\ B_{21} & B_{22} \\ B_{31} & B_{32} \end{bmatrix},$$

then if the sizes of A and B are compatible for multiplication, and if the sizes of all blocks that need to be multiplied are compatible, we can multiply as if the blocks were individual entries,

$$
\begin{aligned}
AB &= \begin{bmatrix} A_{11} & A_{12} & A_{13} \\ A_{21} & A_{22} & A_{23} \\ A_{31} & A_{32} & A_{33} \end{bmatrix} \begin{bmatrix} B_{11} & B_{12} \\ B_{21} & B_{22} \\ B_{31} & B_{32} \end{bmatrix} \\
&= \begin{bmatrix} A_{11}B_{11} + A_{12}B_{21} + A_{13}B_{31} & A_{11}B_{12} + A_{12}B_{22} + A_{13}B_{32} \\ A_{21}B_{11} + A_{22}B_{21} + A_{23}B_{31} & A_{21}B_{12} + A_{22}B_{22} + A_{23}B_{32} \\ A_{31}B_{11} + A_{32}B_{21} + A_{33}B_{31} & A_{31}B_{12} + A_{32}B_{22} + A_{33}B_{32} \end{bmatrix}.
\end{aligned}
$$

Example 2.13 *If*

$$AB = \begin{bmatrix} 1 & 2 & 0 & 0 \\ 3 & 1 & 0 & 0 \\ -1 & 3 & 4 & 2 \\ 2 & 1 & 5 & 1 \end{bmatrix} \begin{bmatrix} 6 & 3 & 0 & 0 \\ 5 & 2 & 0 & 0 \\ 3 & 3 & 4 & 2 \\ 1 & 5 & 2 & 1 \end{bmatrix} = \begin{bmatrix} P & [0] \\ Q & R \end{bmatrix},$$

then

$$[0] = \begin{bmatrix} 0 & 0 \\ 0 & 0 \end{bmatrix},$$

$$P = \begin{bmatrix} 1 & 2 \\ 3 & 1 \end{bmatrix} \begin{bmatrix} 6 & 3 \\ 5 & 2 \end{bmatrix} = \begin{bmatrix} 16 & 7 \\ 23 & 11 \end{bmatrix},$$

$$R = \begin{bmatrix} 4 & 2 \\ 5 & 1 \end{bmatrix} \begin{bmatrix} 4 & 2 \\ 2 & 1 \end{bmatrix} = \begin{bmatrix} 20 & 10 \\ 22 & 11 \end{bmatrix},$$

and

$$
\begin{aligned}
Q &= \begin{bmatrix} -1 & 3 \\ 2 & 1 \end{bmatrix} \begin{bmatrix} 6 & 3 \\ 5 & 2 \end{bmatrix} + \begin{bmatrix} 4 & 2 \\ 5 & 1 \end{bmatrix} \begin{bmatrix} 3 & 3 \\ 1 & 5 \end{bmatrix} \\
&= \begin{bmatrix} 9 & 3 \\ 17 & 8 \end{bmatrix} + \begin{bmatrix} 14 & 22 \\ 16 & 20 \end{bmatrix} \\
&= \begin{bmatrix} 23 & 25 \\ 33 & 28 \end{bmatrix}.
\end{aligned}
$$

Thus

$$AB = \begin{bmatrix} 16 & 7 & 0 & 0 \\ 23 & 11 & 0 & 0 \\ 23 & 25 & 20 & 10 \\ 33 & 28 & 22 & 11 \end{bmatrix}.$$

Exercises

1. Compute the products $E_{12} \begin{bmatrix} 3 & 1 & 2 \\ -1 & 1 & 4 \\ 4 & 5 & 1 \end{bmatrix}$ and $\begin{bmatrix} 3 & 1 & 2 \\ -1 & 1 & 4 \\ 4 & 5 & 1 \end{bmatrix} E_{12}$.

2. Consider the following product of partitioned matrices:

$$\begin{bmatrix} C & D \\ E & F \end{bmatrix} = \begin{bmatrix} 3 & 1 & 2 & 0 & 1 \\ -1 & 4 & 1 & 2 & -1 \\ 1 & 3 & 0 & 1 & 1 \end{bmatrix} \begin{bmatrix} 2 & 0 & -1 & 3 \\ 2 & 1 & 0 & 0 \\ 4 & 3 & 2 & 0 \\ 1 & 1 & 1 & 1 \\ 3 & 2 & 0 & 1 \end{bmatrix}.$$

Where the first matrix is partitioned between the third and fourth columns, and between the second and third rows; and the second matrix is partitioned between the second and third columns, and between the third and fourth rows.

   (a) Are the blocks compatible for block multiplication?
   (b) Write C as a sum of products of blocks from the numerically given matrices, and compute C.
   (c) Same as b for the matrix F.
   (d) What size is the matrix E?

Section 2.3 Systems of Linear Equations

Definition 2.15 *An equation in n unknowns is linear if it is of the form $a_1 x_1 + a_2 x_2 + \cdots + a_n x_n = b$, where the a_i and b are given real* *numbers and the x_i are variables which can take values in R. A solution of this equation is an n-tuple of real numbers $(s_1, s_2, \cdots, s_n)$ such that if we substitute $x_i = s_i$, the equation is satisfied, i.e. $a_1 s_1 + a_2 s_2 + \cdots + a_n s_n = b$. The solution set of this equation is the set of all n-tuples that are solutions.*

Assuming that at least one coefficient a_i is not zero, we can find solutions by choosing the other variables freely and solving for x_i in terms of them.

Example 2.14 *Consider $x + 2y + 3z = 6$. Then $x = 6 - 2y - 3z$. Substituting $y = s$ and $z = t$, we see that $(6 - 2s - 3t, s, t)$ is a solution for all real numbers s and t. In fact this gives all solutions. We could have solved for y and chosen x and z freely, or solved for z and chosen x and y freely.*

Example 2.15 *The solution set of $a_1 x_1 + a_2 x_2 = b$ is a line in R^2.*

Example 2.16 *The solution set of $a_1 x_1 + a_2 x_2 + a_3 x_3 = b$ is a plane in R^3.*

Definition 2.16 *A system of linear equations is a set of one or more linear equations.*

A system of m linear equations in n unknowns can be written as follows:

$$
\begin{aligned}
a_{11} x_1 + a_{12} x_2 + \cdots + a_{1n} x_n &= b_1 \\
a_{21} x_1 + a_{22} x_2 + \cdots + a_{2n} x_n &= b_2 \\
&\ \ \vdots \\
a_{m1} x_1 + a_{m2} x_2 + \cdots + a_{mn} x_n &= b_m.
\end{aligned}
\tag{2.4}
$$

This system of equations can be written in matrix form as $A\mathbf{x} = \mathbf{b}$,

$$
\begin{bmatrix}
a_{11} & a_{12} & \cdots & a_{1n} \\
a_{21} & a_{22} & \cdots & a_{2n} \\
\vdots & \vdots & \ddots & \vdots \\
a_{m1} & a_{m2} & \cdots & a_{mn}
\end{bmatrix}
\begin{bmatrix}
x_1 \\
x_2 \\
\vdots \\
x_n
\end{bmatrix}
=
\begin{bmatrix}
b_1 \\
b_2 \\
\vdots \\
b_m
\end{bmatrix},
$$

in vector form as

$$x_1 \begin{bmatrix} a_{11} \\ a_{21} \\ \vdots \\ a_{m1} \end{bmatrix} + x_2 \begin{bmatrix} a_{12} \\ a_{22} \\ \vdots \\ a_{m2} \end{bmatrix} + \cdots + x_n \begin{bmatrix} a_{1n} \\ a_{2n} \\ \vdots \\ a_{mn} \end{bmatrix} = \begin{bmatrix} b_1 \\ b_2 \\ \vdots \\ b_m \end{bmatrix},$$

or more succinctly as

$$x_1 \mathbf{v}_1 + x_2 \mathbf{v}_2 + \cdots + x_n \mathbf{v}_n = \mathbf{b},$$

where $\mathbf{v}_i$ is the ith column of A.

Remark *Note that $A\mathbf{x} = \mathbf{b}$ has a solution if and only if $\mathbf{b}$ is a linear combination of the columns of A, and the solution tells us the coefficients to use.*

Definition 2.17 *The solution set of a system of m linear equations in n unknowns is a set of n-tuples $(s_1, s_2, \cdots, s_n)$ such that when we substitute $x_i = s_i$, all the equations are satisfied.*

Definition 2.18 *Two systems of equations are equivalent if they have the same solution set.*

Some systems of linear equations are easy to solve.

Example 2.17 *Consider the equations*

$$\begin{array}{rcrcrcl} 3x & + & 2y & - & z & = & 1 \\ 0x & - & 3y & + & 3z & = & 2 \\ 0x & + & 0y & + & 2z & = & 8 \end{array}$$

From the last equation we get $z = 4$. Substituting this in the middle equation gives $y = \frac{10}{3}$, and substituting both of these in the first equation gives $x = -\frac{5}{9}$.

Definition 2.19 *In the system of equations (2.4), the matrix A is called the coefficient matrix of the system. The matrix*

$$[A \ \mathbf{b}] = \begin{bmatrix} a_{11} & a_{12} & \cdots & a_{1n} & b_1 \\ a_{21} & a_{22} & \cdots & a_{2n} & b_2 \\ \vdots & \vdots & \ddots & \vdots & \vdots \\ a_{m1} & a_{m2} & \cdots & a_{mn} & b_m \end{bmatrix}$$

is called the augmented matrix of the system.

The coefficient matrix of the system in Example 2.17 is

$$A = \begin{bmatrix} 3 & 2 & -1 \\ 0 & -3 & 3 \\ 0 & 0 & 2 \end{bmatrix}.$$

The augmented matrix of the system is

$$\begin{bmatrix} A & b \end{bmatrix} = \begin{bmatrix} 3 & 2 & -1 & 1 \\ 0 & -3 & 3 & 2 \\ 0 & 0 & 2 & 8 \end{bmatrix}.$$

The reason the system is so easy to solve is that the coefficient matrix is upper triangular. In this case you can solve for the last unknown first, then the next to last, etc., as we did in the example. This method is called back substitution. (If the coefficient matrix is lower triangular we could do forward substitution.) It is always possible by a sequence of operations that do not change the solution set, to transform a system of linear equations into a system with a coefficient matrix in a form that makes it possible to solve it easily. Let us look at some more examples.

Example 2.18 *Consider the system of equations*

$$\begin{array}{rcrcrcl} x & + & 0y & + & 0z & = & 3 \\ 0x & + & y & + & 0z & = & 4 \\ 0x & + & 0y & + & z & = & 1 \end{array}$$

The augmented matrix of this system is

$$\begin{bmatrix} 1 & 0 & 0 & 3 \\ 0 & 1 & 0 & 4 \\ 0 & 0 & 1 & 1 \end{bmatrix},$$

and the solution is very easy. We can write it down immediately, $x = 3$, $y = 4$, and $z = 1$. The reason this is so easy is that the coefficient matrix is the identity matrix. In this case the solution is immediately obvious and it is unique. Unfortunately not every system can be transformed into one that is this simple.

Example 2.19 *Consider the system*

$$\begin{array}{rcrcrcl} x & + & 0y & + & 2z & = & 4 \\ 0x & + & y & - & 3z & = & 5 \end{array}$$

We can chose z freely, let $z = t$, and solve the second equation for y and the first for x, both in terms of t. Clearly $x = 4 - 2t$ and $y = 5 + 3t$. The augmented matrix of this system is

$$\begin{bmatrix} 1 & 0 & 2 & 4 \\ 0 & 1 & -3 & 5 \end{bmatrix},$$

and the solution set is $\{(4 - 2t, 5 + 3t, t)|t \in R\}$. Geometrically this is the line of intersection of the two planes.

Definition 2.20 *An $m \times n$ matrix is in reduced row echelon form (RREF) if it satisfies the following conditions:*

C1. In each nonzero row, the first nonzero entry is a 1. This 1 is called a leading 1.

C2. The zero rows if any come last.

C3. In every column that contains the leading 1 of some row, all other entries are 0.

C4. The leading 1 in each row occurs to the right of the leading 1 in the row above.

Example 2.20 *The matrix* $\begin{bmatrix} \underline{1} & 0 & 2 & 3 & 0 \\ 0 & \underline{1} & 4 & 5 & 0 \\ 0 & 0 & 0 & 0 & 0 \\ 0 & 0 & 0 & 0 & 0 \end{bmatrix}$ *is in RREF. Leading ones are underlined.*

Example 2.21 *The matrix* $\begin{bmatrix} 0 & 1 & 2 & 0 & 3 \\ 0 & 0 & 4 & 0 & 5 \\ 0 & 0 & 0 & 1 & 0 \\ 0 & 0 & 0 & 0 & 0 \end{bmatrix}$ *is not in RREF. Row two violates C1.*

Example 2.22 *The matrix* $\begin{bmatrix} 0 & 1 & 2 & 3 & 4 \\ 0 & 0 & 0 & 1 & 5 \\ 0 & 0 & 0 & 0 & 0 \end{bmatrix}$ *is not in RREF. C3 is violated.*

Example 2.23 *The matrix*
$$\begin{bmatrix} 0 & 0 & 0 & 0 & 0 \\ 0 & 1 & 2 & 0 & 4 \\ 0 & 0 & 0 & 0 & 0 \\ 0 & 0 & 0 & 1 & 5 \\ 0 & 0 & 0 & 0 & 0 \end{bmatrix}$$ *is not in RREF. C2 is violated.*

Example 2.24 *The matrix* $\begin{bmatrix} 0 & 0 & 0 & 1 & 5 \\ 0 & 1 & 2 & 0 & 4 \end{bmatrix}$ *is not in RREF. It violates C4.*

If the augmented matrix of a system is in RREF, it will be easy to find the solution set of the system. We now show how this is done.

Definition 2.21 *Let A be a matrix in RREF. A column of A that contains a leading 1 is called a leading column, the other columns are non-leading columns.*

Theorem 2.6 *If the augmented matrix of a system of m linear equations in n unknowns is a matrix in RREF, then a solution to the system exists if and only if the last column is not a leading column. In this case the variables corresponding to non-leading columns (non-leading variables) can be chosen freely, and one can solve for the variables corresponding to the leading columns (leading variables).*

Proof. If the last column of the augmented matrix is a leading column, then it represents an equation of the form $0x_1 + 0x_2 + \cdots + 0x_n = 1$, which has no solution, so the whole system has no solution. If there is no leading 1 in the last column, then each leading column corresponds to a variable. Suppose that there is a leading one in column i, row k. Then the equation corresponding to row k contains x_i with coefficient 1, and no other equation contains x_i. Moreover in the equation corresponding to row k all the other leading variables have coefficient 0, so we can solve for x_i in terms of the non-leading variables. An n-tuple will be a solution if and only if each leading variable x_i satisfies the one equation in which it occurs. ∎

Example 2.25 *The matrix*

$$\begin{bmatrix} 1 & 0 & 2 & 3 & 0 \\ 0 & 1 & 4 & 5 & 0 \\ 0 & 0 & 0 & 0 & 0 \\ 0 & 0 & 0 & 0 & 0 \end{bmatrix}$$

is in RREF. If this is the augmented matrix of a system of 4 equations in 4 unknowns, the first four columns correspond to the variables x_1, x_2, x_3, x_4 respectively. Columns one and two are the leading columns. We can chose $x_3 = s$, and $x_4 = t$, and solve for x_1 and x_2. The first row corresponds to the equation $x_1 + 2s + 3t = 0$. and the second row corresponds to the equation $x_2 + 4s + 5t = 0$. Thus the solution set is $\{(-2s - 3t, -4s - 5t, s, t)|s, t \in R\}$.

Now that we know how to solve systems when the augmented matrix is in RREF, we want to find a way to change an arbitrary matrix into a matrix in RREF in such a way that the associated systems of equations are equivalent. To do this we use the following operations.

Definition 2.22 *Given an $m \times n$ matrix A the following changes of A are called Elementary Row Operations:*
  1) Multiply a row vector of A by a nonzero scalar.
  2) Add a multiple of one row of A to another row of A. (If $\mathbf{u}$ and $\mathbf{v}$ represent two different rows of A, replace $\mathbf{u}$ by $\mathbf{u} + k\mathbf{v}$.)
  3) Exchange (transpose) any two rows of A.

This definition is so widely used now that very few people seem to be aware of the fact that an elementary row operation of type 3 can be obtained by a sequence of elementary row operations of the other two types.

Lemma 2.7 *Permuting two rows of A can be accomplished by a sequence of elementary row operations of types 1 and 2.*

Proof.   To exchange row i and row j we apply the following sequence of operations. Add row j to row i, subtract the new row i from row j, add the new row j to row i, and finally multiply the new row j by -1.

The following shows the effect of this on rows i and j:

$$\text{row } i \quad \begin{bmatrix} \mathbf{u} \\ \mathbf{v} \end{bmatrix} \rightarrow \begin{bmatrix} \mathbf{u}+\mathbf{v} \\ \mathbf{v} \end{bmatrix} \rightarrow \begin{bmatrix} \mathbf{u}+\mathbf{v} \\ -\mathbf{u} \end{bmatrix} \rightarrow \begin{bmatrix} \mathbf{v} \\ -\mathbf{u} \end{bmatrix} \rightarrow \begin{bmatrix} \mathbf{v} \\ \mathbf{u} \end{bmatrix}. \ \blacksquare$$
$$\text{row } j$$

Lemma 2.8 *If B comes from A by an elementary row operation, then A comes from B by an elementary row operation of the same kind.*

Proof. Type 1:
$$\begin{bmatrix} \mathbf{u} \\ \mathbf{v} \end{bmatrix} \longrightarrow \begin{bmatrix} c\mathbf{u} \\ \mathbf{v} \end{bmatrix} \longrightarrow \begin{bmatrix} \frac{1}{c}(c\mathbf{u}) \\ \mathbf{v} \end{bmatrix} = \begin{bmatrix} \mathbf{u} \\ \mathbf{v} \end{bmatrix}, \quad \text{as}$$

before we show rows i and j. The other rows are left unchanged.

Type 2:
$$\begin{bmatrix} \mathbf{u} \\ \mathbf{v} \end{bmatrix} \longrightarrow \begin{bmatrix} \mathbf{u}+c\mathbf{v} \\ \mathbf{v} \end{bmatrix} \longrightarrow \begin{bmatrix} \mathbf{u}+c\mathbf{v}-c\mathbf{v} \\ \mathbf{v} \end{bmatrix} = \begin{bmatrix} \mathbf{u} \\ \mathbf{v} \end{bmatrix}.$$

Type 3:
$$\begin{bmatrix} \mathbf{u} \\ \mathbf{v} \end{bmatrix} \longrightarrow \begin{bmatrix} \mathbf{v} \\ \mathbf{u} \end{bmatrix} \longrightarrow \begin{bmatrix} \mathbf{u} \\ \mathbf{v} \end{bmatrix}. \ \blacksquare$$

Lemma 2.9 *Elementary row operations on the augmented matrix $\begin{bmatrix} A & \mathbf{b} \end{bmatrix}$ of a system of equations do not change the solution set of the associated system of linear equations.*

Proof. An operation of type 1 just multiplies one of the equations by a nonzero constant. Clearly any n-tuple $(s_1, s_2, \cdots, s_n)$ that satisfies the original equations will still satisfy the new equations and conversely, since

$$a_{i1}s_1 + a_{i2}s_2 + \cdots + a_{in}s_n = b_i$$

if and only if

$$ca_{i1}s_1 + ca_{i2}s_2 + \cdots + ca_{in}s_n = cb_i.$$

We now consider type 2. If a multiple of row i is added to row j, then any n-tuple $(s_1, s_2, \cdots s_n)$ which satisfies the original equations will satisfy the new equations since only row j is changed and

$$\begin{aligned} &(a_{j1} + ca_{i1})s_1 + (a_{j2} + ca_{i2})s_2 + \cdots + (a_{jn} + ca_{in})s_n \\ = \ &(a_{j1}s_1 + a_{j2}s_2 + \cdots + a_{jn}s_n) + (ca_{i1}s_1 + ca_{i2}s_2 + \cdots + ca_{in}s_n) \\ = \ &b_j + cb_i, \end{aligned}$$

which shows that the equation represented by the new row j is satisfied. Conversely any n-tuple which satisfies the equations represented by the new row j and row i must satisfy the original equations since we can recover the original equations by subtracting c times row i from the new row j. ∎

Definition 2.23 *If the matrix B can be obtained from the matrix A by a finite sequence of elementary row operations, we say that A and B are row equivalent, and write $A \sim B$.*

Theorem 2.7 *For every matrix A there is a matrix B in RREF such that $A \sim B$.*

Proof.   The zero matrix is already a matrix in RREF. If A is not the zero matrix, move to the first column of A which is not all zeros. Suppose this is column p and that a_{jp} is not zero. Multiply row j by $1/a_{jp}$ and and exchange rows j and 1. This produces a leading 1 in the $1p$ position. (If $a_{11} \neq 0$, just multiply the first row by $1/a_{11}$). Now by adding appropriate multiples of row 1 to the other rows we can produce zeros everywhere else in column p. This process is called pivoting on the jp position. This results in a matrix of the following type:

$$
M = \begin{bmatrix} 0 & \cdots & 0 & 1 & * & \cdots & * \\ 0 & \cdots & 0 & 0 & * & \cdots & * \\ \vdots & \ddots & \vdots & \vdots & \vdots & \ddots & \vdots \\ 0 & \cdots & 0 & 0 & * & \cdots & * \end{bmatrix},
$$

where the first $p-1$ columns are zero vectors, the pth column vector is e_1 and the rest could be anything. We shall call the change from A to M a stage one reduction. Next cover up the first row of M and subject the remaining matrix to another stage one reduction. Then cover up the first two rows and subject this matrix to a stage one reduction. The process ends when only zero rows remain or all rows are covered up. The matrix at this stage satisfies C1,C2, and C4, but not necessarily C3. While a column with a leading 1 has zero in all positions below the leading 1, it might not have all zeros above. Some call this stage "row echelon form". To get a matrix in RREF we continue by using the leading 1 in a column to get rid of any nonzero entries above it by adding appropriate multiples of the row containing this leading 1 to the rows above it. ∎

This not only proves the theorem, but gives a procedure for changing A into its RREF. This procedure is so explicitly described that a computer program can be written to do it. In fact many calculators and computers will find the RREF for you. Roundoff error can be a problem if floating point is used. For this reason if you are working with matrices with rational (integer or fraction) entries, it is best to use a program that will compute the entries exactly, and not convert to decimal approximations.

Example 2.26 *The matrix* $\begin{bmatrix} 1 & 2 \\ 0 & 1 \end{bmatrix}$ *is row equivalent to* $\begin{bmatrix} 1 & 3 \\ 0 & 1 \end{bmatrix}$, *and both of these are in row echelon form (but not in RREF).*

We see in the preceding example that a matrix may be row equivalent to more than one row echelon form matrix. We will show in the next section that this cannot happen for the RREF. Every matrix is in fact row equivalent to one and only one matrix in RREF, which is called the RREF of A, or $RREF(A)$. In what follows we use this terminology although strictly speaking we should not use the definite article "the" until we have proved uniqueness.

Example 2.27 *We now do an example showing the step by step reduction of a matrix to RREF. Let*

$$A = \begin{bmatrix} 2 & 4 & 7 & 5 \\ 4 & 2 & 3 & 9 \\ 1 & 1 & 1 & 1 \end{bmatrix}.$$

We could multiply row one by 1/2, but that would give fractions, so if doing this without a computer it is better to exchange row 3 and row 1. Thus

$$A \sim \begin{bmatrix} 1 & 1 & 1 & 1 \\ 4 & 2 & 3 & 9 \\ 2 & 4 & 7 & 5 \end{bmatrix}.$$

Next we subtract 4 times row 1 from row 2 and then subtract 2 times row 1 from row 3. We get

$$\begin{bmatrix} 1 & 1 & 1 & 1 \\ 0 & -2 & -1 & 5 \\ 0 & 2 & 5 & 3 \end{bmatrix}.$$

This completes a stage one reduction. Next we should multiply row 2 by $-1/2$, and then produce zeros below the new leading 1, but we see that we can produce the zero below easily now, so we add the second row to the third, to get

$$\begin{bmatrix} 1 & 1 & 1 & 1 \\ 0 & -2 & -1 & 5 \\ 0 & 0 & 4 & 8 \end{bmatrix}.$$

Now multiply row 2 by $-1/2$ to get

$$\begin{bmatrix} 1 & 1 & 1 & 1 \\ 0 & 1 & \frac{1}{2} & -\frac{5}{2} \\ 0 & 0 & 4 & 8 \end{bmatrix}.$$

Now multiply the last row by $1/4$(divide by 4) to get

$$\begin{bmatrix} 1 & 1 & 1 & 1 \\ 0 & 1 & \frac{1}{2} & -\frac{5}{2} \\ 0 & 0 & 1 & 2 \end{bmatrix}.$$

This is now in row echelon form. To proceed to the RREF we subtract the last row from the first and subtract $1/2$ times the last row from the second to get

$$\begin{bmatrix} 1 & 1 & 0 & -1 \\ 0 & 1 & 0 & -\frac{7}{2} \\ 0 & 0 & 1 & 2 \end{bmatrix},$$

and then subtract the second row from the first, giving

$$\begin{bmatrix} 1 & 0 & 0 & \frac{5}{2} \\ 0 & 1 & 0 & -\frac{7}{2} \\ 0 & 0 & 1 & 2 \end{bmatrix}.$$

This is now the RREF of A. If the original matrix A is viewed as the augmented matrix of a system of 3 equations in 3 unknowns, we see that there is a unique solution, $x_1 = 5/2$, $x_2 = -7/2$, $x_3 = 2$.

Example 2.28 *Let*

$$B = \begin{bmatrix} 1 & 2 & 3 & 6 \\ 2 & 3 & 4 & 9 \\ 3 & 5 & 7 & 13 \end{bmatrix}.$$

We already have a 1 in the upper left hand corner, so we proceed to clear the rest of column 1 by subtracting 2 times row 1 from row 2 and the subtracting 3 times row 1 from row 3. We get

$$\begin{bmatrix} 1 & 2 & 3 & 6 \\ 0 & -1 & -2 & -3 \\ 0 & -1 & -2 & -5 \end{bmatrix}.$$

Next multiply row 2 by −1. *We get*

$$\begin{bmatrix} 1 & 2 & 3 & 6 \\ 0 & 1 & 2 & 3 \\ 0 & -1 & -2 & -5 \end{bmatrix}.$$

Now adding row 2 to row 3 gives

$$\begin{bmatrix} 1 & 2 & 3 & 6 \\ 0 & 1 & 2 & 3 \\ 0 & 0 & 0 & -2 \end{bmatrix}.$$

We can already see at this point that if B is the augmented matrix of a system of 3 equations in 3 unknowns, then there is no solution to the system, since the last row corresponds to the equation $0x_1 + 0x_2 + 0x_3 = -2$. If we are only interested in solving the equations, we can stop here. If we want the RREF of B however, we continue by multiplying the last row by −1/2 *to get*

$$\begin{bmatrix} 1 & 2 & 3 & 6 \\ 0 & 1 & 2 & 3 \\ 0 & 0 & 0 & 1 \end{bmatrix},$$

(the row echelon form). Now using the leading 1 in column 4 to clear the rest of column 4, we get

$$\begin{bmatrix} 1 & 2 & 3 & 0 \\ 0 & 1 & 2 & 0 \\ 0 & 0 & 0 & 1 \end{bmatrix}.$$

It still remains to subtract twice row 2 from row 1 giving

$$\begin{bmatrix} 1 & 0 & -1 & 0 \\ 0 & 1 & 2 & 0 \\ 0 & 0 & 0 & 1 \end{bmatrix}.$$

This is the RREF of B.

Example 2.29 *Let* $C = \begin{bmatrix} 1 & 2 & 3 & 6 \\ 2 & 3 & 4 & 9 \\ 7 & 12 & 17 & 36 \end{bmatrix}$. *We apply elementary row*

operations to bring C into RREF as follows: $C = \begin{bmatrix} 1 & 2 & 3 & 6 \\ 2 & 3 & 4 & 9 \\ 7 & 12 & 17 & 36 \end{bmatrix} \sim$

$\begin{bmatrix} 1 & 2 & 3 & 6 \\ 0 & -1 & -2 & -3 \\ 0 & -2 & -4 & -6 \end{bmatrix} \sim \begin{bmatrix} 1 & 2 & 3 & 6 \\ 0 & 1 & 2 & 3 \\ 0 & -2 & -4 & -6 \end{bmatrix} \sim \begin{bmatrix} 1 & 0 & -1 & 0 \\ 0 & 1 & 2 & 3 \\ 0 & 0 & 0 & 0 \end{bmatrix}$.*If we*

regard C as the augmented matrix of a system of 3 equations in 3 un-knowns, then we see that we can chose x_3 *freely, say* $x_3 = t$, *and then from the first row we get* $x_1 = t$, *from the second row we get* $x_2 = 3 - 2t$. *So one way to describe the solution set is* $\{(t, 3 - 2t, t)|t \in R\}$.

Our method of solving systems of equations by row reduction tries to solve for the variables that come first in terms of variables that come later if possible. If you have a system of equations and there are some variables that you would like to be able to choose freely, solving for the others in terms of these, put the ones you would most like to choose freely last (as x_n, x_{n-1}) and those you would like to be determined first (as x_1, x_2, etc.)

Exercises

1. In Examples 2.21 to 2.24 using elementary row operations, reduce these matrices to RREF.

2. If the following matrix is the RREF of the augmented matrix of a system of equations, find the solution set:

(a) $\begin{bmatrix} 1 & 2 & 0 & -3 & 5 \\ 0 & 0 & 1 & 4 & 7 \\ 0 & 0 & 0 & 0 & 0 \\ 0 & 0 & 0 & 0 & 0 \end{bmatrix}$

(b) $\begin{bmatrix} 1 & 2 & 0 & -3 & 0 \\ 0 & 0 & 1 & 4 & 0 \\ 0 & 0 & 0 & 0 & 1 \\ 0 & 0 & 0 & 0 & 0 \end{bmatrix}$

(c) $\begin{bmatrix} 1 & 0 & -2 & 4 \\ 0 & 1 & 0 & 0 \\ 0 & 0 & 0 & 0 \end{bmatrix}$.

In problems 3 through 5, note that the system of equations you need to solve when trying to express $\mathbf{v}$ as a linear combination of $\mathbf{u}_1, \mathbf{u}_2, \cdots, \mathbf{u}_k$ has augmented matrix $\begin{bmatrix} \mathbf{u}_1 & \mathbf{u}_2 & \cdots & \mathbf{u}_k & \mathbf{v} \end{bmatrix}$.

3. Express the vector $(7, 0, 2)$ as a linear combination of the vectors $(1, 2, -1), (3, 1, 2)$ and $(0, 4, 1)$, if possible.

4. Express the vector $(3, -5, 1)$ as a linear combination of the vectors $(1, 2, 4), (3, 1, 7)$, and $(0, 4, 4)$, if possible.

5. Express $(3, 2, 5)$ as a linear combination of $(1, 3, 0), (2, 1, 4)$, and $(4, 7, 4)$, if possible.

Section 2.4 Uniqueness of the RREF

Definition 2.24 *The row space of a matrix A is $sp\{\mathbf{u}_1, \mathbf{u}_2, \cdots, \mathbf{u}_k\}$, where $\mathbf{u}_1, \mathbf{u}_2, \cdots, \mathbf{u}_k$ are the row vectors of A. We write $rowsp(A)$.*

Definition 2.25 *The column space of a matrix A is $sp\{\mathbf{v}_1, \mathbf{v}_2, \cdots, \mathbf{v}_k\}$, where $\mathbf{v}_1, \mathbf{v}_2, \cdots, \mathbf{v}_k$ are the column vectors of A. We write $colsp(A)$.*

Lemma 2.10 *Elementary row operations do not change the row space of a matrix A.*

Proof. Let A' be the matrix that results from A by a row operation. Let $\mathbf{u}_1, \mathbf{u}_2, \cdots, \mathbf{u}_k$ be the rows of A and $\mathbf{v}_1, \mathbf{v}_2, \cdots, \mathbf{v}_k$ the rows of A' We wish to show $sp\{\mathbf{u}_1, \mathbf{u}_2, \cdots, \mathbf{u}_k\} = sp\{\mathbf{v}_1, \mathbf{v}_2, \cdots, \mathbf{v}_k\}$. Using the Corollary to Theorem 1.6 we need only show that $\mathbf{u}_i \in sp\{\mathbf{v}_1, \mathbf{v}_2, \cdots, \mathbf{v}_k\}$ for all i, and $\mathbf{v}_i \in sp\{\mathbf{u}_1, \mathbf{u}_2, \cdots, \mathbf{u}_k\}$ for all i. Clearly this is true for type 3 row operations, which don't change the row vectors but just rearrange them. When we do an elementary row operation of type one or two, only one row of A is changed, say the jth, so $\mathbf{u}_i = \mathbf{v}_i$ for $i \neq j$. If the row operation is of type 1, then $\mathbf{v}_j = r\mathbf{u}_j \in sp\{\mathbf{u}_1, \mathbf{u}_2, \cdots, \mathbf{u}_k\}$, and $\mathbf{u}_j = (1/r)\mathbf{v}_j \in sp\{\mathbf{v}_1, \mathbf{v}_2, \cdots, \mathbf{v}_k\}$. If the row operation is of type 2, then $\mathbf{v}_j = \mathbf{u}_j + r\mathbf{u}_i \in sp\{\mathbf{u}_1, \mathbf{u}_2, \cdots, \mathbf{u}_k\}$ and $\mathbf{u}_j = \mathbf{v}_j - r\mathbf{u}_i = \mathbf{v}_j - r\mathbf{v}_i \in sp\{\mathbf{v}_1, \mathbf{v}_2, \cdots, \mathbf{v}_k\}$. ∎

Corollary 1 *If A and B are row equivalent they have the same row space.*

Remark *Row operations may change the column space of a matrix as in the following example.*

Example 2.30 *The matrices* $\begin{bmatrix} 1 & 1 & 1 \\ 1 & 1 & 1 \\ 1 & 2 & 1 \end{bmatrix}$ *and* $\begin{bmatrix} 1 & 1 & 1 \\ 0 & 0 & 0 \\ 1 & 2 & 1 \end{bmatrix}$ *have the same row space but not the same column space.*

Definition 2.26 *The row rank of a matrix is the dimension of the row space of A.*

Definition 2.27 *The column rank of a matrix A is the dimension of the column space of A.*

Lemma 2.11 *$A\mathbf{x} = \mathbf{b}$ has a solution if and only if $\mathbf{b}$ is in $colsp(A)$.*

Proof. We know that the column vector $A\mathbf{x}$ is a linear combination of the columns of A (equation 2.2, section 2.1). Thus saying that $A\mathbf{x} = \mathbf{b}$ has a solution is the same as saying that $\mathbf{b}$ is a linear combination of the columns of A, which is the same as saying that $\mathbf{b}$ is in $colsp(A)$. ∎

Lemma 2.12 *$A\mathbf{x} = \mathbf{b}$ has a solution if and only if column $rank(A) =$ column $rank\left(\begin{bmatrix} A & \mathbf{b} \end{bmatrix}\right)$.*

Proof.   This follows immediately from the preceding lemma. ∎

Lemma 2.13 *If a matrix A is row equivalent to a matrix P in RREF, then the nonzero rows of P are a basis of the row space of A.*

Proof.   Since by Lemma 2.10, A and P have the same row space, the nonzero row vectors of P span $\text{rowsp}(A)$, so we need only show that these vectors are linearly independent. Let $\mathbf{u}_1, \mathbf{u}_2, \cdots, \mathbf{u}_k$ be the nonzero row vectors of P. If $c_1\mathbf{u}_1 + c_2\mathbf{u}_2 + \cdots + c_k\mathbf{u}_k = \mathbf{O}$. Since each $\mathbf{u}_i$ has a leading 1, say in the jth component and all the other $\mathbf{u}_i$ are zero in this component, $c_1\mathbf{u}_1 + c_2\mathbf{u}_2 + \cdots + c_k\mathbf{u}_k$ has c_i in the jth component, thus $c_i = 0$. This works for all i. ∎

Theorem 2.8 *Let A and B be two $m \times n$ matrices in RREF and assume A and B have the same row space, then $A = B$.*

Proof.   Let $r = $ row rank(A), select any of the nonzero rows of B and attach it to the matrix A as the last row. This matrix we call A^*. Then row rank$(A^*) = $ row rank(A). Now assume that the leading 1 of the last row of A^* is in column k, a column where A has no leading 1. We will show that this results in a contradiction. If A^* has any leading 1's in any columns to the right of column k, use those leading 1's (by adding scalar multiples of those rows to the last row) to produce 0's in those columns of the last row. This matrix we call A^{**}. Next use the leading 1 of the last row of A^{**} to clear column k, so that column k has zero everywhere except in the last row. Since the last column of A^{**} was zero in all the leading columns of A, these columns will not be affected. Call this new matrix A^{***}. Then except for the rows possibly being in the wrong order, A^{***} is in RREF and clearly has row rank $r + 1$, a contradiction. We conclude that the leading 1 of any row of B must occur in a column where A also has a leading 1. Since this is true of all the rows of B, every leading column of B must be a leading column of A and vice-versa. Thus A and B have their leading 1's in exactly the same columns.

Let $\mathbf{v}$ be a nonzero row of B with leading 1 in column k, and let $\mathbf{u}$ be the row of A which also has a leading 1 in column k. We will show by contradiction that $\mathbf{u} = \mathbf{v}$. Suppose $\mathbf{u} \neq \mathbf{v}$. Then $\mathbf{u} - \mathbf{v}$ has a nonzero entry. Let j be the smallest positive integer such that the jth component of $\mathbf{u} - \mathbf{v}$ is nonzero. We know that j cannot be the index of a leading column since both $\mathbf{u}$ and $\mathbf{v}$ are 0 in all leading columns except for column k, where both are 1. Divide $\mathbf{u} - \mathbf{v}$ by this first nonzero entry. Then $1/c(\mathbf{u} - \mathbf{v})$ is a vector in the row space of A which has a leading 1 in a column that is not a leading column of A. The argument just presented above shows that this is impossible. We conclude that $\mathbf{u} = \mathbf{v}$. Thus the nonzero rows of A and B are identical. ∎

To illustrate the procedures used in the proof consider the following example.

$$\text{Let } A = \begin{bmatrix} 1 & 0 & 6 & 0 & 3 \\ 0 & 1 & 3 & 0 & 5 \\ 0 & 0 & 0 & 1 & 1 \end{bmatrix}, \text{ and } A^* = \begin{bmatrix} 1 & 0 & 6 & 0 & 3 \\ 0 & 1 & 3 & 0 & 5 \\ 0 & 0 & 0 & 1 & 1 \\ 0 & 0 & 1 & 5 & 7 \end{bmatrix}, \text{ then}$$

$$A^{**} = \begin{bmatrix} 1 & 0 & 6 & 0 & 3 \\ 0 & 1 & 3 & 0 & 5 \\ 0 & 0 & 0 & 1 & 1 \\ 0 & 0 & 1 & 0 & 2 \end{bmatrix}, \text{ and } A^{***} = \begin{bmatrix} 1 & 0 & 0 & 0 & -9 \\ 0 & 1 & 0 & 0 & -1 \\ 0 & 0 & 0 & 1 & 1 \\ 0 & 0 & 1 & 0 & 2 \end{bmatrix}.$$

Corollary 1 (Uniqueness of the RREF) *There is only one matrix in RREF that is row equivalent to a given matrix C.*

Proof.   Suppose A and B are both in RREF and both row equivalent to C. Since row operations don't change the row space, A and B have the same row space. Thus $A = B$ by the theorem. ∎

Corollary 2 *If P and Q are $m{\times}n$ matrices, the following are equivalent:*

  1. P and Q are row equivalent.
  2. P and Q have the same row space.
  3. P and Q have the same RREF.

Proof.   We will show $1 \Rightarrow 2 \Rightarrow 3 \Rightarrow 1$. If P and Q are row equivalent, they have the same row space. If P and Q have the same row space then let A be the RREF of P and let B be the RREF of Q. Then A and B have the same row space and by the theorem $A = B$. Clearly if P and Q have the same RREF, call it A, then P and Q are both row equivalent to A and so are row equivalent to each other. ∎

Procedure 2.1 To determine if two sets of vectors S and T span the same subspace of R^n :

1. Let A be the matrix with rows the vectors of S.

2. Let B be the matrix with rows the vectors of T.

3. Find $RREF(A)$ and $RREF(B)$.

4. $Sp\{S\} = sp\{T\}$ if and only if the nonzero rows of $RREF(A)$ and the nonzero rows of $RREF(B)$ are same.

Exercises

1. For each pair of the following matrices, determine whether or not they have the same row space.

$$A = \begin{bmatrix} 1 & 2 & 0 & 5 \\ 0 & 3 & 6 & 9 \\ 0 & 5 & 6 & 14 \end{bmatrix}, \qquad B = \begin{bmatrix} 1 & 1 & 1 & 1 \\ 0 & 3 & 6 & 9 \\ 0 & 5 & 6 & 14 \end{bmatrix},$$

$$C = \begin{bmatrix} -1 & 3 & 6 & 9 \\ 7 & 7 & -6 & 16 \\ -2 & 1 & 6 & 4 \end{bmatrix}, \qquad D = \begin{bmatrix} 1 & 2 & 0 & 5 \\ 0 & 1 & 2 & 3 \\ 0 & 0 & 1 & 1 \end{bmatrix}.$$

2. Determine the row rank of each matrix in problem 1.

3. Use $colsp(A) = rowsp(A^T)$ to determine the column rank of each matrix in problem 1.

Section 2.5 Row Rank Equals Column Rank

We begin with the following very useful result.

Lemma 2.14 *If A and B are row equivalent, their columns satisfy the same linear dependence relations.*

Proof.   A linear dependence relation among the columns of A corresponds to a nonzero solution of $A\mathbf{x} = \mathbf{O}$. Elementary row operations on A do not change the solution set of $A\mathbf{x} = \mathbf{O}$, thus $A\mathbf{x} = \mathbf{O}$ and $B\mathbf{x} = \mathbf{O}$ have the same solution set. It follows that every dependence relation satisfied by the columns of A is also satisfied by the columns of B and vice-versa. ■

Corollary 1 *The columns of A satisfy the same linear dependence relations as the columns of* $RREF(A)$.

Definition 2.28 *The columns of A corresponding to leading columns in* $RREF(A)$ *shall be called the leading columns of A.*

Corollary 2 *The leading columns of A are a basis for the column space of A, and if* $\mathbf{u}_i$ *is a non-leading column of A, then* $\mathbf{u}_i$ *can be expressed as a linear combination of the leading columns of A using as coefficients the entries in column i of* $RREF(A)$.

Example 2.31 *Let the matrix* $A = \begin{bmatrix} \mathbf{u}_1 & \mathbf{u}_2 & \mathbf{u}_3 & \mathbf{u}_4 & \mathbf{u}_5 \end{bmatrix}$ *and*

$$RREF(A) = \begin{bmatrix} \mathbf{v}_1 & \mathbf{v}_2 & \mathbf{v}_3 & \mathbf{v}_4 & \mathbf{v}_5 \end{bmatrix} = \begin{bmatrix} 1 & 0 & -3 & 0 & -1 \\ 0 & 1 & 2 & 0 & 4 \\ 0 & 0 & 0 & 1 & -2 \\ 0 & 0 & 0 & 0 & 0 \end{bmatrix}.$$

It is obvious that $\mathbf{v}_1$, $\mathbf{v}_2$, *and* $\mathbf{v}_4$ *are a basis for the column space of* $RREF(A)$, *and that* $\mathbf{v}_3 = -3\mathbf{v}_1 + 2\mathbf{v}_2$, *and* $\mathbf{v}_5 = -1\mathbf{v}_1 + 4\mathbf{v}_2 - 2\mathbf{v}_4$. *Thus we can conclude that* $\mathbf{u}_1$, $\mathbf{u}_2$, *and* $\mathbf{u}_4$ *are a basis of* $colsp(A)$, *and that* $\mathbf{u}_3 = -3\mathbf{u}_1 + 2\mathbf{u}_2$, *and* $\mathbf{u}_5 = -1\mathbf{u}_1 + 4\mathbf{u}_2 - 2\mathbf{u}_4$. *The same relations hold for the columns of any other matrix which has the same RREF as A.*

Example 2.32 *Let*

$$A = \begin{bmatrix} \mathbf{u}_1 & \mathbf{u}_2 & \mathbf{u}_3 & \mathbf{u}_4 \end{bmatrix} = \begin{bmatrix} 1 & 2 & 0 & 2 \\ -2 & -5 & 5 & 6 \\ 0 & -3 & 15 & 18 \\ 0 & -2 & 10 & 8 \\ 3 & 6 & 0 & 6 \end{bmatrix},$$

then

$$RREF(A) = \begin{bmatrix} \mathbf{v}_1 & \mathbf{v}_2 & \mathbf{v}_3 & \mathbf{v}_4 \end{bmatrix} = \begin{bmatrix} 1 & 0 & 10 & 0 \\ 0 & 1 & -5 & 0 \\ 0 & 0 & 0 & 1 \\ 0 & 0 & 0 & 0 \\ 0 & 0 & 0 & 0 \end{bmatrix}.$$

Clearly $\mathbf{v}_1$, $\mathbf{v}_2$, *and* $\mathbf{v}_4$ *are a basis of the column space of* $RREF(A)$, *moreover it is easy to see that* $\mathbf{v}_3 = 10\mathbf{v}_1 - 5\mathbf{v}_2$. *Thus we can conclude that* $\mathbf{u}_1$, $\mathbf{u}_2$, *and* $\mathbf{u}_4$ *are a basis of the column space of* A, *and that* $\mathbf{u}_3 = 10\mathbf{u}_1 - 5\mathbf{u}_2$.

Remark **(Warning)** *The leading columns of the RREF do not form a basis of* $colsp(A)$. $Colsp(RREF(A)) \neq colsp(A)$, *since elementary row operations on* (A) *may change* $colsp(A)$. *For a basis of* $colsp(A)$ *we must use the leading columns of* A.

Theorem 2.9 *For any matrix* A, *row* $rank(A) = $ *column* $rank(A) = $ *number of leading ones in* $RREF(A)$.

Proof.   Since by Lemma 2.13 the nonzero rows of the $RREF(A)$ are a basis of the row space of A, we know that the row rank of A equals the number of leading ones in $RREF(A)$. By Corollary 2 to Lemma 2.14 the columns of A corresponding to leading columns of $RREF(A)$ are a basis of $colsp(A)$, so the column rank of A also equals the number of leading ones in $RREF(A)$. ■

In view of this result we no longer need distinguish between row rank and column rank. We make the following definition.

Definition 2.29 *The rank of a matrix* A *equals the number of leading ones in* $RREF(A)$.

Since we now have row rank equals column rank equals rank, we can replace the terms row rank and column rank with rank in all previous results. In particular we can restate Lemma 2.12 as follows:

Lemma 2.15 $A\mathbf{x} = \mathbf{b}$ *has a solution if and only if* $rank(A) = rank\left(\begin{bmatrix} A & \mathbf{b} \end{bmatrix}\right).$

Procedure 2.2 To find the rank of a matrix A:

1. Find $RREF(A)$

2. Count the number of leading ones.

3. This gives the $rank(A)$ which is also the dimension of the row space and the dimension of the column space.

Sometimes one can find the rank of a matrix just by inspection.

Example 2.33 Let $A = \begin{bmatrix} 1 & 2 & 1 \\ 2 & 4 & 2 \\ -3 & -6 & -3 \end{bmatrix}$. It is obvious that the second and third rows are multiples of the first, therefore the rank of this matrix is 1. (In this case the same applies to the columns.)

Example 2.34 Let $A = \begin{bmatrix} 1 & 2 & 2 \\ 3 & 1 & 2 \\ 2 & 4 & 4 \end{bmatrix}$. The first two rows are linearly independent and the third row is a multiple of (twice) the first. Therefore the rank of this matrix is 2. This is not obvious from the columns, but we now know that the dimension of the column space of A is 2. Since the first two columns of A are linearly independent, the third column must be a linear combination of the first two.

Theorem 2.10 $Rowsp(AB) \subset rowsp(B)$, with equality if A is invertible.

Proof.   The rows of AB are linear combinations of the rows of B, so the first part is obvious. If A is invertible, we have $B = A^{-1}(AB)$, so that $rowsp(B) = rowsp\{A^{-1}(AB)\} \subset rowsp(AB)$, thus we have containment both ways so $rowsp(B) = rowsp(AB)$. ■

Theorem 2.11 $Colsp(AB) \subset colsp(A)$, with equality if B is invertible.

Proof.   The proof is similar to the previous theorem, and is left as an exercise. ∎

Theorem 2.12 $Rank(AB) \leq \min\{rank(A), rank(B)\}$. *and multiplication on either side by an invertible matrix does not change the rank.*

Proof.   Since $rowsp(AB) \subset rowsp(B)$, $rank(AB) \leq rank(B)$. Since $colsp(AB) \subset colsp(A)$, $rank(AB) \leq rank(A)$. This proves the first part of the theorem. If A is a square invertible matrix, then we have by Theorem 2.10 that $rowsp(AB) = rowsp(B)$, so $rank(AB) = rank(B)$. If B is a square invertible matrix then by Theorem 2.11, $colsp(AB) = colsp(A)$, so $rank(AB) = rank(A)$. ∎

If you want to find a basis for $W = sp\{u_1, u_2, \cdots u_n\}$, there are two ways to do it:

Procedure 2.3a

1. Put the given vectors as the columns of a matrix A.

2. Find $RREF(A)$.

3. The leading ones in the RREF tell you which of the columns in the original matrix to keep for your basis (Keep the leading columns of A).

Procedure 2.3b

1. Put the given vectors as the rows of a matrix C.

2. Find RREF(C).

3. The nonzero rows of the RREF(C) are then a basis of W.

The advantage of the first method is that it allows you to select a basis from the given vectors. The advantage of the second way is that it gives a very "nice" basis, one that allows you to see what kind of vectors are in W.

Example 2.35 *Let* $W = sp\{u_1, u_2, u_3, u_4\}$, *where* $u_1 = (1, -2, 0, 0, 3)$, $u_2 = (2, -5, -3, -2, 6)$, $u_3 = (0, 5, 15, 10, 0)$, *and* $u_4 = (2, 6, 18, 8, 6)$. *Find a basis for* W. *Using the first method, we put the vectors as the columns of a matrix* A. *Let*

$$A = \begin{bmatrix} u_1 & u_2 & u_3 & u_4 \end{bmatrix}$$

$$= \begin{bmatrix} 1 & 2 & 0 & 2 \\ -2 & -5 & 5 & 6 \\ 0 & -3 & 15 & 18 \\ 0 & -2 & 10 & 8 \\ 3 & 6 & 0 & 6 \end{bmatrix}$$

$$\sim \begin{bmatrix} 1 & 0 & 10 & 0 \\ 0 & 1 & -5 & 0 \\ 0 & 0 & 0 & 1 \\ 0 & 0 & 0 & 0 \\ 0 & 0 & 0 & 0 \end{bmatrix}.$$

Then $W = colsp(A)$. *The leading ones in columns 1,2,and 4 tell us that the set* $\{u_1, u_2, u_4\}$ *is a basis of* W.

Using the second method, we put the vectors u_i *as rows and let* $C = A^T$. *Then*

$$C = \begin{bmatrix} u_1 \\ u_2 \\ u_3 \\ u_4 \end{bmatrix} = \begin{bmatrix} 1 & -2 & 0 & 0 & 3 \\ 2 & -5 & -3 & -2 & 6 \\ 0 & 5 & 15 & 10 & 0 \\ 2 & 6 & 18 & 8 & 6 \end{bmatrix},$$

$$RREF(C) = \begin{bmatrix} 1 & 0 & 0 & -2 & 3 \\ 0 & 1 & 0 & -1 & 0 \\ 0 & 0 & 1 & 1 & 0 \\ 0 & 0 & 0 & 0 & 0 \end{bmatrix},$$

and the vectors $(1, 0, 0, -2, 3)$, $(0, 1, 0, -1, 0)$, $(0, 0, 1, 1, 0)$ *form a basis of* W. *Thus every vector in* W *is of the form*

$$a(1, 0, 0, -2, 3) + b(0, 1, 0, -1, 0) + c(0, 0, 1, 1, 0)$$

which simplifies to

$$(a, b, c, -2a - b + c, 3a).$$

We see that every vector in W must have its 5th component 3 times its first component. Thus the vector $(1, 2, 3, 1, 1)$ cannot be in W. The relation between the 4th component and the first three is a bit more complicated, namely $x_4 = -2x_1 - x_2 + x_3$. Thus $(1, 2, 3, 1, 3)$ is not in W either. But $(1, 2, 3, -1, 3)$ is in W (with $a = 1, b = 2, c = 3$). In fact W can now be described as the solution set of the homogeneous equations

$$2x_1 + x_2 - x_3 + x_4 = 0$$

and

$$-3x_1 + x_5 = 0.$$

Here are some hints on when to use columns and when to use rows:

1. If you are interested in the vectors themselves and relations among them, or in picking a basis from among the given vectors themselves, then write them as the columns of a matrix and row reduce.

2. If you are interested in the subspace W and what kind of vectors are in W, characterizing W by homogeneous equations, or comparing W with other subspaces, put the spanning set as rows of a matrix and row reduce.

3. If you are only interested in $\dim(W)$ put the spanning set as either rows or columns of a matrix, and $\dim(W)$ will equal the rank of the matrix.

Exercises

1. Let $A = \begin{bmatrix} 12 & 24 & 0 & 6 & 8 \\ 6 & 12 & 3 & 0 & 16 \\ 18 & 36 & 6 & 3 & 36 \end{bmatrix}$. Then

$$RREF(A) = \begin{bmatrix} 1 & 2 & 0 & \frac{1}{2} & \frac{2}{3} \\ 0 & 0 & 1 & -1 & 4 \\ 0 & 0 & 0 & 0 & 0 \end{bmatrix}.$$

If the columns of A are called u_1, u_2, u_3, u_4, u_5

(a) Find a basis of $colsp(A)$.

(b) Write each of the other columns as linear combinations of the basis found in answer to part a.

(c) What is the rank of (A)?

2. If $A = [u_1, u_2, u_3, u_4, u_5, u_6, u_7]$ and

$$RREF(A) = \begin{bmatrix} 1 & 2 & 0 & 3 & 1 & 0 & 3 \\ 0 & 0 & 1 & -2 & 2 & 0 & -1 \\ 0 & 0 & 0 & 0 & 0 & 1 & 2 \\ 0 & 0 & 0 & 0 & 0 & 0 & 0 \end{bmatrix},$$

answer the same three questions as in problem 1.

3. If $A = \begin{bmatrix} 5 & 1 & 1 & -2 & 0 & 21 \\ 6 & 2 & 0 & -2 & 1 & 30 \\ 12 & 3 & 2 & -4 & 1 & 54 \\ 8 & 2 & 1 & -3 & 1 & 37 \end{bmatrix} = [u_1, u_2, u_3, u_4, u_5, u_6],$ then

$$RREF(A) = \begin{bmatrix} 1 & 0 & 0 & -1 & 0 & 4 \\ 0 & 1 & 0 & 2 & 0 & 2 \\ 0 & 0 & 1 & 1 & 0 & -1 \\ 0 & 0 & 0 & 0 & 1 & 2 \end{bmatrix},$$

answer the same questions as in problem 1.

4. Let $C = \begin{bmatrix} 2 & 2 & 6 \\ 1 & 2 & 4 \\ 3 & 3 & 9 \end{bmatrix}$. Find rank$(C)$.

5. Let $A = \begin{bmatrix} 1 & 3 & 1 & -2 & -3 \\ 1 & 4 & 3 & -1 & -4 \\ 2 & 3 & -4 & -7 & -3 \\ 3 & 8 & 1 & -7 & -8 \end{bmatrix}.$

(a) Find a basis for the row space of A.

(b) Find a basis for the column space of A.

(c) Find the rank of A.

(d) Find a system of linear homogeneous equations that characterize $rowsp(A)$. (See Example 2.35, second method.)

6. Let $A = \begin{bmatrix} 1 & 0 & 0 & 0 \\ 0 & 1 & 2 & 3 \\ 0 & 0 & 1 & 2 \end{bmatrix}$. Find a linear homogeneous equation

that characterizes $rowsp(A)$. (See Example 2.35, second method.) Which of the following vectors are in $rowsp(A)$? $(1, 3, 0, 2)$, $(2, -1, 1, 3)$, $(1, 1, 2, 1)$, $(-1, 2, 1, 0)$.

7. Let W equal the subspace of R^4 spanned by the vectors $(3, -1, 2, 2)$, $(0, -4, -1, -2)$, $(-6, 2, -4, -4)$, $(-1, -4, -2, -1)$, $(7, -7, 3, 6)$.

   (a) What is the dimension of W?

   (b) Select a basis of W from among the given vectors and express the other vectors as linear combinations of these basis vectors.

Section 2.6 Null Space of A

We have discussed solving the linear system of equations $Ax = b$ by reducing the augmented matrix $[A \; b]$ to RREF using elementary row operations. The special case $Ax = O$ is of interest. We can solve it in the same way and we need not even carry the zero column of the augmented matrix along, since if a matrix has a zero column, that zero column will remain unchanged by elementary row operations.

Definition 2.30 *A system of linear equations is called homogeneous if the constant term in each equation is 0, i.e. the matrix form is $Ax = O$.*

We note that a system of homogeneous linear equations always has a solution, for one can choose all the variables equal to zero and this will be a solution. This is called the trivial solution. We will be interested in finding non-trivial solutions.

Lemma 2.16 *A non-trivial solution of $Ax = O$ exists if and only if the columns of A are linearly dependent.*

Proof.   Because of equation 2.2, a nontrivial solution of $Ax = O$ gives a dependence relation on the columns of A and vice-versa.   ∎

Lemma 2.17 $Ax = O$ *has only the trivial solution if and only if the RREF of A has a leading* 1 *in every column.*

Proof.   If there is a column of $RREF(A)$ with no leading 1, this variable can be chosen freely, so there is a non-trivial solution. If there is a leading 1 in every column of $RREF(A)$, all the variables are determined and must be 0.   ∎

Lemma 2.18 *The columns of A are linearly independent if and only if $RREF(A)$ has a leading* 1 *in every column.*

Proof.   Follows immediately from Lemmas 2.19 and 2.20.   ∎

Theorem 2.13 *If A is an $m \times n$ matrix, the solution set of $Ax = O$ is a subspace of R^n.*

Proof. Let $S =$ the solution set of $Ax = O$. We have already seen that the S is not empty, since it contains the zero vector. We need only check that S is closed under vector addition and scalar multiplication. Let **u** and **v** be in S. Then $Au = O$ and $Av = O$. Now $A(u+v) = Au + Av = O + O = O$. Also $A(ru) = (Ar)u = (rA)u = r(Au) = rO = O$.
∎

Definition 2.31 *The solution set of $Ax = O$ is called the null space of A, written $nullsp(A)$.*

Definition 2.32 *If S and T are sets of vectors, and if every vector in T is the sum of a vector in S and a fixed vector **v**, then we say T is a translation of S.*

Theorem 2.14 *Let **b** be a vector in $colsp(A)$, and let $T =$ the solution of set of $Ax = b$. Then $T = nullsp(A) + \{v\}$, where **v** is any fixed solution of $Ax = b$.*

Proof. Let u be a solution of $Ax = b$. Then $A(u - v) = Au - Av = b - b = O$, so that $u - v \in nullsp(A)$. Let $u - v = w$, then $u = w + v \in nullsp(A) + \{v\}$. Conversely let z be an element of $nullsp(A) + \{v\}$. Then $z = n + v$ for some vector n in nullsp(A), but then $Az = A(n + v) = An + Av = O + b = b$, so z is a solution of $Ax = b$. ∎

Remark *If* $b \neq O$, T *is not a subspace, since if* $Au = b$ *and* $Av = b$ *then* $A(u + v) = Au + Av = b + b = 2b$, *so* $u + v \notin T$. *However by the Theorem 2.14,* T *is a translation of* $nullsp(A)$.

Definition 2.33 *The nullity of* A *is the dimension of* $nullsp(A)$.

Lemma 2.19 *If* A *is an* $m \times n$ *matrix,* $nullity(A) \leq n - rank(A)$.

Proof. Let u be a vector in the null space of A. Then $Au = O$. This means that the dot product of u with each row of A is 0. Thus the dot product of u with every vector in the row space of A is 0. (See properties of the dot product in Section 2.1.) Let $S = rowsp(A)$ and $T = nullsp(A)$. If u is in $S \sqcap T$ then $u \cdot u = 0$, so $u = O$. Thus $S \sqcap T = \{O\}$, and using Theorem 1.12 we conclude that $\dim(S + T) = \dim(S) + \dim(T)$. Since A is $m \times n$, S and T are both subspaces of R^n, so we have $\dim(S + T) \leq n$. Thus $\dim(S) + \dim(T) \leq n$. Since $\dim(T) = nullity(A)$ and $\dim(S) = rank(A)$, we have $rank(A) + nullity(A) \leq n$. ∎

Remark *The previous lemma is a partial result on the way to the important theorem that* $rank(A) + nullity(A) = n$, *which we will soon prove.*

We now explain how to find a basis of the null space of a matrix A. We start with an example.

Example 2.36 *Suppose we are looking for a basis for the null space of*
A *and* $RREF(A) = \begin{bmatrix} 1 & 0 & -5 & 1 \\ 0 & 1 & 2 & -3 \\ 0 & 0 & 0 & 0 \\ 0 & 0 & 0 & 0 \end{bmatrix}$. *Then the augmented matrix*

of $A\mathbf{x} = \mathbf{O}$ *will reduce to*

$$\begin{bmatrix} 1 & 0 & -5 & 1 & 0 \\ 0 & 1 & 2 & -3 & 0 \\ 0 & 0 & 0 & 0 & 0 \\ 0 & 0 & 0 & 0 & 0 \end{bmatrix}.$$

We see that x_3 and x_4 can be chosen freely. Let $x_3 = s$, and $x_4 = t$. From the first row of the matrix we read off the equation $x_1 - 5x_3 + x_4 = 0$, or $x_1 = 5s - t$. From the second row we read the equation $x_2 + 2x_3 - 3x_4 = 0$, or $x_2 = -2s + 3t$. The general solution vector is

$$\begin{bmatrix} x_1 \\ x_2 \\ x_3 \\ x_4 \end{bmatrix} = \begin{bmatrix} 5s - t \\ -2s + 3t \\ s \\ t \end{bmatrix}.$$

We can write

$$\begin{bmatrix} 5s - t \\ -2s + 3t \\ s \\ t \end{bmatrix} = \begin{bmatrix} 5s \\ -2s \\ s \\ 0 \end{bmatrix} + \begin{bmatrix} -t \\ 3t \\ 0 \\ t \end{bmatrix}$$

$$= s \begin{bmatrix} 5 \\ -2 \\ 1 \\ 0 \end{bmatrix} + t \begin{bmatrix} -1 \\ 3 \\ 0 \\ 1 \end{bmatrix}.$$

The two vectors $\begin{bmatrix} 5 \\ -2 \\ 1 \\ 0 \end{bmatrix}$ *and* $\begin{bmatrix} -1 \\ 3 \\ 0 \\ 1 \end{bmatrix}$ *are a basis for* $nullsp(A)$.

Another way to obtain these vectors is to look at $\{I - RREF(A)\}$. *We have*

$$\{I - RREF(A)\} = \begin{bmatrix} 1 & 0 & 0 & 0 \\ 0 & 1 & 0 & 0 \\ 0 & 0 & 1 & 0 \\ 0 & 0 & 0 & 1 \end{bmatrix} - \begin{bmatrix} 1 & 0 & -5 & 1 \\ 0 & 1 & 2 & -3 \\ 0 & 0 & 0 & 0 \\ 0 & 0 & 0 & 0 \end{bmatrix}$$

$$= \begin{bmatrix} 0 & 0 & 5 & -1 \\ 0 & 0 & -2 & 3 \\ 0 & 0 & 1 & 0 \\ 0 & 0 & 0 & 1 \end{bmatrix}.$$

We see that the nonzero columns are a basis of the $nullsp(A)$.

Remark (**Warning**) *This trick will only work if A is square and if the leading ones of the RREF are on the diagonal. However we can modify this trick to work in all cases.*

Definition 2.34 *A matrix H is said to be in Hermite form if it satisfies the following four conditions:*

1. *H is square.*
2. *Each diagonal entry h_{ii} is either 0 or 1.*
3. *If $h_{ii} = 1$, the rest of column i is all zeros.*
4. *If $h_{ii} = 0$, then the ith row of H is the zero vector.*

If A is a square matrix, then by permuting the rows of $RREF(A)$ so that the leading ones are moved to the main diagonal, one obtains a matrix H in Hermite form. It's just as easy to change a matrix in Hermite form to its unique RREF by rearranging the rows. Moreover H is uniquely determined by $RREF(A)$. The matrix H is called the Hermite form of A.

If A is not square, suppose A is $m \times n$ and $m < n$. Find $RREF(A)$ and add $n - m$ zero rows. If $m > n$, then $RREF(A)$ must have at least $m - n$ zero rows, since there can only be at most n leading ones. In this case we omit $m - n$ zero rows. In either case we get a square matrix, uniquely determined by $RREF(A)$. Now if necessary rearrange the rows of A so that the leading ones are on the diagonal as before. By this method one obtains a unique matrix H having the same row space as A.

Definition 2.35 *The matrix obtained from $RREF(A)$ by the above procedure we will call the Hermite form of A, or $Hermite(A)$.*

Remark *Given A, $Hermite(A)$ is uniquely determined. This follows from the uniqueness of the RREF. If A is square, $Hermite(A)$ differs* *from $RREF(A)$ only by a rearrangement of the rows, and in this case the Hermite Normal Form is in the literature, in fact it predates the use of the RREF. In case A is $m \times n$, with $m \neq n$, $Hermite(A)$ is $n \times n$, and moreover the null spaces of $RREF(A)$ and $Hermite(A)$ are the same, since adding or deleting zero rows amounts to adding or deleting equations that simply say $0 = 0$. Thus A, $RREF(A)$, and $Hermite(A)$ all have the same null space as well as the same row space.*

Lemma 2.20 *If H is a matrix in Hermite Form, then $HH = H$.*

Proof.   Let $H = [h_{ij}]$, and let $\mathbf{u}_j$ be the row vectors of H. Then

$$
HH \;=\; H \begin{bmatrix} \mathbf{u}_1 \\ \mathbf{u}_2 \\ \vdots \\ \vdots \\ \mathbf{u}_n \end{bmatrix}
$$

$$
=\; \begin{bmatrix} h_{11}\mathbf{u}_1 + h_{12}\mathbf{u}_2 + \cdots + h_{1n}\mathbf{u}_n \\ \vdots \\ h_{j1}\mathbf{u}_1 + h_{j2}\mathbf{u}_2 + \cdots + h_{jn}\mathbf{u}_n \\ \vdots \\ h_{n1}\mathbf{u}_1 + h_{n2}\mathbf{u}_2 + \cdots + h_{nn}\mathbf{u}_n \end{bmatrix}.
$$

If $h_{jj} = 0$, then the jth row of H is the zero vector and so the jth row of HH is also the zero vector. If $h_{jj} = 1$, then for $k \neq j$ either $h_{jk} = 0$, in which case $h_{jk}\mathbf{u}_k = \mathbf{O}$, or $h_{jk} \neq 0$, in which case we must have $h_{kk} = 0$, and hence $\mathbf{u}_k = \mathbf{O}$, so that in either case $h_{jk}\mathbf{u}_k = \mathbf{O}$ for $k \neq j$. Thus the jth row of HH is just $1\mathbf{u}_j = \mathbf{u}_j$. Thus the zero rows of H are reproduced in HH and so are the nonzero rows of H. ∎

Lemma 2.21 *Let $H = Hermite(A)$. The non-zero columns of $H - I$ are a basis of $nullsp(A)$.*

Proof.   Since $HH = H$ we have $H(H - I) = HH - H = [0]$, so the columns of $H - I$ are all in $nullsp(H)$. The nonzero columns of $H - I$ are linearly independent. To see this recall that H is upper triangular and so is I, therefore $H - I$ is upper triangular. Thus no column which is nonzero on the main diagonal can be a linear combination of preceding columns. The nonzero columns of $H - I$ come from columns of H that had zero on the main diagonal, and so the nonzero columns of $H - I$ all have a -1 on the main diagonal. Thus they are linearly independent. That the set of nonzero columns of $H - I$ is a basis for $nullsp(A)$ now follows because there are $n - rank(A)$ of them and dimension of $nullsp(A) \leq n - rank(A)$ by Lemma 2.19. ∎

Example 2.37 *Let A be a matrix with*

$$RREF(A) = \begin{bmatrix} 1 & 2 & 0 & -3 \\ 0 & 0 & 1 & 4 \\ 0 & 0 & 0 & 0 \end{bmatrix}.$$

To find $H = Hermite(A)$ we add a zero row to $RREF(A)$ and rearrange the rows. This gives

$$H = \begin{bmatrix} 1 & 2 & 0 & -3 \\ 0 & 0 & 0 & 0 \\ 0 & 0 & 1 & 4 \\ 0 & 0 & 0 & 0 \end{bmatrix}.$$

Then

$$H - I = \begin{bmatrix} 0 & 2 & 0 & -3 \\ 0 & -1 & 0 & 0 \\ 0 & 0 & 0 & 4 \\ 0 & 0 & 0 & -1 \end{bmatrix},$$

and the nonzero columns of this matrix are a basis of the nullspace of A.

Remark   *Clearly we could use the nonzero rows of $I - H$ instead of $H - I$ as the basis of $nullsp(A)$, if we prefer, but $H - I$ is easier, since we don't have to change the signs of all the entries of H.*

Procedure 2.4 To find a basis for $nullsp(A)$:

1. Find $RREF(A)$.

2. Add or delete zero rows to change $RREF(A)$ into a square matrix.

3. Rearrange the rows if necessary so that the leading ones are on the diagonal (the matrix you now have is the Hermite form of A, call it H).

4. Subtract the identity matrix from H (form $H - I$).

5. The nonzero columns of the resulting matrix are a basis of $nullsp(A)$.

Remark   *If you have a linear combination of the columns of A which equals $\mathbf{O}$ (dependence relation), the coefficients give an n-tuple in $nullsp(A)$.*

Using this fact you can sometimes get the nullspace quickly. For example if $A = \begin{bmatrix} 1 & 0 & 2 \\ 2 & 1 & 4 \\ 1 & 0 & 2 \end{bmatrix}$, the last column is twice the first, so $\begin{bmatrix} 2 \\ 0 \\ -1 \end{bmatrix}$ is in the null space. Clearly $rank(A) = 2$, thus $nullity(A) = 1$, so $\begin{bmatrix} 2 \\ 0 \\ -1 \end{bmatrix}$ is a basis of nullsp(A).

Remark   *You can find a basis of $rowsp(A)$, $colsp(A)$, and $nullsp(A)$ with just one row reduction. Once you have $RREF(A)$, the nonzero rows are a basis of $rowsp(A)$, the leading columns of A are a basis of $colsp(A)$, and the nonzero columns of $Hermite(A) - I$ are a basis of $nullsp(A)$.*

Theorem 2.15 *If A is $m \times n$, $rank(A) + nullity(A) = n$.*

Proof.   Since the number of nonzero column vectors of $H - I = n - rank(A)$, and we have proved in the preceding theorem that these nonzero column vectors are a basis of $nullsp(A)$, it follows that $nullity(A) = n - rank(A)$. ∎

Corollary 1 *If A is an $m \times n$ matrix, $\{basis\ of\ rowsp(A)\} \sqcup \{basis\ of\ nullsp(A)\}$ is a basis of R^n.*

Proof. Let $S = rowsp(A)$ and let $T = nullsp(A)$. Since $S \sqcap T = \{\mathbf{O}\}$
(see proof of Lemma 2.19) we have $\dim(S+T) = \dim(S)+\dim(T) = n$,
by Theorem 2.15. Thus $S+T = R^n$. Thus $\{\text{basis of } rowsp(A)\} \sqcup \{\text{basis}$
of $nullsp(A)\}$ is a spanning set of R^n containing n vectors, and hence
is a basis of R^n. ∎

The preceding corollary gives a better way to extend a linearly independent set in R^n to a basis of R^n than the method of Procedure
1.2.

Procedure 2.5 To extend a linearly independent set of vectors in
R^n to a basis of R^n:

1. Put the linearly independent set as the rows of a matrix A.

2. Find a basis of $nullsp(A)$.

3. The original set of vectors and the basis of $nullsp(A)$ together
   form a basis of R^n.

Exercises

For the matrices in problems 1 through 4, find the Hermite form of A
and a basis for $nullsp(A)$:

1. $A = \begin{bmatrix} 1 & 0 & -1 & 0 \\ 0 & 1 & 4 & 0 \\ 0 & 0 & 0 & 1 \\ 0 & 0 & 0 & 0 \end{bmatrix}$.

2. $A = \begin{bmatrix} 1 & 2 & 0 \\ 0 & 0 & 1 \\ 0 & 0 & 0 \end{bmatrix}$

3. $A = \begin{bmatrix} 1 & 0 & 0 & -4 \\ 0 & 1 & 0 & 2 \\ 0 & 0 & 1 & 1 \end{bmatrix}$

4. $A = \begin{bmatrix} 1 & 0 & 0 \\ 0 & 1 & 0 \\ 0 & 0 & 1 \\ 0 & 0 & 0 \end{bmatrix}$.

5. Prove that if $Ax = O$ has only the trivial solution and A is square, then $RREF(A) = I = Hermite(A)$.

6. Let $A = \begin{bmatrix} 12 & 24 & 0 & 6 & 8 \\ 6 & 12 & 3 & 0 & 16 \\ 18 & 36 & 6 & 3 & 36 \end{bmatrix}$. Then

$$RREF(A) = \begin{bmatrix} 1 & 2 & 0 & \frac{1}{2} & \frac{2}{3} \\ 0 & 0 & 1 & -1 & 4 \\ 0 & 0 & 0 & 0 & 0 \end{bmatrix}.$$

Find a basis for $nullsp(A)$.

7. If

$$RREF(A) = \begin{bmatrix} 1 & 2 & 0 & 3 & 1 & 0 & 3 \\ 0 & 0 & 1 & -2 & 2 & 0 & -1 \\ 0 & 0 & 0 & 0 & 0 & 1 & 2 \\ 0 & 0 & 0 & 0 & 0 & 0 & 0 \end{bmatrix},$$

find a basis for $nullsp(A)$.

8. If $A = \begin{bmatrix} 5 & 1 & 1 & -2 & 0 & 21 \\ 6 & 2 & 0 & -2 & 1 & 30 \\ 12 & 3 & 2 & -4 & 1 & 54 \\ 8 & 2 & 1 & -3 & 1 & 37 \end{bmatrix}$, then

$$RREF(A) = \begin{bmatrix} 1 & 0 & 0 & -1 & 0 & 4 \\ 0 & 1 & 0 & 2 & 0 & 2 \\ 0 & 0 & 1 & 1 & 0 & -1 \\ 0 & 0 & 0 & 0 & 1 & 2 \end{bmatrix}.$$

Find a basis for $nullsp(A)$.

9. Let $A = \begin{bmatrix} 1 & 3 & 1 & -2 & -3 \\ 1 & 4 & 3 & -1 & -4 \\ 2 & 3 & -4 & -7 & -3 \\ 3 & 8 & 1 & -7 & -8 \end{bmatrix}$.

(a) Find a basis for the row space of A.

(b) Find a basis for the column space of A.

(c) Find a basis for the null space of A.

10. Let $A = \begin{bmatrix} 1 & 2 & 2 \\ 3 & 1 & 2 \\ 2 & 4 & 4 \end{bmatrix}$, find a basis for $nullsp(A)$.

Section 2.7 Elementary Matrices

Definition 2.36 *An elementary matrix is an $n \times n$ matrix that can be obtained from the $n \times n$ identity matrix by a single elementary row operation.*

Lemma 2.22 *Left multiplication of any $m \times n$ matrix A by an $m \times m$ elementary matrix E results in a matrix that comes from A by the same elementary row operation that was performed on I to get E.*

Proof. Let $\mathbf{u}_1, \mathbf{u}_2, \cdots, \mathbf{u}_m$ be the rows of A. If row j of I is multiplied by a number c, we get the matrix E with 1 on the diagonal except for the jj position which has c, and zeros elsewhere. Then EA will have the same rows as A, except for row j, which will be multiplied by c. For example

$$EA = \begin{bmatrix} 1 & 0 & 0 & 0 \\ 0 & c & 0 & 0 \\ 0 & 0 & 1 & 0 \\ 0 & 0 & 0 & 1 \end{bmatrix} \begin{bmatrix} \mathbf{u}_1 \\ \mathbf{u}_2 \\ \mathbf{u}_3 \\ \mathbf{u}_4 \end{bmatrix} = \begin{bmatrix} \mathbf{u}_1 \\ c\mathbf{u}_2 \\ \mathbf{u}_3 \\ \mathbf{u}_4 \end{bmatrix},$$

using the result that the rows of EA are linear combinations of the rows of A, with coefficients for row i of EA coming from row i of E. If E comes from the identity matrix by adding c times row j to row i, we have $E = I + cE_{ij}$, so that E looks like the $m \times m$ identity matrix with one extra nonzero entry c in the ij position. In this case the rows of EA will be the rows of A except for row i, which will be $\mathbf{u}_i + c\mathbf{u}_j$. For example

$$EA = \begin{bmatrix} 1 & 0 & c & 0 \\ 0 & 1 & 0 & 0 \\ 0 & 0 & 1 & 0 \\ 0 & 0 & 0 & 1 \end{bmatrix} \begin{bmatrix} \mathbf{u}_1 \\ \mathbf{u}_2 \\ \mathbf{u}_3 \\ \mathbf{u}_4 \end{bmatrix} = \begin{bmatrix} \mathbf{u}_1 + c\mathbf{u}_3 \\ \mathbf{u}_2 \\ \mathbf{u}_3 \\ \mathbf{u}_4 \end{bmatrix}.$$

Last if E results from the $m \times m$ identity by exchanging two rows, say row i and j, then row i of E has all zeros except for a one in the ij position, so row i of EA is $\mathbf{u}_j$, row j of E has all zeros except for a one in the ji position, so row j of EA is $\mathbf{u}_i$. The remaining rows of E are just like the $m \times m$ identity, so the remaining rows of EA are just equal to the corresponding rows of A. For example

$$EA = \begin{bmatrix} 1 & 0 & 0 & 0 \\ 0 & 0 & 1 & 0 \\ 0 & 1 & 0 & 0 \\ 0 & 0 & 0 & 1 \end{bmatrix} \begin{bmatrix} \mathbf{u}_1 \\ \mathbf{u}_2 \\ \mathbf{u}_3 \\ \mathbf{u}_4 \end{bmatrix} = \begin{bmatrix} \mathbf{u}_1 \\ \mathbf{u}_3 \\ \mathbf{u}_2 \\ \mathbf{u}_4 \end{bmatrix}.$$

■

Lemma 2.23 *Each elementary matrix is invertible and its inverse is an elementary matrix of the same kind.*

Proof.   If E is an elementary matrix that comes from I (the identity matrix) by multiplying row j by the nonzero number c, and if E' is the elementary matrix that comes from I by multiplying row j by $1/c$, then $E'E = I$, since E' multiplies row j of E by $1/c$. If E is of the second type, that is if E comes from I by adding c times row j to row i, then let E' be the matrix that comes from I by adding $-c$ times row j to row i. Then $E'E = I$. If E comes from I by exchanging two rows, then clearly a second exchange of the same two rows gets back the identity matrix, so that $EE = I$. ■

Theorem 2.16 *If A is an $n \times n$ matrix the following are equivalent:*
   1. *A has a left inverse*
   2. *$A\mathbf{x} = \mathbf{O}$ has only the trivial (zero) solution*
   3. *$RREF(A) = I$*
   4. *$A = product$ of elementary matrices*
   5. *A is invertible.*

Proof. We will show $1 \Rightarrow 2 \Rightarrow 3 \Rightarrow 4 \Rightarrow 5 \Rightarrow 1$. Assume that A has a left inverse. Then there exists an $n \times n$ matrix B such that $BA = I$. Now if $Ax = O$ we can multiply both sides on the left by B to get $B(Ax) = BO = O$. Using the associative law for matrix multiplication gives $(BA)x = O$. But $BA = I$, so $Ix = O$, whence $x = O$. Thus the only solution of $Ax = O$ is $x = O$. Now assume $Ax = O$ has only the trivial solution. Then every column of $RREF(A)$ must be a leading column, since non-leading columns result in free variables, which would give nonzero solutions. So every column of $RREF(A)$ has a leading 1. Since A has the same number of rows and columns, each row of A has a leading 1, and these must occur on the diagonal, since no column is skipped. Since the columns with leading ones are zeros elsewhere, this shows that $RREF(A)$ is just the $n \times n$ identity matrix. This proves $2 \Rightarrow 3$. To see that $3 \Rightarrow 4$, assume that $RREF(A) = I$. We know there is a sequence of elementary row operations that change A into $RREF(A)$, and each of these elementary row operations can be done by multiplying A on the left by an elementary matrix. Let $E_1, E_2, \cdots, E_k$ be a sequence of elementary matrices such that $E_k E_{k-1} \cdots E_2 E_1 A = I$. Multiplying both sides on the left by $E_1^{-1} E_2^{-1} \cdots E_k^{-1}$, we see that $A = E_1^{-1} E_2^{-1} \cdots E_k^{-1}$. Since the inverse of an elementary matrix is an elementary matrix, this is a product of elementary matrices. To see that $4 \Rightarrow 5$, recall that a product of invertible matrices is invertible (Corollary to Theorem 2.5) so that if A is equal to a product of elementary matrices then A is invertible. Last, $5 \Rightarrow 1$ is obvious since when we say A is invertible, we mean A has a two-sided inverse (Definition 2.13), so clearly A has a left inverse. ■

Suppose A is an invertible $n \times n$ matrix, and let $E_1, E_2, \cdots, E_k$ be the elementary matrices corresponding to a sequence of elementary row operations that change A into $RREF(A)$. Then $RREF(A) = E_k E_{k-1} \cdots E_2 E_1 A$. Now if A is invertible, $RREF(A) = I$, so we have $E_k E_{k-1} \cdots E_2 E_1 A = I$. Multiply both sides on the right by A^{-1}. This gives $E_k E_{k-1} \cdots E_2 E_1 A A^{-1} = I A^{-1}$, or $E_k E_{k-1} \cdots E_2 E_1 I = A^{-1}$. This last equation says that **the same sequence of elementary row operations that change A into I will change I into A^{-1}.** Based on this we have the following method for finding A^{-1}.

Procedure 2.6 To find the inverse of an $n \times n$ matrix A:

1. Form an $n \times 2n$ matrix by putting A and the $n \times n$ identity matrix

I side by side, $\begin{bmatrix} A & I \end{bmatrix}$.

2. Row reduce as you do when finding $RREF(A)$ until the first n columns are in RREF.

3. If $RREF(A) = I$, then A is invertible and A^{-1} will appear in columns $n + 1$ to $2n$.

4. If A is not invertible, you will discover this because $RREF(A)$ will not be I.

For a square matrix such as A, $RREF(A)$ is either the identity matrix or else has at least one zero row. In row reducing $\begin{bmatrix} A & I \end{bmatrix}$, if a zero row appears in the A part at any time, you know that A is not invertible. Otherwise you can continue until the A part becomes I, then you will have $\begin{bmatrix} A & I \end{bmatrix} \sim \begin{bmatrix} I & A^{-1} \end{bmatrix}$.

Remark   *If A is an invertible matrix, $A\mathbf{x} = \mathbf{b}$ implies that $\mathbf{x} = A^{-1}\mathbf{b}$, so we can solve this system of equations by inverting the coefficient matrix and multiplying A^{-1} times $\mathbf{b}$. It is usually more efficient to row reduce the augmented matrix $\begin{bmatrix} A & \mathbf{b} \end{bmatrix}$, but if you are solving $A\mathbf{x} = \mathbf{b}$ for many different vectors $\mathbf{b}$ with the same coefficient matrix A, it may be worthwhile to find A^{-1}.*

Procedure 2.7 For a 2×2 matrix $A = \begin{bmatrix} a & b \\ c & d \end{bmatrix}$ with $ad - bc \neq 0$, the inverse is

$$\frac{1}{ad - bc} \begin{bmatrix} d & -b \\ -c & a \end{bmatrix} = \begin{bmatrix} \frac{d}{ad-bc} & \frac{-b}{ad-bc} \\ \frac{-c}{ad-bc} & \frac{a}{ad-bc} \end{bmatrix},$$

which you can check by multiplying the two matrices.

We have shown in Theorem 2.16 that if A has a left inverse then A is invertible. Because of Theorem 2.4 the left inverse must equal the two sided inverse. Thus we have proved:

Theorem 2.17 *If A and B are $n \times n$ matrices with $BA = I$ then $AB = I$ also. In other words a left inverse of A must be a two-sided inverse of A.*

If C is a right inverse of A, i.e. if $AC = I$, then A is a left inverse of C and so by the preceding theorem we have $CA = I$, whence A is invertible and $C = A^{-1}$. So we have proved:

Theorem 2.18 *If A is an $n \times n$ matrix, a right inverse of A must also be a two-sided inverse of A.*

Here are some more properties of A that are equivalent to those in Theorem 2.16.

Theorem 2.19 *If A is an $n \times n$ matrix, the following are equivalent.*
1. $RREF(A) = I$
2. $Rank(A) = n$
3. $Rowsp(A) = R^n$
4. *Rows of A are a basis of R^n*
5. $Colsp(A) = R^n$
6. *Columns of A are a basis of R^n*
7. $A\mathbf{x} = \mathbf{b}$ *has a unique solution for all* $\mathbf{b} \in R^n$
8. $A\mathbf{x} = \mathbf{b}$ *has a solution for all* $\mathbf{b} \in R^n$
9. $A\mathbf{x} = \mathbf{b}$ *has a unique solution for at least one* $\mathbf{b} \in R^n$.

Proof. We will prove $1 \Rightarrow 2 \Rightarrow 3 \Rightarrow 4 \Rightarrow 5 \Rightarrow 6 \Rightarrow 7 \Rightarrow 8 \Rightarrow 1$, and then prove $1 \Longleftrightarrow 9$.

If $RREF(A) = I$ then $rank(A) = n$, by definition of rank. If $rank(A) = n$, then we have dimension $rowsp(A) = n$, but $rowsp(A)$ is a subspace of R^n, and since they have the same dimension, we have $rowsp(A) = R^n$. If $rowsp(A) = R^n$, then since the n rows of A are a spanning set of R^n, a space of dimension n, by Corollary 6 to Theorem 1.10 (Replacement Theorem), the rows of A are a basis of R^n. Assumming the rows of A are a basis of R^n, then the dimension of $rowsp(A) = n$, whence dimension of $colsp(A) = n$ also, so $colsp(A) = R^n$. If $colsp(A) = R^n$, then since the n columns of A are a spanning set of R^n, they must be a basis of R^n using again Corollary 6 of the Replacement Theorem. If the columns of A are a basis of R^n, then $A\mathbf{x} = \mathbf{b}$ has a unique solution for all $\mathbf{b} \in R^n$, because each vector can be written uniquely as a linear combination of basis vectors. Clearly $7 \Rightarrow 8$. To see that $8 \Rightarrow 1$, let $\mathbf{u}_i$ be the solution to $A\mathbf{x} = \mathbf{e}_i$, (where $\mathbf{e}_i$ are columns of the identity matrix). Then $A \begin{bmatrix} \mathbf{u}_1 & \mathbf{u}_2 & \cdots & \mathbf{u}_n \end{bmatrix} = I$, so A is invertible, and hence $RREF(A) = I$. Last to show that $1 \iff 9$. If $RREF(A) = I$, we have $A\mathbf{x} = \mathbf{b}$ has a unique solution for every $\mathbf{b}$, and so for at least one $\mathbf{b}$. If $RREF(A) \neq I$, then since A is square, there must be at least one zero row, hence at least one column with no leading one. Thus we have at least one free variable, and whenever a solution exists to $A\mathbf{x} = \mathbf{b}$, many solutions exist. ■

Coding and decoding

One of the most guarded secrets of any country concerns the coding and decoding of secret messages. Its necessity in times of war is obvious. "Bodyguard of Lies" by Anthony C. Brown is a fascinating book about code breaking by the Allies during World War II and what a vital part of winning the war against the Axis this was. It reads like a spy novel of the highest caliber, but all based on historical truth.

The agencies which are in charge of these activities like to stay out of the limelight, but much of this information did come out, albeit many years after the events. In the United States coding and decoding is the business of the National Security Agency, which occupies a large complex in the D.C. area. Its budget is never revealed to the general public and only a few select members of Congress get this information. Mathematicians have been involved in various aspects of these activities and one of the most prominent was Alan Turing who designed a monster computer during the war whose sole purpose was breaking the German codes.

A very simple way to code a message is to assign to each letter of the alphabet a different letter, for example if we assign to the letter s the letter b, then every time an s occurs in the message we replace it with b. Here is a message that has been coded in this way: GTJ GMHNYWJ OLGT GTJ OHMWU LP GTKG GTJMJ LP BHG K HBJ GH HBJ ZHMMJPXHBUJBZJ YJGOJJB THMPJP KBU THMPJP' KPPJP. AHTB DUB BJNSKBB In this message GTH occurs twice and stands for the same three letter word each time. Have some fun and try to decode this message. It isn't very hard. Certain words of short length occur frequently and there are letters which turn up more often than others and that also gives a clue. (ETAOINSHRDLU is probably a list in order of frequency of the most common letters in English.)

The first way to make decoding of the above type of code more difficult is to omit the separation of words. If that were done with the coded message above, decoding would have become much more difficult, though no great challenge to an expert, because certain sequences of 2 and 3 letters would still be plain, and also the frequency of a given letter could still be observed. A more interesting idea is to assign a number to each letter of the alphabet, for example $a = 1, b = 2, \cdots, z = 26$. However since 3 stands for c each time 3 appears in the message, frequencies and sequences can still be used as a clue to which number stands for which letter. If we now change the numbers three at a time by multiplying by an invertible 3x3 matrix, then a 3 in the final string of numbers no longer means there was a c in that position in the message. For example we could use the matrix

$$A = \begin{bmatrix} 1 & 0 & 1 \\ 1 & 1 & 0 \\ 1 & 1 & 1 \end{bmatrix}.$$

We encode the message "come home" as follows. First substituting numbers for the letters we get "3 15 13 5 8 15 13 5". Then multiply

$$\begin{bmatrix} 3 & 15 & 13 \end{bmatrix} \begin{bmatrix} 1 & 0 & 1 \\ 1 & 1 & 0 \\ 1 & 1 & 1 \end{bmatrix} = \begin{bmatrix} 31 & 28 & 16 \end{bmatrix},$$

$$\begin{bmatrix} 5 & 8 & 15 \end{bmatrix} \begin{bmatrix} 1 & 0 & 1 \\ 1 & 1 & 0 \\ 1 & 1 & 1 \end{bmatrix} = [28\ 23\ 20],$$

and

$$\begin{bmatrix} 13 & 5 & 0 \end{bmatrix} \begin{bmatrix} 1 & 0 & 1 \\ 1 & 1 & 0 \\ 1 & 1 & 1 \end{bmatrix} = \begin{bmatrix} 18 & 5 & 13 \end{bmatrix}.$$

The coded message now reads "31 28 16 28 23 20 18 5 13". Note that the first time 28 appears it stands for o, and the second time it appears it stands for e. The first m in the message is represented by 16, while the second m is represented by 18. Thus frequencies no longer give away certain letters, and sequences no longer give away certain frequent short words. However if you know what matrix is being used, you can use A^{-1} to undo the scrambling. In this case

$$A^{-1} = \begin{bmatrix} 1 & 1 & -1 \\ -1 & 0 & 1 \\ 0 & -1 & 1 \end{bmatrix},$$

and

$$\begin{bmatrix} 31 & 28 & 16 \end{bmatrix} \begin{bmatrix} 1 & 1 & -1 \\ -1 & 0 & 1 \\ 0 & -1 & 1 \end{bmatrix} = \begin{bmatrix} 3 & 15 & 13 \end{bmatrix},$$

the first three numbers or the original sequence.

While this method of coding is not foolproof, it is harder to decode than the first method. We scrambled by writing triples as rows and multiplying by A on the right. We could have written triples as columns and multiplied by A on the left. Either way is OK, but you have to know which is being used and do it the same way when decoding.

Finding coordinates with respect to a given basis

Let $B = \{v_1, v_2, \cdots, v_n\}$ be a basis for R^n, and u a vector in R^n. Each vector of u can be written uniquely as a linear combination

$$u = a_1 v_1 + a_2 v_2 + \cdots a_n v_n.$$

Writing u and the v_i as columns,we can write this in matrix notation as

$$u = \begin{bmatrix} v_1 & v_2 & \cdots & v_n \end{bmatrix} \begin{bmatrix} a_1 \\ a_2 \\ \vdots \\ a_n \end{bmatrix}.$$

We call the column vector $\begin{bmatrix} a_1 \\ a_2 \\ \vdots \\ a_n \end{bmatrix}$ the coordinate vector of $\mathbf{u}$ with respect

to the basis B, and use the notation

$$[\mathbf{u}]_B = \begin{bmatrix} a_1 \\ a_2 \\ \vdots \\ a_n \end{bmatrix}.$$

Then we can rewrite the previous equation as

$$\mathbf{u} = \begin{bmatrix} \mathbf{v}_1 & \mathbf{v}_2 & \cdots & \mathbf{v}_n \end{bmatrix} [\mathbf{u}]_B.$$

Since the vectors $\mathbf{v}_1, \mathbf{v}_2, \cdots, \mathbf{v}_n$ form a basis of R^n, the matrix $\begin{bmatrix} \mathbf{v}_1 & \mathbf{v}_2 & \cdots & \mathbf{v}_n \end{bmatrix}$ is invertible, and we see that

$$[\mathbf{u}]_B = \begin{bmatrix} \mathbf{v}_1 & \mathbf{v}_2 & \cdots & \mathbf{v}_n \end{bmatrix}^{-1} \mathbf{u}.$$

Thus we can find the coordinates of $\mathbf{u}$ with respect to the basis B simply by multiplying the vector $\mathbf{u}$ by the inverse of the matrix whose columns are the vectors in the basis B. If we pervert notation slightly and let $B = \begin{bmatrix} \mathbf{v}_1 & \mathbf{v}_2 & \cdots & \mathbf{v}_n \end{bmatrix}$ as well as $B = \{\mathbf{v}_1, \mathbf{v}_2, \cdots, \mathbf{v}_n\}$, then we can write

$$[\mathbf{u}]_B = B^{-1}\mathbf{u}$$

. Let $B' = \begin{bmatrix} \mathbf{v}'_1 & \mathbf{v}'_2 & \cdots & \mathbf{v}'_n \end{bmatrix}$ be a second basis of R^n, then

$$\mathbf{u} = B'[\mathbf{u}]_{B'} = B[\mathbf{u}]_B,$$

so we have an equation relating $[\mathbf{u}]_B$ and $[\mathbf{u}]_{B'}$, and we can solve for either one in terms of the other and the two bases. Note if we solve for $[\mathbf{u}]_B$ we get $[\mathbf{u}]_B = B^{-1}B'[\mathbf{u}]_{B'}$. The matrix $P = B^{-1}B'$ is sometimes called the transition matrix from the basis B' to the basis B, because it changes coordinates form basis B' coordinates to basis B coordinates. One can see that since the columns of P are B^{-1} times the columns of B', and since multiplying a vector by B^{-1} gives the coordinates of that vector with respect to B, the columns of P are the coordinates of the B' basis vectors with respect to the B basis. However all of this is easily derived if one just remembers the matrix equation

$$B[\mathbf{u}]_B = \mathbf{u}$$

where B is the matrix whose columns are the vectors in the basis B.

Exercises

1. Let $C = \begin{bmatrix} 2 & 2 & 6 \\ 1 & 2 & 4 \\ 3 & 3 & 9 \end{bmatrix}$. Find rank$(C)$.

2. Find the inverse of each of the following matrices, if it exists:

(a) $A = \begin{bmatrix} 1 & 1 & 1 \\ 1 & 1 & -2 \\ 2 & 1 & 1 \end{bmatrix}$

(b) $B = \begin{bmatrix} 1 & -1 & 0 \\ 1 & 4 & 0 \\ 0 & 2 & 2 \end{bmatrix}$

(c) $C = \begin{bmatrix} 1 & 1 & -1 \\ 2 & 5 & -1 \\ 2 & 0 & -3 \end{bmatrix}$

(d) $D = \begin{bmatrix} 1 & 3 & 5 \\ -1 & -2 & 1 \\ 2 & 7 & 6 \end{bmatrix}$

(e) $F = \begin{bmatrix} 3 & 2 \\ 5 & 1 \end{bmatrix}$.

3. Do the triples $(1, 2, 2)$, $(1, 5, 0)$, $(-1, -1, -3)$ form a basis of R^3? (Compare with 2c.)

4. 4. Assuming the same method and same matrix is being used as in the example in the section on coding above, decode the message "38 19 33 32 28 19 44 30 39".

5. Find the coordinates of the vector $(4, 4, 8)$ with respect to the basis $B = \{(1, 2, -1), (2, 1, -3), (1, 0, 1)\}$.

6. Let

$$B = \{(1, 1, 2), (0, 1, 3), (0, 2, 3)\}$$

$$B' = \{(0, 1, 1), (2, 3, 0), (1, 0, 3)\}.$$

Each of these sets is a basis of R^3. If

$$[\mathbf{w}]_B = \begin{bmatrix} 1 \\ 1 \\ 2 \end{bmatrix},$$

find $\mathbf{w}$ and find $[\mathbf{w}]_{B'}$.

Section 2.8 Equivalence

In section 2.3 we defined elementary row operations. We can define elementary column operations similarly.

Definition 2.37 *Given an $m \times n$ matrix A, the following changes are called elementary column operations:*

1. *Multiply a column of A by a nonzero scalar.*
2. *Add a multiple of one column of A to another column of A.*
3. *Exchange any two columns of A.*

Clearly an elementary column operation on A can be performed by performing an elementary row operation on A^T and then transposing the result. Just as an elementary row operation can be performed by multiplying on the left by an elementary matrix, an elementary column operation can be performed by multiplying on the right by an elementary matrix. The transpose of an elementary matrix is an elementary matrix, and if $EA = B$, then $A^T E^T = B^T$. Since B comes from A by an elementary row operation, B^T comes from A^T by an elementary column operation.

Example 2.38 *Let* $A = \begin{bmatrix} 1 & 2 & 3 \\ 4 & 5 & 6 \\ 7 & 8 & 9 \end{bmatrix}$ *and* $E = \begin{bmatrix} 1 & 2 & 0 \\ 0 & 1 & 0 \\ 0 & 0 & 1 \end{bmatrix}$. *Then*

$$EA = \begin{bmatrix} 1 & 2 & 0 \\ 0 & 1 & 0 \\ 0 & 0 & 1 \end{bmatrix} \begin{bmatrix} 1 & 2 & 3 \\ 4 & 5 & 6 \\ 7 & 8 & 9 \end{bmatrix} = \begin{bmatrix} 9 & 12 & 15 \\ 4 & 5 & 6 \\ 7 & 8 & 9 \end{bmatrix} = B, \text{ and } A^T E^T =$$

$$\begin{bmatrix} 1 & 4 & 7 \\ 2 & 5 & 8 \\ 3 & 6 & 9 \end{bmatrix} \begin{bmatrix} 1 & 0 & 0 \\ 2 & 1 & 0 \\ 0 & 0 & 1 \end{bmatrix} = \begin{bmatrix} 9 & 4 & 7 \\ 12 & 5 & 8 \\ 15 & 6 & 9 \end{bmatrix} = B^T.$$

Definition 2.38 *Two* $m \times n$ *matrices* A *and* B *are equivalent if there exists a sequence of row and column operations that transforms* A *into* B.

Lemma 2.24 *Let* A *and* B *be* $m \times n$ *matrices.* A *is equivalent to* B *if and only if* $B = PAQ$ *for invertible matrices* P *(size* $m \times m$*) and* Q *(size* $n \times n$*).*

Proof.   This follows from the fact that both elementary row operations and elementary column operations can be done by multiplication by elementary matrices, and the fact that a matrix P is invertible if and only if P is a product of elementary matrices (Theorem 2.16). ∎

Theorem 2.20 *Every* $m \times n$ *matrix* A *is equivalent to an* $m \times n$ *matrix of the form* $\begin{bmatrix} I_{r \times r} & [0] \\ [0] & [0] \end{bmatrix}$

Proof.   First transform A by elementary row operations to B, a matrix in RREF (B is the RREF form of A, but the uniqueness of the RREF is not needed for this proof). Then transpose B and use elementary row operations on B^T to transform it to a matrix C in RREF. Then C^T is of the required form and C^T is equivalent to A. ∎

Example 2.39 *If* A *is a matrix with*

$$RREF(A) = B = \begin{bmatrix} 1 & 0 & 2 & 3 & 0 \\ 0 & 1 & 4 & 5 & 0 \\ 0 & 0 & 0 & 0 & 1 \\ 0 & 0 & 0 & 0 & 0 \end{bmatrix},$$

then

$$B^T = \begin{bmatrix} 1 & 0 & 0 & 0 \\ 0 & 1 & 0 & 0 \\ 2 & 4 & 0 & 0 \\ 3 & 5 & 0 & 0 \\ 0 & 0 & 1 & 0 \end{bmatrix}.$$

$$C = RREF(B^T) = \begin{bmatrix} 1 & 0 & 0 & 0 \\ 0 & 1 & 0 & 0 \\ 0 & 0 & 1 & 0 \\ 0 & 0 & 0 & 0 \\ 0 & 0 & 0 & 0 \end{bmatrix},$$

and

$$C^T = \begin{bmatrix} 1 & 0 & 0 & 0 & 0 \\ 0 & 1 & 0 & 0 & 0 \\ 0 & 0 & 1 & 0 & 0 \\ 0 & 0 & 0 & 0 & 0 \end{bmatrix}.$$

This last matrix is equivalent to A.

We now give a second and independent proof that

$$row\ rank(A) = column\ rank(A)$$

for any matrix A. (See Theorem 2.9)

Proof.   Let P and Q be invertible matrices such that

$$PAQ = \begin{bmatrix} I_{r \times r} & [0] \\ [0] & [0] \end{bmatrix}.$$

Let $\mathbf{u}_1, \mathbf{u}_2, \cdots, \mathbf{u}_m$ be the columns of P^{-1}. Then

$$\begin{aligned} AQ &= P^{-1} \begin{bmatrix} I_{r \times r} & [0] \\ [0] & [0] \end{bmatrix} \\ &= \begin{bmatrix} \mathbf{u}_1 & \mathbf{u}_2 & \cdots & \mathbf{u}_m \end{bmatrix} \begin{bmatrix} I_{r \times r} & [0] \\ [0] & [0] \end{bmatrix} \\ &= \begin{bmatrix} \mathbf{u}_1 & \cdots & \mathbf{u}_r & [0] \end{bmatrix}. \end{aligned}$$

Clearly this matrix has column rank r, since the columns of P^{-1} are a linearly independent set. We also know that *column rank* $AQ =$ *column rank*(A), since by Theorem 2.11 $colsp(AQ) = colsp(A)$ if Q is invertible. (This part of Theorem 2.11 is equivalent to the assertion that elementary column operations don't change the column space.) Thus

$$colrank(A) = colrank(AQ) = r.$$

Let $\mathbf{v}_1, \mathbf{v}_2, \cdots, \mathbf{v}_n$ be the rows of Q^{-1}. Then

$$PA = \left[\begin{array}{cc} I_{r\times r} & [0] \\ {[0]} & [0] \end{array} \right] Q^{-1}$$

$$= \left[\begin{array}{cc} I_{r\times r} & [0] \\ {[0]} & [0] \end{array} \right] \left[\begin{array}{c} \mathbf{v}_1 \\ \mathbf{v}_2 \\ \vdots \\ \mathbf{v}_n \end{array} \right]$$

$$= \left[\begin{array}{c} \mathbf{v}_1 \\ \vdots \\ v_r \\ {[0]} \end{array} \right].$$

This matrix has row rank r. Now $rowrank(PA) = rowrank(A)$, since elementary row operations don't change the row space. Thus

$$rowrank(A) = rowrank(PA) = r,$$

and we have

$$colrank(A) = r = rowrank(A).$$

■

Theorem 2.21 *Let A and B be $m \times n$ matrices. Then A is equivalent to B if and only if A and B have the same rank.*

Proof.   Suppose that A and B have the same rank. Now

$$PAQ = \left[\begin{array}{cc} I_{r\times r} & [0] \\ {[0]} & [0] \end{array} \right]$$

and

$$P'BQ' = \begin{bmatrix} I_{s\times s} & [0] \\ [0] & [0] \end{bmatrix}$$

for some invertible P, Q, P' and Q'. Since multiplication by an invertible matrix on either side does not change the rank (Theorem 2.12), we have $s = rank(B)$ and $r = rank(A)$, thus $s = r$. Hence

$$\begin{aligned} A &= P^{-1}\begin{bmatrix} I_{r\times r} & [0] \\ [0] & [0] \end{bmatrix}Q^{-1} \\ &= P^{-1}P'BQ'Q^{-1}, \end{aligned}$$

so A is equivalent to B. Conversely if A is equivalent to B, then $B = GAR$, where G and R are invertible, but then A and B must have the same rank, again using Theorem 2.12. ■

Exercises

1. Let $A = \begin{bmatrix} 2 & 1 \\ 1 & 2 \end{bmatrix}$, and $B = \begin{bmatrix} 3 & 1 \\ 1 & 3 \end{bmatrix}$. Find invertible matrices P and Q such that $B = PAQ$.

2. Find necessary and sufficient conditions on a and b for the matrices $\begin{bmatrix} a & 1 \\ 1 & a \end{bmatrix}$ and $\begin{bmatrix} b & 1 \\ 1 & b \end{bmatrix}$ to be equivalent.

3. Let A and B be $n \times n$ matrices of rank k. Show that in the equation $B = PAQ$, with P, Q invertible, the matrices P and Q are never unique.

4. If $rank(AB) = rank(A)$, prove that $colsp(AB) = colsp(A)$.

Chapter 3

Determinants

Section 3.1 Mappings and Permutations

We consider mappings from a finite set S into itself. For simplicity's sake we call the elements of S $1, 2, 3, \cdots, n$.
$$S = \{1, 2, 3, \cdots, n\}.$$
The mappings from S into S will be denoted by capital letters and the image of 1 under the mapping P will be denoted by $1P$ rather than by $P(1)$ which is the functional notation favored by some. Our way does away with writing so many brackets and also permits the reading of consecutive mappings from left to right. Clearly there are n^n distinct mappings from S into S. Our real interest however focuses on only those mappings which are 1 : 1 and onto.

Definition 3.1 P *is* 1 : 1 *if* $xP = yP$ *only when* $x = y$.

Definition 3.2 P *is onto in case the set* $\{xP\}$ *is the whole set S.*

For a finite set S a mapping P of S into S is 1 : 1 if and only if P is onto. This is sometimes called the pigeon hole principle.

Definition 3.3 *A* 1 : 1 *mapping from S onto S is called a permutation on S, and the set of permutations on S when S has n elements is called the symmetric group, denoted by S_n.*

The number of distinct elements in S_n is $n!$ (the product of all the integers starting with 1 and including n, $n! = 1 \cdot 2 \cdot 3 \cdot 4 \cdots (n-1) \cdot n$. To see this consider how we would define a permutation P in S_n. There are n possibilities for $1P$, but after $1P$ is chosen, there are only $n-1$ possibilities for $2P$, since no repeats are allowed. Then there are only $n-2$ choices for $3P$, etc. When $n-1$ choices have been made, nP must be that number which has not yet been chosen, and so there is only one possibility. The total number of possibilities is the product of the number of possibilities at each stage.

Our next objective is to introduce a method of combining two permutations P and Q to obtain a new permutation which we will call $P * Q$. The new permutation $P * Q$ is defined when we know the images it gives for every element in S. We define $P * Q$ by the equation

$$x(P * Q) = (xP)Q.$$

For example if $S = \{1, 2, 3\}$, define P by $1P = 2$, $2P = 3$, and $3P = 1$. Define Q by $1Q = 2$, $2Q = 1$, and $3Q = 3$. Then

$$1(P * Q) = (1P)Q = 2Q = 1,$$

$$2(P * Q) = (2P)Q = 3Q = 3,$$

$$3(P * Q) = (3P)Q = 1Q = 2.$$

Note that $P * Q$ is again a permutation on S.

Property 1. The product $*$ of two permutations on a finite set S is again a permutation on S.

Property 1 is often described by saying S_n is closed under the product $*$.

Property 2 There exists a mapping I known as the identity mapping and defined by $xI = x$ for all x in S, and I has the property that $P * I = I = I * P$.

Property 3 For each P in S_n there exists a P^{-1} with the property $P * P^{-1} = I = P^{-1} * P$.

To find this mapping suppose that $xP = y$. Then define P^{-1} by defining $yP^{-1} = x$. Since P is $1:1$ and onto, P^{-1} is a permutation on S. Then $y(P^{-1} * P) = xP = y$ and $x(P * P^{-1}) = yP^{-1} = x$, so that property 3 follows.

Property 4 $(P * Q) * R = P * (Q * R)$, or $*$ is associative.

The last property is obvious since P, Q, and R are mappings. A set which is closed under an operation $*$ and which satisfies the four properties just mentioned for all P, Q, R is called a GROUP.

Theorem 3.1 S_n *is a group with n! elements.*

Definition 3.4 *A group is called commutative in case* $P * Q = Q * P$ *for all P and Q in the group.*

We show that S_3 is not commutative. Let $1P = 2$, $2P = 1$, $3P = 3$. Let $1Q = 3$, $2Q = 2$, $3Q = 1$. Then

$$1(P * Q) = 2Q = 2,$$

$$2(P * Q) = 1Q = 3,$$

$$3(P * Q) = 3Q = 1.$$

On the other hand

$$1(Q * P) = 3P = 3,$$

$$2(Q * P) = 2P = 1,$$

$$3(Q * P) = 1P = 2.$$

Clearly $P * Q$ and $Q * P$ are not the same.

The following result indicates how universal S_n and its subgroups are in the universe of finite groups.

Caley's Theorem. *Every finite group with n elements is isomorphic to (indistinguishable from) a subgroup of* S_n.

A proof of this result would take us too far out of our way.

Within S_n for $n > 1$ there exists a specific subgroup with $\frac{n!}{2}$ elements known as the alternating group and designated by A_n. To distinguish the elements which lie in A_n from those that don't we need to explain the disjoint cycle notation and define odd and even permutations.

Given a permutation P on a set S of n elements, the disjoint cycle representation of P is arrived at as follows: start with any x in S and begin with a round bracket and an x. The next entry in the cycle will be xP, if x is not fixed by P. If x is fixed by P, we close the cycle and write (x). If x is not fixed by P, we continue the cycle, with each entry being the image under P of the preceding entry, so we get $(x, xP, xPP, xPPP, \cdots)$. At some point there is a repeat. The first repeat will be $xPPP \cdots P = x$, and at this point we do not write x, but close the brackets. For example suppose that $n = 5$ and $1P = 2$, $2P = 4$, $3P = 1$, $4P = 5$, $5P = 3$. Then the disjoint cycle notation for P is $P = (12453)$. If $n = 7$ and if Q agrees with P on the first five numbers and $6Q = 7$, $7Q = 6$, then the disjoint cycle notation for Q is $Q = (12453)(67)$. The disjoint cycle is a short way of writing a permutation. It also gives additional information regarding the alternating subgroup, which we will explain shortly. Observe that in the disjoint cycle notation a cycle might start with any number in that cycle, so that a given permutation P could be written in several ways which don't look identical. For example $P = (12453) = (31245) = (45312)$, etc. Also the order in which the different cycles are written can be different, as in $Q = (12453)(67) = (67)(12453)$. However all disjoint cycle representations of P would agree on the number of cycles, the number of elements in each cycle, and which elements go together in one cycle. Sometimes it is convenient to omit all those cycles which have only one element in them. We shall follow this convention, except for the identity permutation, since $I = (1)(2)(3) \cdots (n)$ is a disjoint cycle representation of I, nothing would be left if we omit all cycles of length one. Once a permutation is written in disjoint cycle form it is easy to explain whether it gets put in A_n (the even permutations) or not (the odd permutations).

Definition 3.5 *If P is any permutation in disjoint cycle form, then $N(P)$ is defined as the number of elements in S not fixed by P minus the number of disjoint cycles of length greater than 1 in a disjoint cycle form of P.*

Note that $N(I) = 0 - 0 = 0$. Let $R = (12453)(67)$. Then $N(R) = 7 - 2 = 5$. Let $Q = (12)(37)(46)(59)$. Then $N(Q) = 8 - 4 = 4$. Also note that if n is the number of elements in S, and if q is the number of cycles in P (with cycles of length one included), then $N(P) = n - q$, for the number of fixed elements equals the number of cycles of length one, so these cancel each other.

Definition 3.6 *A permutation P is called even if $N(P)$ is divisible by 2 and odd otherwise.*

We need to find out if the parity (even or oddness) of two permutations P and Q determines the parity of $P * Q$.

Definition 3.7 *A transposition is a permutation consisting of one cycle of length two, such as (xy).*

In S_3 there are exactly three transpositions, namely (12), (13), and (23). In S_n with $n > 1$ the number of transpositions is $n(n-1)/2$. All transpositions are odd permutations, since $N\{(xy)\} = 2 - 1 = 1$.

Theorem 3.2 *Let P be any permutation and T be any transposition in S_n. Then P and $P * T$ always have opposite parity.*

Proof.   Suppose $T = (xy)$, with x and y distinct elements in S. Write out P in disjoint cycle form and this time include all cycles of length one if there are any. Suppose first that x and y occur in the same cycle of P. Then $P = (xa_1 \cdots a_s y b_1 \cdots b_t)(c_1 \cdots c_u) \cdots (q_1 \cdots q_v)$. Then $P * T = (xa_1 \cdots a_s)(y b_1 \cdots b_t)(c_1 \cdots c_u) \cdots (q_1 \cdots q_v)$, and it is easy to see that $N(P * T) + 1 = N(P)$, since $P * T$ and P both have all the elements of S in disjoint cycles but $P * T$ has one more cycle than P. Thus P and $P * T$ have opposite parity. We are left with the case where x and y occur in different cycles in P. Suppose now that $P = (xa_1 \cdots a_s)(y b_1 \cdots b_t)(c_1 \cdots c_u) \cdots (q_1 \cdots q_v)$ and $T = (xy)$. Then $P * T = (xa_1 \cdots a_s y b_1 \cdots b_t)(c_1 \cdots c_u) \cdots (q_1 \cdots q_v)$. Again both P and $P * T$ have all the elements of S in cycles, but P has one more cycle than $P * T$, therefore $N(P * T) = N(P) + 1$, so $P * T$ and P have opposite parity. ■

Lemma 3.1 *Every permutation can be written as a product of transpositions.*

Proof.   Note that $(12 \cdots r) = (12)(13) \cdots (1r)$. From this we can write any cycle as a product of transpositions. Thus if we write a permutation P in disjoint cycle form and then write each cycle as a product of transpositions, we will have P as a product of transpositions. For example $(13469)(237) = (13)(14)(16)(19)(23)(27)$. ∎

There is more than one way to write a permutation as a product of transpositions. Thus $(1234) = (12)(13)(14) = (23)(13)(23)(13)(14)$. However we have the following result.

Lemma 3.2 *If any permutation P is written as a product of r transpositions and again as a product of s transpositions, then $N(P)$, r and s always have the same parity.*

Proof.   Suppose that $P = T_1 * T_2 * \cdots * T_r$ where the T_i are all transpositions. Since each transposition is its own inverse in S_n it follows that $P * T_r * T_{r-1} * \cdots * T_1 = I$. Since I is even, we see that $N(P)$ after r changes of parity must be even. Thus $N(P)$ and r have the same parity. ∎

In the example $P = (1234)$ we note that $N(P) = 3$ and we also saw that P could be written as the product of three transpositions and also as a product of five transpositions. Clearly three is the smallest number of transpositions whose product equals P.

Theorem 3.3 *The following multiplication table is valid for all permutations in S_n.*

*	even	odd
even	even	odd
odd	odd	even

Proof.   We write arbitrary P and Q in S_n as products of transpositions as follows: $P = T_1 * T_2 * \cdots * T_r$, and $Q = U_1 * U_2 * \cdots * U_s$. Then parities of P, Q, and $P * Q$ are the same as the parities of r, s, and $r + s$ respectively. That is precisely the table we have, since $2p + 2q = 2(p + q)$, showing that the sum of two even integers is even, and $2p + (2q + 1) = 2(p + q) + 1$, showing that the sum of an even and an odd integer is odd. Finally $(2p + 1) + (2q + 1) = 2(p + q + 1)$, shows that the sum of two odd integers is always even. ∎

Corollary 1 *The alternating group A_n of all even permutations of S_n is a subgroup of S_n.*

Proof. First I, the identity permutation is an even permutation, thus $I \in A_n$. Also if P and Q are even permutations then $P * Q$ is even so that A_n is closed under $*$. If P is an even permutation they represent it as a product of an even number of transpositions. Then P^{-1} can be written as the product of the same transpositions in reverse order. Since the number of these is again even, this shows that P^{-1} is even, hence $P^{-1} \in A_n$. This is enough to show that A_n is a subgroup of S_n. ∎

Lemma 3.3 *For $n > 1$, A_n has $\frac{n!}{2}$ elements.*

Proof. Let $\{P_1, P_2, \cdots P_k\}$ be the set of all even permutations, A_n. Let T be a fixed transposition of S_n. Form the set $\{P_1 * T, P_2 * T, \cdots P_k * T\}$. As a result of Theorem 3.3, every permutation in the second set is odd, and we will show that every odd permutation of S_n is included. If Q is any odd permutation, then $Q * T$ is even, so $Q * T = P_j$ for some j, and hence $P_j * T = (Q * T) * T = Q$ is in the second set. Also the k odd permutations in the second set are all distinct, for if $P_i * T = P_j * T$, then multiplying this equation on the right by T and using again that $T * T = I$, we get $P_i = P_j$. This shows that in S_n there are the same number of odd permutations as even permutations. Thus the number of even permutations in S_n is $\frac{n!}{2}$. ∎

Remark (Historical note) *The alternating groups have played an important role in what has become known as Galois Theory. Évariste Galois (1811-1832) was a French mathematician who died very young, under mysterious circumstances. He established the groundwork for solving a number of classic problems, by introducing groups in a novel way and many regard him as the father of modern abstract algebra. His work remained unpublished and unpublicized until long after his death. Only after others had obtained the same results were his papers discovered by the mathematical community at large, thus adding to the mystery.*

Exercises

1. If $1P = 3$, $2P = 4$, $3P = 2$, $4P = 1$, $5P = 6$, and $6P = 5$, is a permutation in S_6, write P in disjoint cycle form. Find $N(P)$. Is P even or odd? Write P as a product of transpositions.

2. If P is as in problem 1, find P^{-1} and write it in disjoint cycle form.

3. If $P = (132)(45)$, find $P * P$.

4. If $P = (12746)(385)$ and $Q = (16)(3825)$, write $P * Q$ and $Q * P$ in disjoint cycle form. Calculate the parity of each of the four permutations and verify that Theorem 3.3 works for them.

5. Suppose the permutation P consists of two disjoint cycles having length r and s. Calculate the least power of P that equals the identity in terms of r and s. Start first with some specific numbers, for example $r = 4$ and $s = 6$, before answering the question in general.

6. For what values of n is S_n a commutative group? Same question for A_n. Give reasons for your answers.

7. Determine the cycle structure of all the elements of A_4.

   For the permutations in problem 4, determine the inverse of each, and write it in disjoint cycle form.

8. What is the smallest number of transpositions whose product equals $(123 \cdots n)$? Explain.

9. There are n persons lining up for a meal in a cafeteria. Explain in how many different ways they can line up.

10. If there are $k = 2n$ persons forming partners for lab work, explain in how many different ways this can be done.

11. Let S be a finite set. Prove that a mapping from S into S is onto if and only if it is 1:1.

12. Give an example of a function from $Z+$, the set of all positive integers into itself which is

(a) 1:1 but not onto.

(b) onto but not 1:1.

Section 3.2 Determinants

Definition 3.8 *If P is a permutation then $sgn(P) = 1$ if P is even and $sgn(P) = -1$ if P is odd.*

Definition 3.9 *If $A = [a_{ij}]$ is an nxn matrix then the determinant of A denoted by the symbol $|A|$ or $det(A)$ is defined by*

$$|A| = \sum_{P \in S_n} sgn(P) a_{1P,1} a_{2P,2} \cdots a_{nP,n}.$$

By $a_{1P,1}$ we mean the entry in A in column 1 and row $1P$. We include the commas for clarity, but they could be omitted. The formula says that to compute the determinant of A we are to choose an entry from column 1, any row (the row we chose becomes $1P$), then choose an entry from column 2, any row except the row chosen in column one, continue in this way until we have chosen one and only one entry from each column in turn, without repeating any row. Now multiply these n numbers together and multiply by 1 or -1 depending on whether the row subscripts come from the column subscripts by an even or odd permutation P. This gives one term in the summation. One must repeat this over all possible ways of picking one and only one entry from each column without repeating any row, in other words repeat over all possible permutations of the row subscripts, multiplying each term by 1 or -1 depending on whether the row subscripts come from the column subscripts by an even or odd permutation. There will be $n!$ terms in this sum. Once we develop some rules for evaluating determinants we will not have to use the defining formula very often, but in the beginning it is all we have to work with.

For 2×2 matrices, there are $2! = 2$ terms. We have

$$\begin{vmatrix} a_{11} & a_{12} \\ a_{21} & a_{22} \end{vmatrix} = a_{11}a_{22} - a_{21}a_{12}.$$

The first term comes from the identity permutation and the second term comes from the permutation $1P = 2$, $2P = 1$, a transposition, so $sgn(P) = -1$. So we have the familiar formula for the determinant of a 2×2 matrix

$$\begin{vmatrix} a & b \\ c & d \end{vmatrix} = ad - bc,$$

the product of the main diagonal minus the product of the other diagonal. Of course this only works for 2×2 matrices.

For 3×3 matrices there are $3! = 6$ terms. Let us look at the definition of determinant in the case $n = 3$. The 6 permutations in S_3 written in disjoint cycle form are $I, (123), (132), (13), (12),$ and (23). Using these permutations in order, the definition gives

$$\begin{vmatrix} a_{11} & a_{12} & a_{13} \\ a_{21} & a_{22} & a_{23} \\ a_{31} & a_{32} & a_{33} \end{vmatrix}$$
$$= a_{11}a_{22}a_{33} + a_{21}a_{32}a_{13} + a_{31}a_{12}a_{23} - a_{31}a_{22}a_{13} - a_{21}a_{12}a_{33} - a_{11}a_{32}a_{23}.$$

The first three permutations are even, so these terms get multiplied by $+1$, while the last three permutations are odd, so these terms get multiplied by -1.

For 4×4 matrices, we have $4! = 24$ terms, 12 with $+1$ and 12 with -1. For 5×5 matrices, we have $5! = 120$ terms and we definitely don't want to write them out! Fortunately for some special matrices the determinant is easy to compute.

Lemma 3.4 *If A is an upper triangular matrix, say*

$$A = \begin{bmatrix} a_{11} & a_{12} & \cdots & a_{1n} \\ 0 & a_{22} & \cdots & a_{2n} \\ \vdots & 0 & \ddots & \vdots \\ 0 & \cdots & 0 & a_{nn} \end{bmatrix},$$

then

$$|A| = a_{11}a_{22} \cdots a_{nn}.$$

Proof. Note that $A = [a_{ij}]$ with $a_{ij} = 0$ whenever $i > j$. Consider the product $a_{1P,1}a_{2P,2}\cdots a_{nP,n}$. If for even one integer j we have $jP > j$, then the whole product is 0, so the only surviving terms will have $jP \leq j$. But the sum $1P + 2P + \cdots + nP = 1 + 2 + \cdots + n$ (since P only permutes the integers 1 to n the sum is still the same). Now we have $jP \leq j$ for all nonzero terms, but if strict inequality occurs for some k, say $kP < k$, we would have $1P + 2P + \cdots + nP < 1 + 2 + \cdots + n$. Thus we must have $kP = k$ for all k, so that $P = I$, $sgn(P) = 1$, and $|A| = a_{11}a_{22}\cdots a_{nn}$. ∎

Corollary 1 $|I| = 1$.

Each elementary row operation has a predictable effect on the determinant, and once we learn what that is we can calculate determinants by using row operations to change them to upper triangular matrices. That every square matrix is row equivalent to an upper triangular matrix follows from the fact that every matrix is row equivalent to a matrix in RREF, and matrices in RREF are upper triangular.

Lemma 3.5 $\left|A^T\right| = |A|$.

Proof. We recall that the columns of A^T are the rows of A, so that if $B = A^T$ we have $b_{ij} = a_{ji}$. Thus

$$\left|A^T\right| = \sum_{P \in S_n} sgn(P)a_{1,1P}a_{2,2P}\cdots a_{n,nP}.$$

Rearrange the order of the a_{ij} in each term so that the column subscripts are in the natural order, $1, 2, \cdots, n$. Note that we want to put first $a_{j,jP}$ such that $jP = 1$, so we want $j = 1P^{-1}$. Similarly we want to put second $a_{k,kP}$ such that $kP = 2$, or in other words $k = 2P^{-1}$, and so on. Thus

$$\left|A^T\right| = \sum_{P \in S_n} sgn(P)a_{1P^{-1},1}a_{2P^{-1},2}\cdots a_{nP^{-1},n}.$$

Now $sgn(P) = sgn(P^{-1})$ since P and P^{-1} always have the same parity, so we have

$$\left|A^T\right| = \sum_{P \in S_n} sgn(P^{-1})a_{1P^{-1},1}a_{2P^{-1},2}\cdots a_{nP^{-1},n}.$$

As P goes through all permutations on $\{1, 2, \cdots, n\}$ so does P^{-1}, since $P = (P^{-1})^{-1}$ every permutation is the inverse of some permutation. Thus $\left|A^T\right| = |A|$. ∎

Corollary 1 *If A is a lower triangular matrix then $|A| = a_{11}a_{22}\cdots a_{nn}$.*

Proof.   Follows from Lemmas 3.4 and 3.5 since the transpose of a lower triangular matrix is upper triangular. ∎

Lemma 3.6 *If B is the $n \times n$ matrix obtained from A by switching rows i and j, then $|B| = -|A|$.*

Proof.   Let $T = (ij)$. Then $b_{k,r} = a_{kT,r}$. Thus we have

$$
\begin{aligned}
|B| &= \sum_{P \in S_n} sgn(P)b_{1P,1}b_{2P,2}\cdots b_{nP,n} \\
&= \sum_{P \in S_n} sgn(P)a_{1P*T,1}a_{2P*T,2}\cdots a_{nP*T,n}.
\end{aligned}
$$

Since T is a transposition $sgn(P*T) = -sgn(P)$. Thus

$$
|B| = \sum_{P \in S_n} -sgn(P*T)a_{1P*T,1}a_{2P*T,2}\cdots a_{nP*T,n}.
$$

As P ranges over all permutations in S_n so does $P*T$. Thus $|B| = -|A|$. ∎

Corollary 1 *Permuting the rows of A by a permutation P multiplies $|A|$ by $sgn(P)$.*

Corollary 2 *Permuting the columns of A by a permutation P multiplies $|A|$ by $sgn(P)$.*

Proof.   Follows from Lemmas 3.5 and 3.6. ∎

Corollary 3 *If a matrix has two identical rows (or columns) then $|A| = 0$.*

Proof.   Suppose rows i and j of A are identical. Switch them. Lemma 3.6 implies that $|B| = -|A|$. But $B = A$. Thus $|A| = -|A|$. Consequently $2|A| = 0$, Thus $|A| = 0$. To get the result for columns, use Lemma 3.5. ∎

Lemma 3.7 *If B is obtained from A by multiplying a fixed row (column) of A by a real number c, then $|B| = c|A|$.*

Proof.   This can be seen from the formula that defines $|A|$, since each term (product of n elements) in the sum contains one element from each row, and so each term in $|B|$ will be c times the corresponding term in $|A|$. ∎

Lemma 3.8 *Denote the row vectors of A by $\mathbf{u}_1, \mathbf{u}_2, \cdots, \mathbf{u}_n$. Suppose that $\mathbf{u}_i$ is equal to a sum of two vectors, $\mathbf{u}_i = \mathbf{v} + \mathbf{w}$. Then*

$$|A| = \begin{vmatrix} \mathbf{u}_1 \\ \vdots \\ \mathbf{v} + \mathbf{w} \\ \vdots \\ \mathbf{u}_n \end{vmatrix} = \begin{vmatrix} \mathbf{u}_1 \\ \vdots \\ \mathbf{v} \\ \vdots \\ \mathbf{u}_n \end{vmatrix} + \begin{vmatrix} \mathbf{u}_1 \\ \vdots \\ \mathbf{w} \\ \vdots \\ \mathbf{u}_n \end{vmatrix}.$$

Proof.   Let $\mathbf{u}_j = (a_{j1}, a_{j2}, \cdots, a_{jn})$ and let $\mathbf{v} = (b_1, b_2, \cdots, b_n)$ and $\mathbf{w} = (c_1, c_2, \cdots, c_n)$. Then $a_{ik} = b_k + c_k$. In the definition of $|A|$ replace a_{ik} by $b_k + c_k$ and use the distributive rule for real numbers. This will give the desired result. ∎

Lemma 3.9 *If B is obtained from A by adding c times row j to row i, with $i \neq j$, then $|B| = |A|$. (Same result is true for columns.)*

Proof.   Again let $\mathbf{u}_1, \mathbf{u}_2, \cdots, \mathbf{u}_n$ be the rows of A. Then we have

$$|B| = \begin{vmatrix} \mathbf{u}_1 \\ \vdots \\ \mathbf{u}_i + c\mathbf{u}_j \\ \vdots \\ \mathbf{u}_n \end{vmatrix} = \begin{vmatrix} \mathbf{u}_1 \\ \vdots \\ \mathbf{u}_i \\ \vdots \\ \mathbf{u}_n \end{vmatrix} + \begin{vmatrix} \mathbf{u}_1 \\ \vdots \\ c\mathbf{u}_j \\ \vdots \\ \mathbf{u}_n \end{vmatrix} = \begin{vmatrix} \mathbf{u}_1 \\ \vdots \\ \mathbf{u}_i \\ \vdots \\ \mathbf{u}_n \end{vmatrix} + c \begin{vmatrix} \mathbf{u}_1 \\ \vdots \\ \mathbf{u}_j \\ \vdots \\ \mathbf{u}_n \end{vmatrix}$$

using Lemmas 3.8 and 3.7. Now the first of the two determinants on the right is just $|A|$, while the second is zero because two rows are equal, namely $\mathbf{u}_j$ in row i and also in row j. Thus $|B| = |A|$. ∎

We now know the effect on $|A|$ of elementary row operations. Multiplying a row or column by a nonzero number multiplies $|A|$ by that number, adding a multiple of one row (column) to another row (column) does not change $|A|$, and exchanging two rows (columns) multiplies $|A|$ by -1. Since the determinant of a triangular matrix is easy to compute, we can compute determinants by using either row or column operations to change A into a triangular matrix. Since some row and column operations do change $|A|$ we have to keep track of these so that we can relate $|A|$ to the determinant of the triangular matrix we get. It is useful to note that while elementary row operations on A do change $|A|$, they do not change it from zero to nonzero or from nonzero to zero. Thus if $|A| = 0$, any matrix B that is row equivalent to A also has determinant zero, while if $|A| \neq 0$, any matrix that is row equivalent to A also has nonzero determinant.

Lemma 3.10 *If A has a zero row (or column), then $|A| = 0$.*

Proof.   In the definition of $|A|$, every term (product of n elements) in the sum has one element from each row (and one from each column), thus if any row (or column) is all zeros, every term will be zero. ∎

From this Lemma and the remarks above, we see that if at any stage in the row reduction of A we get a zero row, we can stop and conclude that $|A| = 0$. In fact we have a stronger result.

Theorem 3.4 *If A is an $n \times n$ matrix, then $|A| = 0$ if and only if $rank(A) < n$.*

Proof.   Let $Q = RREF(A)$. Since Q is square, if every row has a leading one then $Q = I$, otherwise Q has at least one zero row. If $|A| = 0$, we must have $|Q| = 0$ since A and Q are row equivalent. Thus $Q \neq I$, so Q has a zero row and hence $rank(A) < n$. (Recall $rank(A)$ equals the number of leading ones in $RREF(A) = Q$). Conversely if $rank(A) < n$, then Q has a zero row, whence $|Q| = 0$ and so $|A| = 0$ also. ∎

Corollary 1 *A is invertible if and only if $|A| \neq 0$.*

Proof.   From the above theorem,$|A| \neq 0$ if and only if $rank(A) = n$. From Theorem 2.19, $rank(A) = n$ is equivalent to $RREF(A) = I$. Then from Theorem 2.16, $RREF(A) = I$ is equivalent to A is invertible. ∎

We can add $|A| \neq 0$ to the long list of things that are equivalent to $RREF(A) = I$. Looking back at Theorems 2.16 and 2.19 we see that just knowing that $|A| \neq 0$ gives us a lot of information about the matrix A. For example knowing whether $|A|$ is zero or not determines whether or not $A\mathbf{x} = \mathbf{O}$ has nonzero solutions. It also determines whether or not the rows of A are a basis of R^n. $|A| \neq 0 \Longleftrightarrow$ *the* rows of A are a basis for R^n was proved for $n = 2$ in Theorem 1.3, so we have now generalized that result to R^n.

Definition 3.10 *An $n \times n$ matrix A is singular if $|A| = 0$. If $|A| \neq 0$, A is nonsingular.*

Remark   *The preceding corollary now can be stated: A is invertible if and only if A is nonsingular.*

Lemma 3.11 *If A is any $n \times n$ matrix and E any elementary $n \times n$ matrix then $|EA| = |E||A|$.*

Proof.   Since E comes from the identity matrix I by an elementary row operation, we can use our results about the effect of row operations on determinants. If E comes from I by multiplying a row by c, the $|E| = c$, but also $|EA| = c|A|$, so the result holds. If E comes from I by adding a multiple of row j to row i, $i \neq j$, then $|E| = 1$, and we also know that $|EA| = |A|$, so again the result holds. If E comes from I by switching two rows, then $|E| = -1$, however in this case EA comes from A by switching two rows so that $|EA| = -|A|$, and again the result holds. ∎

Theorem 3.5 *If A and B are any $n \times n$ matrices, then $|AB| = |A||B|$.*

Proof.   If A has rank $< n$, then so does AB, since $rank(AB) \leq rank(A)$ by Theorem 2.12, and thus Theorem 3.4 implies that $|AB| = 0$. Since $rank(A) < n$, we also have $|A| = 0$. Thus no matter what B is, both $|AB|$ and $|A||B|$ are zero and the equation holds. If $\text{rank}(A) = n$, then by Theorems 2.19 and 2.16 we know that A equals a product of elementary matrices, $A = E_1 E_2 \cdots E_t$. Then repeated use of the previous lemma gives $|AB| = |E_1 E_2 \cdots E_t B| = |E_1||E_2 \cdots E_t B| = |E_1||E_2||E_3 \cdots E_t B| = \cdots = |E_1||E_2||E_3|\cdots|E_t||B|$. This is valid for all B, so using $B = I$ we get $|A| = |E_1||E_2||E_3|\cdots|E_t|$, and substituting this in the previous equation gives $|AB| = |A||B|$. ∎

Expansion by Minors

If A is an $n \times n$ matrix, the minor $M_{ij}(A)$ is the $(n-1) \times (n-1)$ submatrix obtained from A by deleting the ith row and j^{th} column of A. When there is no doubt what matrix is referred to we just write M_{ij}. The following equation relates $|A|$ to $|M_{j1}|$.

$$|A| = a_{11}|M_{11}| - a_{21}|M_{21}| + a_{31}|M_{31}| - \cdots (-1)^{n+1} a_{n1}|M_{n1}|.$$

This formula follows from the definition of determinant by considering the sum over all permutations in the following way. First take all permutations P with $1P = 1$, the sum of all these terms will give $a_{11}|M_{11}|$. Then take all permutations with $1P = 2$. The sum of these terms will give $-a_{21}|M_{21}|$, and so on. We omit the details. The above formula is called expansion by minors around the first column of A. By interchanging columns, we can bring any column into the first position. If one does this very carefully, one can deduce the following formula:

$$|A| = a_{1j}(-1)^{1+j}|M_{1j}| + a_{2j}(-1)^{2+j}|M_{2j}| + \cdots (-1)^{n+j} a_{nj}|M_{nj}|.$$

This is the formula for expansion by minors around the jth column. By passing to the transpose and remembering that $\left|A^T\right| = |A|$, one can establish the formula for expansion by minors around the ith row.

$$|A| = a_{i1}(-1)^{i+1}|M_{i1}| + a_{i2}(-1)^{i+2}|M_{i2}| + \cdots (-1)^{i+n} a_{in}|M_{in}|.$$

Example 3.1 Let $A = \begin{bmatrix} 1 & 2 & 3 \\ 4 & 5 & 6 \\ 7 & 8 & 9 \end{bmatrix}$. We use expansion by minors around the first column to get

$$|A| = \begin{vmatrix} 1 & 2 & 3 \\ 4 & 5 & 6 \\ 7 & 8 & 9 \end{vmatrix}$$

$$= 1 \begin{vmatrix} 5 & 6 \\ 8 & 9 \end{vmatrix} - 4 \begin{vmatrix} 2 & 3 \\ 8 & 9 \end{vmatrix} + 7 \begin{vmatrix} 2 & 3 \\ 5 & 6 \end{vmatrix}$$
$$= (45 - 48) - 4(18 - 24) + 7(12 - 15) = -3 + 24 - 21 = 0.$$

Using expansion by minors around column three gives:

$$|A| = \begin{vmatrix} 1 & 2 & 3 \\ 4 & 5 & 6 \\ 7 & 8 & 9 \end{vmatrix}$$

$$= 3 \begin{vmatrix} 4 & 5 \\ 7 & 8 \end{vmatrix} - 6 \begin{vmatrix} 1 & 2 \\ 7 & 8 \end{vmatrix} + 9 \begin{vmatrix} 1 & 2 \\ 4 & 5 \end{vmatrix}$$
$$= 3(32 - 35) - 6(8 - 14) + 9(5 - 8) = -9 + 36 - 27 = 0.$$

We could also have seen that the determinant of the matrix in the previous example is zero by using Lemma 3.9. Subtracting row two from row three and then subtracting row one from row two gives:

$$\begin{vmatrix} 1 & 2 & 3 \\ 4 & 5 & 6 \\ 7 & 8 & 9 \end{vmatrix} = \begin{vmatrix} 1 & 2 & 3 \\ 4 & 5 & 6 \\ 3 & 3 & 3 \end{vmatrix} = \begin{vmatrix} 1 & 2 & 3 \\ 3 & 3 & 3 \\ 3 & 3 & 3 \end{vmatrix} = 0,$$

since two rows are equal.

Row reduction to triangular form is usually the most efficient way to calculate determinants. However if there are a lot of zeros, expansion by minors can be useful. Often a combination of the two methods is used. If you are working without a calculator, you try to take advantage of any properties of the particular matrix you are dealing with, rather than following one method consistently. There are now good computer and calculator programs that will find determinants for you. However if your program uses floating point, roundoff error may be a problem. Also matrices can get too big for the computer to handle by a fixed method, even with the large memory capacity and fast computation times of today's computers. Programmers try to write programs that will tell the computer to look for and take advantage of special properties, like patterns or lots of zeros, but this is difficult. Here is an example of a class of matrices whose determinants we can find by taking advantage of their pattern.

Example 3.2 *Let A be the $n \times n$ matrix which has one everywhere except for the main diagonal where it has $n + 1$.*

$$A = \begin{bmatrix} n+1 & 1 & 1 & \cdots & 1 \\ 1 & n+1 & 1 & \cdots & 1 \\ 1 & 1 & n+1 & \ddots & \vdots \\ \vdots & \vdots & \ddots & \ddots & 1 \\ 1 & 1 & \cdots & 1 & n+1 \end{bmatrix}.$$

To calculate $|A|$, start by adding all other rows to the first row. This does not change $|A|$, so

$$|A| = \begin{vmatrix} 2n & 2n & 2n & \cdots & 2n \\ 1 & n+1 & 1 & \cdots & 1 \\ 1 & 1 & n+1 & \ddots & \vdots \\ \vdots & \vdots & \ddots & \ddots & 1 \\ 1 & 1 & \cdots & 1 & n+1 \end{vmatrix}.$$

We can factor out the $2n$ from row one, since multiplying a row by a real number multiplies the determinant by that number. Thus

$$|A| = 2n \begin{vmatrix} 1 & 1 & 1 & \cdots & 1 \\ 1 & n+1 & 1 & \cdots & 1 \\ 1 & 1 & n+1 & \ddots & \vdots \\ \vdots & \vdots & \ddots & \ddots & 1 \\ 1 & 1 & \cdots & 1 & n+1 \end{vmatrix}.$$

Next subtract row one from all other rows.

$$|A| = 2n \begin{vmatrix} 1 & 1 & 1 & \cdots & 1 \\ 0 & n & 0 & \cdots & 0 \\ 0 & 0 & n & \ddots & \vdots \\ \vdots & \vdots & \ddots & \ddots & 0 \\ 0 & 0 & \cdots & 0 & n \end{vmatrix}.$$

Now the matrix is upper triangular, so the determinant is the product of the diagonal elements and we get:$|A| = 2n(n^{n-1}) = 2n^n$.

Definition 3.11 *If $A = [a_{ij}]$, then the adjoint of A, written $adj(A)$, is defined by $adj(A) = [C_{ij}]^T$, where $C_{ij} = (-1)^{i+j} |M_{ij}|$.*

Remark *C_{ij} is what a_{ij} is multiplied by in the expansion by minors of $|A|$ around any the ith row or j^{th} column. For this reason C_{ij} is sometimes called the cofactor of a_{ij}.*

Lemma 3.12 $A \cdot adj(A) = |A| \, I = \begin{bmatrix} |A| & 0 & 0 & \cdots & 0 \\ 0 & |A| & 0 & \cdots & 0 \\ \vdots & 0 & \ddots & \ddots & \vdots \\ 0 & \vdots & \ddots & |A| & 0 \\ 0 & 0 & \cdots & 0 & |A| \end{bmatrix}.$

Proof. The diagonal elements of $A \cdot adj(A)$ are all $|A|$. This follows directly from the formula for expansion by minors. To see that the off diagonal elements are all zero, given positive integers i and j with $i \neq j$, form a new matrix A^* which agrees with A in all rows except row j, but row j of A^* has row i of A. Thus A^* has two identical rows, row i and row j. Thus $|A^*| = 0$. The minors of A^* for elements in row j are exactly the same as the minors of A for row j, since A^* differs from A only in row j. Expanding $|A^*|$ by minors around row j results in $0 = |A^*| = a_{i1}C_{j1} + a_{i2}C_{j2} + \cdots + a_{in}C_{jn}$. But this sum is the ij^{th} entry of $A \cdot adj(A)$. ∎

Corollary 1 *If $|A| \neq 0$, then $A^{-1} = \frac{1}{|A|} adj(A)$.*

Proof. It follows from the lemma that $\frac{1}{|A|} adj(A)$ is a right inverse of A. Since a right inverse must also be a left inverse, the corollary is proved. ∎

Remark *The method for finding the inverse of a 2×2 matrix in Procedure 2.7 is a special case of this corollary. For larger matrices row reducing $[A \; I]$ is usually a more efficient method of computing A^{-1}.*

Cramer's Rule

Consider the system of linear equations whose matrix form is $A\mathbf{x} = \mathbf{b}$. If the coefficient matrix A is invertible, Cramer's rule gives a formula for the solution. This is of more theoretical interest than practical use, since row reduction of the augmented matrix $[A\ \mathbf{b}]$ to RREF is a more efficient method of solving the system. However Cramer's rule does allow one to solve for one of the unknowns without finding the others.

Suppose you wish to solve $A\mathbf{x} = \mathbf{b}$, and assume $|A| \neq 0$. We know from previous theorems that this means that A is invertible and their is a unique solution for any $\mathbf{b}$. Let the columns of A be $\mathbf{u}_1, \mathbf{u}_2, \cdots, \mathbf{u}_n$. So that

$$A = \begin{bmatrix} \mathbf{u}_1 & \mathbf{u}_2 & \cdots & \mathbf{u}_n \end{bmatrix}.$$

Let

$$A_k = \begin{bmatrix} \mathbf{u}_1 & \mathbf{u}_2 & \cdots & \mathbf{u}_{k-1} & \mathbf{b} & \mathbf{u}_{k+1} & \cdots & \mathbf{u}_n \end{bmatrix},$$

be the matrix obtained from A by replacing the kth column by $\mathbf{b}$. Then

$$x_k = \frac{|A_k|}{|A|}$$

gives the solution of $A\mathbf{x} = \mathbf{b}$.

Proof.   We may write the system of equations in the form

$$x_1\mathbf{u}_1 + x_2\mathbf{u}_2 + \cdots + x_n\mathbf{u}_n = \mathbf{b}.$$

We know a unique solution exists because we are assuming $|A| \neq 0$. Let $x_1, x_2, \cdots x_n$ be the solution. Then substituting for $\mathbf{b}$ in A_k we get

$$|A_k| = \begin{vmatrix} \mathbf{u}_1 & \cdots & \mathbf{u}_{k-1} & x_1\mathbf{u}_1 + x_2\mathbf{u}_2 + \cdots + x_n\mathbf{u}_n & \mathbf{u}_{k+1} & \cdots & \mathbf{u}_n \end{vmatrix}.$$

Now we can write $|A_k|$ as a sum of determinants

$$|A_k| = \begin{vmatrix} \mathbf{u}_1 & \cdots & x_1\mathbf{u}_1 & \cdots & \mathbf{u}_n \end{vmatrix} + \cdots + \begin{vmatrix} \mathbf{u}_1 & \cdots & x_k\mathbf{u}_k & \cdots & \mathbf{u}_n \end{vmatrix} + \cdots + \begin{vmatrix} \mathbf{u}_1 & \cdots & x_n\mathbf{u}_n & \cdots & \mathbf{u}_n \end{vmatrix}.$$

Factoring out the scalars gives

$$|A_k| \;=\; x_1 \left| \begin{array}{ccccc} \mathbf{u}_1 & \cdots & \mathbf{u}_1 & \cdots & \mathbf{u}_n \end{array} \right| + \cdots + x_k \left| \begin{array}{ccccc} \mathbf{u}_1 & \cdots & \mathbf{u}_k & \cdots & \mathbf{u}_n \end{array} \right| +$$
$$\cdots + x_n \left| \begin{array}{ccccc} \mathbf{u}_1 & \cdots & \mathbf{u}_n & \cdots & \mathbf{u}_n \end{array} \right|$$

Now we observe that all but one of the determinants in this sum are zero because of a repeated row. The only term that survives is $x_k \left| \begin{array}{ccccc} \mathbf{u}_1 & \cdots & \mathbf{u}_k & \cdots & \mathbf{u}_n \end{array} \right|$. Thus we have

$$|A_k| \;=\; x_k \left| \begin{array}{ccccc} \mathbf{u}_1 & \cdots & \mathbf{u}_k & \cdots & \mathbf{u}_n \end{array} \right|$$
$$= x_k |A|,$$

and solving for x_k gives the desired result. ∎

Vandermonde Matrices

Another class of matrices whose special character makes their determinants easy to find are the Vandermonde matrices. Such a matrix is of the form

$$A = \begin{bmatrix} 1 & x_1 & x_1^2 & x_1^3 & \cdots & x_1^{n-1} \\ 1 & x_2 & x_2^2 & x_2^3 & \cdots & x_2^{n-1} \\ \vdots & \vdots & \vdots & \vdots & & \vdots \\ 1 & x_n & x_n^2 & x_n^3 & \cdots & x_n^{n-1} \end{bmatrix}.$$

Using mathematical induction, it can be shown that the formula for the determinant of A is

$$|A| = \prod_{j>i}(x_j - x_i).$$

This means multiply together all differences obtained by taking an element in the second column and subtracting from it an element above it. A Vandermonde matrix with x_i all distinct will have nonzero determinant and will thus be invertible.

We give a proof for the 3×3 case.

$$|A| = \begin{vmatrix} 1 & a & a^2 \\ 1 & b & b^2 \\ 1 & c & c^2 \end{vmatrix} = (b-a)(c-a)(c-b).$$

Proof.   Subtract a times column two from column three, then subtract a times column one from column two. This gives

$$|A| = \begin{vmatrix} 1 & 0 & 0 \\ 1 & b-a & b^2 - ab \\ 1 & c-a & c^2 - ac \end{vmatrix}$$

$$= \begin{vmatrix} 1 & 0 & 0 \\ 1 & b-a & b(b-a) \\ 1 & c-a & c(c-a) \end{vmatrix}.$$

Expanding by minors around the first row gives

$$|A| = \begin{vmatrix} b-a & b(b-a) \\ c-a & c(c-a) \end{vmatrix}$$

$$= (b-a)(c-a) \begin{vmatrix} 1 & b \\ 1 & c \end{vmatrix}$$

$$= (b-a)(c-a)(c-b).$$

■

The same steps will reduce the $n \times n$ case to the $(n-1) \times (n-1)$ case. First Starting at the right, from each column subtract x_1 times the column to the left of it. Then expand around row 1. You will get just one $(n-1) \times (n-1)$ matrix. Factor out $x_k - x_1$ from row $k-1$ of this matrix and you will have an $(n-1) \times (n-1)$ Vandermonde matrix and you can use the induction hypothesis.

Example 3.3 *Let* $A = \begin{bmatrix} 1 & 2 & 4 & 8 \\ 1 & 3 & 9 & 27 \\ 1 & 5 & 25 & 125 \\ 1 & -2 & 4 & -8 \end{bmatrix}$. *This is a Vandermonde matrix with* $x_1 = 2, x_2 = 3, x_3 = 5,$ *and* $x_4 = -2$. *Using the formula for the determinant of a Vandermonde matrix we have*

$$|A| = (3-2)(5-2)(-2-2)(5-3)(-2-3)(-2-5)$$
$$= 1 \cdot 3(-4)2(-5)(-7) = -840.$$

Dominant Diagonal
Definition 3.12 *An* $n \times n$ *matrix* $A = [a_{ij}], n > 1,$ *is said to have dominant diagonal if*

$$a_{jj} > \sum_{k=1, k \neq j}^{n} |a_{jk}|.$$

For a 1×1 matrix, dominant diagonal will just mean $a_{11} > 0$.

Theorem 3.6 *If A has dominant diagonal then $|A| > 0$.*

Proof.   The proof is by induction on n. If $n = 1$, then $a_{11} > 0$ and thus $|A| > 0$. Assume the result is true for $(n-1) \times (n-1)$ matrices that have dominant diagonal. Let A be an $n \times n$ matrix with dominant diagonal. Expand A by minors around the first column of A. Thus $|A| = a_{11} |M_{11}| - a_{21} |M_{21}| + \cdots (-1)^{n+1} a_{n1} |M_{n1}|$, where M_{j1} are the minors of A obtained by crossing out row j and column 1. Since M_{11} will again have dominant diagonal, it follows from the induction hypothesis that $|M_{11}| > 0$. We shall form a new matrix $B = [b_{ij}]$ obtained from A by either adding column j ot column 1 or subtracting column j from column 1 for all $j > 1$. In all events this does not change the determinant, so that $|B| = |A|$. In expanding $|B|$ by minors around the first column we observe that the last $n-1$ columns of B and A are identical, and therefore so are the minors of the first column. Thus we have

$|A| = |B| = b_{11} |M_{11}| - b_{21} |M_{21}| + \cdots (-1)^{n+1} b_{n1} |M_{n1}|$. Now it is time to specify which B we want. This is determined solely on the basis of whether $(-1)^{j+1} |M_{j1}|$ is positive or negative. If positive, then add column j to column 1 and if negative, then subtract column j from column 1. Since the diagonal elements are dominant this guarantees that all the n terms in the expansion of $|B|$ above are nonnegative. But we know that $|M_{11}| > 0$, and $b_{11} > 0$, so that $b_{11} |M_{11}| > 0$. Therefore $|A| > 0$. This completes the induction. ■

Corollary 1

$$|A| \geq \left\{ a_{11} - \sum_{j=2}^{n} |a_{1j}| \right\} \left\{ a_{22} - \sum_{j=3}^{n} |a_{2j}| \right\} \cdots \{a_{nn}\},$$

and

$$2^{n-1} a_{11} a_{22} \cdots a_{nn} \geq \left\{ a_{11} + \sum_{j=2}^{n} |a_{1j}| \right\} \left\{ a_{22} + \sum_{j=3}^{n} |a_{2j}| \right\} \cdots \{a_{nn}\} \geq |A|.$$

Proof. By induction on n. All the inequalities are trivially true for $n = 1$. Assume true for matrices of size $(n-1) \times (n-1)$ with dominant diagonal. It follows from the proof of the preceding theorem that

$$|A| \geq \left\{ a_{11} - \sum_{j=2}^{n} |a_{1j}| \right\} |M_{11}|,$$

and since M_{11} has dominant diagonal, we can use the induction hypothesis. This gives the lower bound on $|A|$. We now go after the upper bound on $|A|$. In the proof of the theorem above we added or subtracted column j from column 1 to obtain a matrix B from A. Now we will get another matrix C in the same way, only this time we will fix it up so that $c_{k1}(-1)^{k+1} |M_{k1}|$ is nonpositive instead of nonnegative. This will mean that if we omit all but the first term in the expansion of $|C| = |A|$ around the first column we will have $|A| \leq c_{11} |M_{11}| \leq (a_{11} + |a_{12}| + |a_{13}| + \cdots + |a_{1n}|) |M_{11}| \leq 2a_{11} |M_{11}|$. Applying the induction hypothesis to $|M_{11}|$, the desired result follows.

Example 3.4 Let $A = \begin{bmatrix} 5 & 0 & 1 & -1 \\ 3 & 6 & 0 & 1 \\ -2 & -3 & 7 & -1 \\ 1 & 2 & -4 & 8 \end{bmatrix}$. Then A has dominant diagonal. One may wish to verify that $|A| = 1,561$. The corollary gives the following estimates $|A| \geq (3)(5)(6)(8) = 720$, and $|A| \leq (7)(7)(8)(8) \leq 2^3(5)(6)(7)(8)$, or $|A| \leq 3,136 \leq 13,440$.

Exercises

1. For what values of a is $\begin{vmatrix} a & 1 & 2 \\ -1 & 3 & 5 \\ 2 & 4 & 1 \end{vmatrix} = 0$?

2. Evaluate

   (a) $\begin{vmatrix} 1 & 2 & 3 \\ 1 & 4 & 9 \\ 1 & 8 & 27 \end{vmatrix}$

(b) $\begin{vmatrix} 1 & 2 & 1 \\ 2 & 1 & 0 \\ -1 & -3 & 2 \end{vmatrix}$

3. Evaluate

(a) $\begin{vmatrix} 5 & 1 & 1 & 1 \\ 1 & 5 & 1 & 1 \\ 1 & 1 & 5 & 1 \\ 1 & 1 & 1 & 5 \end{vmatrix}$

(b) $\begin{vmatrix} 7 & 1 & 1 & 1 \\ 1 & 7 & 1 & 1 \\ 1 & 1 & 7 & 1 \\ 1 & 1 & 1 & 7 \end{vmatrix}$.

4. Evaluate $\begin{vmatrix} e^x & e^x & e^x \\ xe^x & e^x + xe^x & 2e^x + xe^x \\ x^2e^x & x^2e^x + 2xe^x & x^2e^x + 4xe^x + 2e^x \end{vmatrix}$.

5. (a) Are the row vectors of the matrices in problems 2,3,and 4 linearly independent?

   (b) Same question for the column vectors.

6. Suppose the row vectors v_j of an $n \times n$ matrix A are nonzero and satisfy $v_i \cdot v_j = 0$ for $i \neq j$. Prove that $|A| \neq 0$. (Hint: consider AA^T.)

7. Use the corollary to Lemma 3.12 to invert the following matrices:

   (a) $\begin{bmatrix} 2 & 1 \\ 3 & 2 \end{bmatrix}$

   (b) $\begin{bmatrix} 1 & 1 & 1 \\ 2 & 1 & 0 \\ 1 & 2 & 1 \end{bmatrix}$.

8. If A is an $n \times n$ matrix with a everywhere on the main diagonal and b everywhere else, find $|A|$.

9. If A is $n \times n$ and invertible, find the following in terms of $|A|$:

(a) $|A^{-1}|$

(b) $|cA|$

(c) $\left|A^k\right|$

(d) $|adj(A)|$.

10. Evaluate
$$
\begin{vmatrix}
1 & 1 & -1 & 2 & 1 \\
2 & -1 & 4 & 3 & 1 \\
1 & 3 & 1 & 0 & 1 \\
0 & 2 & 4 & 2 & 1 \\
1 & 0 & -1 & 0 & 1
\end{vmatrix}.
$$

11. Solve the following system of equations for y using Cramer's rule

$$
\begin{array}{rcrcrcr}
-x & - & 2y & + & z & = & 3 \\
-x & + & 7y & - & 3z & = & 12 \\
2x & - & 4y & + & 8z & = & 5
\end{array}
$$

Recall that $Ax = b$ can also be solved by row reducing $\begin{bmatrix} A & b \end{bmatrix}$, or by finding A^{-1} (when $|A| \neq 0$), in which case the solution is $x = A^{-1}b$. Which of the three methods is least work? Check your Cramer's Rule answer by one of the other methods.

12. Solve the following system of equations for each of the unknowns using Cramer's rule:

$$
\begin{array}{rcrcrcrcr}
w & + & x & + & y & + & z & = & 4 \\
2w & + & 4x & + & 8y & + & 16z & = & 28 \\
3w & + & 9x & + & 27y & + & 81z & = & 120 \\
w & - & x & + & y & - & z & = & 0
\end{array}
$$

13. Solve the following system for x and y in terms of $a, b, c, d, r,$ and s, using Cramer's rule, assuming $ad - bc \neq 0$:

$$
\begin{array}{rcl}
ax + by & = & r \\
cx + dy & = & s
\end{array}
$$

14. Find $|A|$ for the following Vandermonde matrix A, using the for-
mula for Vandermonde determinants. $A = \begin{bmatrix} 1 & 3 & 9 & 27 \\ 1 & 7 & 49 & 343 \\ 1 & 2 & 4 & 8 \\ 1 & -4 & 16 & -64 \end{bmatrix}$.

Check your answer by finding the determinant some other way.

15. Let $A = \begin{bmatrix} 6 & 0 & -1 & 3 \\ 1 & 8 & 2 & 3 \\ 2 & 0 & 5 & 1 \\ 1 & 1 & 1 & 4 \end{bmatrix}$.

   (a) Verify that A has dominant diagonal and is therefore invertible.

   (b) Find upper and lower bounds for $|A|$ as done in example 3.4.

   (c) Find $|A|$.

Section 3.3 Determinants of Block Triangular Matrices

Theorem 3.7 *Let* $M = \begin{bmatrix} A & C \\ [0] & B \end{bmatrix}$ *be a square matrix with square blocks A and B on the diagonal. Let M be $n \times n$, A $r \times r$, and B $s \times s$, with $n = r + s$, and assume there is an $s \times r$ block of zeros below A. Then $|M| = |A| \, |B|$.*

Proof. Since $M = [m_{ij}]$ has the property $m_{ij} = 0$ whenever $i > r$ and $j \le r$, then in

$$|M| = \sum_{P \in S_n} sgn(P) m_{1P,1} m_{2P,2} \cdots m_{nP,n}$$

we need only consider those permutations P which the map subset $\{1, 2, \cdots r\}$ onto itself, for if $jP > r$ while $j \le r$, we have $m_{jP,j} = 0$, and then whole term going with P is 0. Such a permutation must then map $\{r+1, r+2, \cdots, n\}$ onto itself also. Thus P can be written as a product of a permutation P_1 which fixes $\{r+1, r+2, \cdots, n\}$ and agrees with P on $\{1, 2, \cdots r\}$, and a permutation P_2 which fixes $\{1, 2, \cdots r\}$ and agrees with P on $\{r+1, r+2, \cdots, n\}$. Thus $sgn(P) = sgn(P_1)sgn(P_2)$. Let P_1' be P_1 restricted to $\{1, 2, \cdots r\}$, and let P_2' be P_2 restricted to $\{r+1, r+2, \cdots, n\}$. Then $sgn(P_1') = sgn(P_1)$ and $sgn(P_2') = sgn(P_2)$ and we have

$$sgn(P)m_{1P,1}m_{2P,2} \cdots m_{nP,n}$$
$$= sgn(P_1')a_{1P_1,1} \cdots a_{rP_1,r}sgn(P_2')b_{(r+1)P_2,r+1} \cdots b_{nP_2,n}.$$

This implies $|M| = |A|\,|B|$. ∎

Corollary 1 *If* $M = \begin{bmatrix} A_1 & & & & \\ 0 & A_2 & & * & \\ 0 & 0 & \ddots & & \\ \vdots & \vdots & & \ddots & \ddots \\ 0 & 0 & \cdots & 0 & A_s \end{bmatrix}$, *is an* $n \times n$ *upper triangular block matrix with* s *square blocks on the diagonal then* $|M| = |A_1|\,|A_2| \cdots |A_s|$.

Proof. The proof is by induction on s. We have proved the result for $s = 2$. Now If we let $A = \begin{bmatrix} A_1 & & & & \\ 0 & A_2 & & * & \\ 0 & 0 & \ddots & & \\ \vdots & \vdots & & \ddots & \ddots \\ 0 & 0 & \cdots & 0 & A_{s-1} \end{bmatrix}$ and $B = A_s$.

Then $M = \begin{bmatrix} A & * \\ [0] & B \end{bmatrix}$, so by the previous theorem $|M| = |A|\,|B|$, but $|B| = |A_s|$, and by the induction hypothesis $|A| = |A_1|\,|A_2| \cdots |A_{s-1}|$, so $|M| = |A_1|\,|A_2| \cdots |A_s|$ as desired. ∎

Exercises
Find the determinants of the following matrices.

1. $\begin{bmatrix} 1 & 2 & 0 & 0 \\ 3 & 4 & 0 & 0 \\ 9 & 10 & 5 & 6 \\ 1 & 12 & 7 & 8 \end{bmatrix}.$

2. $\begin{bmatrix} 1 & 3 & 4 & 2 \\ 3 & 1 & 2 & 4 \\ 0 & 0 & 3 & 5 \\ 0 & 0 & 5 & 3 \end{bmatrix}.$

3. $\begin{bmatrix} 1 & 2 & 3 & 4 & 4 & 4 \\ 3 & 2 & 1 & 4 & 4 & 4 \\ 4 & 5 & 1 & 4 & 4 & 4 \\ 0 & 0 & 0 & 2 & 3 & 4 \\ 0 & 0 & 0 & 4 & 3 & 2 \\ 0 & 0 & 0 & 5 & 6 & 2 \end{bmatrix}.$

4. $\begin{bmatrix} 1 & 2 & 3 & 3 & 3 & 3 \\ 2 & 1 & 4 & 4 & 4 & 4 \\ 0 & 0 & 1 & 2 & 5 & 6 \\ 0 & 0 & 2 & 1 & 5 & 6 \\ 0 & 0 & 0 & 0 & 2 & 1 \\ 0 & 0 & 0 & 0 & 1 & 2 \end{bmatrix}.$

Chapter 4

Eigenvalues, Eigenvectors, and Diagonalization

Section 4.1 Eigenvalues and Eigenvectors

Definition 4.1 *A mapping T from R^n into R^n is called a linear transformation on R^n if for all $\mathbf{u}, \mathbf{v} \in R^n$ and all $c \in R$ we have*

    i) $T(\mathbf{u} + \mathbf{v}) = T(\mathbf{u}) + T(\mathbf{v})$ *and*

    ii) $T(c\mathbf{u}) = cT(\mathbf{u})$.

In this section we will assume all vectors are column vectors. Thus $\mathbf{u}$ will be a column vector and $\mathbf{u}^T$ will be a row vector. In particular we denote the natural basis vectors of R^n by $\mathbf{e}_j$, the column n-tuple with all zeros except a one in the jth row.

For any $n \times n$ matrix A if we define a mapping by $T(\mathbf{u}) = A\mathbf{u}$, then this will be a linear transformation from R^n to R^n, since for any vectors $\mathbf{u}, \mathbf{w} \in R^n$ we have $T(\mathbf{u} + \mathbf{w}) = A(\mathbf{u} + \mathbf{w}) = A\mathbf{u} + A\mathbf{w} = T(\mathbf{u}) + T(\mathbf{w})$, and $T(c\mathbf{u}) = A(c\mathbf{u}) = cA\mathbf{u} = cT(\mathbf{u})$. Moreover all linear transformations from R^n to R^n are of this form, for if $T(\mathbf{e}_j) = \mathbf{v}_j$, then using properties i and ii freely

$$T \begin{bmatrix} c_1 \\ c_2 \\ \vdots \\ c_n \end{bmatrix} = T(c_1 \mathbf{e}_1 + c_2 \mathbf{e}_2 + \cdots + c_n \mathbf{e}_n)$$

$$= c_1 T(\mathbf{e}_1) + c_2 T(\mathbf{e}_2) + \cdots + c_n T(\mathbf{e}_n)$$

$$= c_1 \mathbf{v}_1 + c_2 \mathbf{v}_2 + \cdots + c_n \mathbf{v}_n$$

$$= \begin{bmatrix} \mathbf{v}_1 & \mathbf{v}_2 & \cdots & \mathbf{v}_n \end{bmatrix} \begin{bmatrix} c_1 \\ c_2 \\ \vdots \\ c_n \end{bmatrix}$$

$$= A \begin{bmatrix} c_1 \\ c_2 \\ \vdots \\ c_n \end{bmatrix} .$$

where $A = \begin{bmatrix} \mathbf{v}_1 & \mathbf{v}_2 & \cdots & \mathbf{v}_n \end{bmatrix}$. Thus $T(\mathbf{u}) = A\mathbf{u}$ for any column vector $\mathbf{u}$. We have proved the following:

Theorem 4.1 *The linear transformations from R^n to R^n are exactly the mappings $T : \mathbf{u} \to A\mathbf{u}$ where A is any $n \times n$ matrix, and the relation between T and A is $T(\mathbf{u}) = A\mathbf{u}$ if and only if the images of the natural basis vectors are the columns of A, i.e. $T(\mathbf{e}_j) = \mathbf{v}_j$, where $\mathbf{v}_j$ is the jth column of A.*

Definition 4.2 *A number k is an eigenvalue of a linear transformation T if there exists a nonzero vector $\mathbf{v} \in R^n$ such that $T(\mathbf{v}) = k\mathbf{v}$. The vector $\mathbf{v}$ is called an eigenvector of T for the eigenvalue k.*

If A is the matrix associated with T we have $T(\mathbf{v}) = A\mathbf{v}$, and so solving $T(\mathbf{v}) = k\mathbf{v}$ is equivalent to solving $A\mathbf{v} = k\mathbf{v}$. The eigenvalues and eigenvectors of T are also called eigenvalues and eigenvectors of the matrix A. "Eigen" is a German word but is widely used in this context. The closest English translation is "characteristic", thus eigenvectors are sometimes called characteristic vectors, and eigenvalues are sometimes called characteristic values.

The equation $A\mathbf{v} = k\mathbf{v}$ is equivalent to $(A - kI)\mathbf{v} = \mathbf{O}$, so k is an eigenvalue of A if and only if this equation has nonzero solutions. The eigenvectors of A are the nonzero vectors in the null space of $(A - kI)$. Now $(A - kI)\mathbf{v} = \mathbf{O}$ has nonzero solutions if and only if $|A - kI| = 0$ (the determinant of the coefficient matrix equals zero). We can use this to find all the eigenvalues. Once we have found an eigenvalue k, we know how to find the null space of $(A - kI)$, so we can find the eigenvectors of A for k. We state these observations as a theorem.

Theorem 4.2 *The eigenvalues of A are those numbers k for which $|A - kI| = 0$. If k is an eigenvalue of A, the eigenvectors of A for k are the nonzero solutions of $(A - kI)\mathbf{v} = \mathbf{O}$.*

Definition 4.3 *If k is an eigenvalue of A, the eigenspace of A for the eigenvalue k is the null space of $(A - kI)$, that is all solutions of $(A - kI)\mathbf{v} = \mathbf{O}$.*

The eigenspace of A for k is a subspace of R^n of dimension $\{n - rank(A - kI)\}$, and consists of all the eigenvectors of A for k and the zero vector.

Example 4.1 *Let* $A = \begin{bmatrix} -1 & 2 & 0 \\ 1 & 2 & 1 \\ 0 & 2 & -1 \end{bmatrix}$. *Then*

$$|A - kI| = \begin{vmatrix} -1 - k & 2 & 0 \\ 1 & 2 - k & 1 \\ 0 & 2 & -1 - k \end{vmatrix}.$$

The eigenvalues of A will be the values of k which make this determinant 0. If we subtract column one from column three, we don't change the determinant, so

$$|A - kI| = \begin{vmatrix} -1 - k & 2 & 1 + k \\ 1 & 2 - k & 0 \\ 0 & 2 & -1 - k \end{vmatrix}.$$

Now add row one to row three to get

$$|A - kI| = \begin{vmatrix} -1 - k & 2 & 1 + k \\ 1 & 2 - k & 0 \\ -1 - k & 4 & 0 \end{vmatrix}.$$

Expanding around column three gives

$$|A - kI| = (1 + k) \begin{vmatrix} 1 & 2 - k \\ -1 - k & 4 \end{vmatrix}$$
$$= (1 + k)(4 - (2 - k)(-1 - k))$$
$$= (1 + k)(-k^2 + k + 6)$$
$$= -(1 + k)(k^2 - k - 6)$$
$$= -(k + 1)(k - 3)(k + 2).$$

Thus the eigenvalues are $k = -1, 3,$ *and* -2. *To find the eigenvectors that go with* $k = -1$, *we look for the null space of the matrix*

$$(A - (-1)I) = \begin{bmatrix} 0 & 2 & 0 \\ 1 & 3 & 1 \\ 0 & 2 & 0 \end{bmatrix}.$$

$$RREF\,(A - (-1)I) = \begin{bmatrix} 1 & 0 & 1 \\ 0 & 1 & 0 \\ 0 & 0 & 0 \end{bmatrix},$$

which is also the Hermite form H, *so forming*

$$(H - I) = \begin{bmatrix} 0 & 0 & 1 \\ 0 & 0 & 0 \\ 0 & 0 & -1 \end{bmatrix},$$

we see that $\begin{bmatrix} 1 \\ 0 \\ -1 \end{bmatrix}$ *is a basis of this null space of* $(A - (-1)I)$, *and hence of the eigenspace of* A *for* $k = -1$. *To find the eigenvectors that go with* $k = 3$, *we look for the null space of the matrix*

$$(A - 3I) = \begin{bmatrix} -4 & 2 & 0 \\ 1 & -1 & 1 \\ 0 & 2 & -4 \end{bmatrix}.$$

The RREF of this matrix is

$$RREF(A - 3I) = \begin{bmatrix} 1 & 0 & -1 \\ 0 & 1 & -2 \\ 0 & 0 & 0 \end{bmatrix},$$

this is also the Hermite form H, so

$$(H - I) = \begin{bmatrix} 0 & 0 & -1 \\ 0 & 0 & -2 \\ 0 & 0 & -1 \end{bmatrix},$$

and $\begin{bmatrix} -1 \\ -2 \\ -1 \end{bmatrix}$ *is a basis of the eigenspace for* $k = 3$. *Last for* $k = -2$,

$$(A - (-2)I) = \begin{bmatrix} 1 & 2 & 0 \\ 1 & 4 & 1 \\ 0 & 2 & 1 \end{bmatrix},$$

with

$$RREF(A) = \begin{bmatrix} 1 & 0 & -1 \\ 0 & 1 & \frac{1}{2} \\ 0 & 0 & 0 \end{bmatrix}.$$

A basis for this eigenspace is $\begin{bmatrix} -1 \\ \frac{1}{2} \\ -1 \end{bmatrix}$.

Remark   *When you are looking for the eigenvectors of A for an eigenvalue c and you row reduce* $(A - cI)$ *to its RREF, you will always get at least one zero row. If you do not, it means that either c is not really an eigenvalue of A or you have made a mistake in the row reduction. This is because if c really is an eigenvalue of A, then* $(A - cI)\mathbf{v} = \mathbf{O}$ *has nonzero solutions, so* $RREF(A - cI)$ *cannot be the identity matrix.*

Remark   **(Warning)** *When you are looking for eigenvalues of A, do not perform row or column operations on A before subtracting the variable k from the diagonal elements. If you do, you are likely to change the eigenvalues. As in the example above, you can perform row and column operations on* $(A - kI)$, *keeping track of their effect on* $|A - kI|$. *It is not true that if B comes from A by elementary row operations A and B always have the same eigenvalues. (See exercises 8 and 9 at the end of this section.)*

Remark   *When you have found an eigenvector* **u** *of the matrix A for the value c, it is easy to check your answer. Just multiply A times* **u** *to be sure that A**u** = c**u***.

If a matrix is triangular (upper or lower) the eigenvalues are easy to find, they are just the diagonal entries. This is because the determinant of a triangular matrix is the product of the diagonal elements.

Example 4.2 *Consider the matrix* $A = \begin{bmatrix} 2 & 4 & 1 \\ 0 & 2 & -1 \\ 0 & 0 & 3 \end{bmatrix}$. *Then* $|A - kI| =$

$\begin{vmatrix} 2-k & 4 & 1 \\ 0 & 2-k & -1 \\ 0 & 0 & 3-k \end{vmatrix} = (2-k)(2-k)(3-k)$, *and the eigenvalues are*

$k = 2$ *and* 3. *It is left as an exercise for the reader to find the eigenvectors and check them.*

Definition 4.4 *The characteristic polynomial of an* $n \times n$ *matrix A is* $(-1)^n |A - kI| = |kI - A| = p(k)$.

Definition 4.5 *If* $p(k)$ *is the characteristic polynomial of A, the equation* $p(k) = 0$, *is the characteristic equation of A.*

Let us examine the characteristic polynomial of A in some detail. When expanding the determinant $|kI - A|$, powers of k as high as n and $n - 1$ appear only in the product $(k - a_{11})(k - a_{22}) \cdots (k - a_{nn})$. Since

$$(k-a_{11})(k-a_{22}) \cdots (k-a_{nn}) = k^n - (a_{11}+a_{22}+\cdots+a_{nn})k^{n-1}+\cdots+t_1 k+t_0,$$

$p(k)$ must have degree n and must be of the form

$$p(k) = k^n - (a_{11} + a_{22} + \cdots + a_{nn})k^{n-1} + \cdots + s_1 k + s_0.$$

Substituting $k = 0$ into the expression above gives $p(0) = s_0$. Then using the definition of $p(k)$ we have

$$s_0 = p(0) = |0I - A| = |-A|.$$

Definition 4.6 *Let A be an $n \times n$ matrix, $A = [a_{ij}]$, then $tr(A) = a_{11} + a_{22} + \cdots + a_{nn}$. This is called the trace of A.*

We have proved the following theorem.

Theorem 4.3 *If $p(k) = k^n + s_{n-1}k^{n-1} + \cdots s_1 k + s_0$, is the characteristic polynomial of an $n \times n$ matrix A, then $s_{n-1} = -tr(A)$ and $s_0 = (-1)^n |A|$. For $n = 2$ we have $p(k) = k^2 - tr(A)k + |A|$.*

For $n > 2$ calculating $p(k)$ takes time and there are many opportunities for making mistakes. Fortunately there are computer programs that can find $p(k)$ for reasonable n. However $tr(A)$ is very easy to calculate and always a good check on $p(k)$.

Example 4.3 *Let $A = \begin{bmatrix} 1 & 2 \\ 3 & 4 \end{bmatrix}$, find $p(k)$. Solution: $tr(A) = 5$, and $|A| = 4 - 6 = -2$, so*

$$p(k) = k^2 - 5k - 2.$$

The eigenvalues of A are

$$c_1 = \frac{5 + \sqrt{33}}{2}, \text{ and } c_2 = \frac{5 - \sqrt{33}}{2}.$$

One can also calculate $|kI - A|$ directly.

$$\begin{vmatrix} k-1 & -2 \\ -3 & k-4 \end{vmatrix} = (k-1)(k-4) - 6$$

$$= k^2 - 5k - 2.$$

Remark *Although $p(k)$ is defined as $|kI - A|$, it is more convenient to calculate $|A - kI| = (-1)^n p(k)$, since in the latter we do not have to change the signs of all the entries of A, but just subtract k form the diagonal entries. If n is even these are the same, if n is odd, one is the negative of the other. To find the eigenvalues of A we are going to solve $p(k) = 0$, and $-p(k) = 0$ has the same solutions.*

Example 4.4 *Let* $B = \begin{bmatrix} 0 & 1 \\ -1 & 0 \end{bmatrix}$. *Then* $tr(B) = 0$ *and* $|B| = 1$.
Thus $p(k) = k^2 + 1$. *The roots of this equation are complex. B has no real eigenvalues.*

Example 4.5 *Let* $C = \begin{bmatrix} 0 & -c_0 \\ 1 & -c_1 \end{bmatrix}$, $tr(C) = -c_1$ *and* $|C| = c_0$. *Thus* $p(k) = k^2 + c_1 k + c_0$.

Example 4.6 *Let* $D = \begin{bmatrix} 2 & 0 & 1 \\ 0 & 3 & 0 \\ 1 & 0 & 2 \end{bmatrix}$. *Then*

$$|D - kI| = \begin{vmatrix} 2-k & 0 & 1 \\ 0 & 3-k & 0 \\ 1 & 0 & 2-k \end{vmatrix}.$$

Expanding by minors around row 2 gives

$$|D - kI| = (3 - k)\begin{vmatrix} 2-k & 1 \\ 1 & 2-k \end{vmatrix}$$
$$= (3 - k)(k^2 - 4k + 3)$$
$$= (3 - k)(k - 3)(k - 1).$$

Thus $p(k) = (k - 1)(k - 3)^2$. *The eigenvalues of D are 1 and 3. Three is a root of multiplicity two, or a double root. We continue this example by computing the eigenspaces of D. For* $k = 3$, *we have*

$$(D - 3I) = \begin{bmatrix} -1 & 0 & 1 \\ 0 & 0 & 0 \\ 1 & 0 & -1 \end{bmatrix}.$$

$$RREF\,(D - 3I) = \begin{bmatrix} 1 & 0 & -1 \\ 0 & 0 & 0 \\ 0 & 0 & 0 \end{bmatrix},$$

which is also the Hermite form H of $(D - 3I)$. *Thus*

$$(H - I) = \begin{bmatrix} 0 & 0 & -1 \\ 0 & -1 & 0 \\ 0 & 0 & -1 \end{bmatrix},$$

giving $\begin{bmatrix} 0 \\ 1 \\ 0 \end{bmatrix}$ *and* $\begin{bmatrix} 1 \\ 0 \\ 1 \end{bmatrix}$ *as a basis of the eigenspace of D for* $k = 3$.

Now for $k = 1$,

$$(D - 1I) = \begin{bmatrix} 1 & 0 & 1 \\ 0 & 2 & 0 \\ 1 & 0 & 1 \end{bmatrix}.$$

Clearly this matrix has rank 2, so the null space will have dimension 1. It is also easy to see that $\begin{bmatrix} 1 \\ 0 \\ -1 \end{bmatrix}$ *is in the null space of this matrix.*

Thus $\begin{bmatrix} 1 \\ 0 \\ -1 \end{bmatrix}$ *is a basis of the eigenspace of D for* $k = 1$. *(You could also have used the Hermite form of* $(D - 1I)$ *as above to find a basis of this eigenspace.)*

Definition 4.7 *An* $n \times n$ *matrix A is said to be similar to an* $n \times n$ *matrix B if there exists an invertible* $n \times n$ *matrix P such that* $B = P^{-1}AP$. *We write* $A \approx B$ *to indicate that A is similar to B.*

Lemma 4.1 *(1)* $A \approx A$
(2) $A \approx B$ *implies* $B \approx A$
(3) $A \approx B$ *and* $B \approx C$ *implies* $A \approx C$.

Proof.   Left to the reader.

Definition 4.8 *Any relation which satisfies (1), (2), and (3) is called an equivalence relation.*

The relation of similarity separates all $n \times n$ matrices into mutually exclusive equivalence classes, so that all matrices similar to a fixed A belong to one class and no other.

Example 4.7 *The identity matrix I of size* $n \times n$ *is similar only to itself, since* $P^{-1}IP = P^{-1}P = I$. *In fact if A is a scalar matrix, A commutes with all* $n \times n$ *matrices thus we also have* $P^{-1}AP = P^{-1}PA = A$, *and so each scalar matrix is similar only to itself.*

Theorem 4.4 *If A is similar to B, then B and A have the identical characteristic polynomials, and hence have the same trace, same determinant, and same eigenvalues. In addition they have the same rank.*

Proof.   Let $B = P^{-1}AP$. Then $(B - kI) = (P^{-1}AP - kI)$. However $kI = kIP^{-1}P = P^{-1}kIP$, so that $P^{-1}AP - kI = P^{-1}AP - P^{-1}kIP = P^{-1}(A - kI)P$. Consequently $|B - kI| = |P^{-1}(A - kI)P|$. But since $|CD| = |C| |D|$, we get

$$
\begin{aligned}
|B - kI| &= \left|P^{-1}(A - kI)P\right| \\
&= \left|P^{-1}\right| |A - kI| |P| \\
&= \left|P^{-1}\right| |P| |A - kI| \\
&= |A - kI|.
\end{aligned}
$$

Thus we have shown that $|B - kI| = |A - kI|$, so that B and A have identical characteristic polynomials. That B and A have the same rank follows from the fact that multiplication by an invertible matrix does not change the rank (Theorem 2.12). ∎

Does the converse of Theorem 4 hold? In other words if A and B are two matrices having the same rank and same characteristic polynomial does this guarantee that $A \approx B$? The answer is no and we present an example to show this.

Example 4.8 *Let* $A = \begin{bmatrix} 2 & 0 & 0 \\ 1 & 2 & 0 \\ 0 & 0 & 2 \end{bmatrix}$, $B = \begin{bmatrix} 2 & 0 & 0 \\ 1 & 2 & 0 \\ 0 & 1 & 2 \end{bmatrix}$. *Suppose that* $A \approx B$. *Then* $B = P^{-1}AP$. *Therefore as we have just seen in the proof of the preceding theorem,* $(A - 2I)$ *and* $(B - 2I)$ *are similar. But*

$(A - 2I) = \begin{bmatrix} 0 & 0 & 0 \\ 1 & 0 & 0 \\ 0 & 0 & 0 \end{bmatrix}$ *and* $(B - 2I) = \begin{bmatrix} 0 & 0 & 0 \\ 1 & 0 & 0 \\ 0 & 1 & 0 \end{bmatrix}$. *Now* $(A - 2I)$

has rank one, while $(B - 2I)$ *has rank two, a contradiction. Consequently A and B are not similar. On the other hand A and B have the same rank and the same characteristic polynomial, namely* $(k-2)^3$. *We see that a necessary condition for* $A \approx B$ *is that* $(A - kI) \approx (B - kI)$ *for all k, but even this condition added to the other two is not enough* *to guarantee that A and B are similar. The problem of finding necessary and sufficient conditions for the similarity of two matrices has been solved but is beyond the scope of this book.*

Definition 4.9 *An $n \times n$ matrix A is said to be diagonalizable if there exists an invertible matrix P such that*

$$P^{-1}A\,P = \begin{bmatrix} d_1 & 0 & 0 & \cdots & 0 \\ 0 & d_2 & 0 & \cdots & 0 \\ 0 & 0 & d_3 & \ddots & \vdots \\ \vdots & \vdots & \ddots & \ddots & 0 \\ 0 & 0 & \cdots & 0 & d_n \end{bmatrix} = D,$$

where D is a diagonal matrix, i.e. the only nonzero entries of D are on the main diagonal.

Theorem 4.5 *If A is diagonalizable, with D as above, then the characteristic polynomial of A must be $p(k) = (k - d_{11})(k - d_{22}) \cdots (k - d_{nn})$ and $d_{11}, d_{22}, \cdots, d_{nn}$ must be the eigenvalues of A.*

Proof. Since A and D are similar, they have the same characteristic polynomial and same eigenvalues, by Theorem 4.4. Since $(D - kI)$ is diagonal its determinant is the product of the diagonal elements. ∎

Remark **(1)** *If the characteristic polynomial of A has complex roots, then A is not diagonalizable since the characteristic polynomial of a diagonal matrix with real numbers as entries has no complex roots.*

Remark **(2)** *We have restricted ourselves to vectors and matrices with real numbers as entries, and we have used the real numbers as our scalars, thus when we say a matrix A is diagonalizable, we mean using only real numbers in P and D we have $P^{-1}AP = D$. To be more precise, we should say in this case that A is diagonalizable over the real numbers. If we allow complex entries in P and D and $P^{-1}AP = D$, we could say that A is diagonalizable over the complex numbers.*

Remark **(3)** *If we use the complex numbers as our scalars and as entries in our matrices and vectors, the results we have proved prior* *to Remark 1 are still true. This is because the complex numbers under addition and multiplication behave very much like the real numbers. In particular they satisfy the axioms of a field. When working with n-tuples of complex numbers, one slight change is required, and that is in the definition of length. Instead of length$(\mathbf{v}) = \sqrt{\mathbf{v} \cdot \mathbf{v}}$, we use length$(\mathbf{v}) = \sqrt{\bar{\mathbf{v}} \cdot \mathbf{v}}$, where $\bar{\mathbf{v}}$ is the n-tuple whose entries are the complex conjugates of the entries in $\mathbf{v}$.*

There are many applications where we want to diagonalize a matrix A if possible. It is not always possible, even when the eigenvalues are all real. The matrices in Example 4.8, $A = \begin{bmatrix} 2 & 0 & 0 \\ 1 & 2 & 0 \\ 0 & 0 & 2 \end{bmatrix}$, $B = \begin{bmatrix} 2 & 0 & 0 \\ 1 & 2 & 0 \\ 0 & 1 & 2 \end{bmatrix}$ cannot be similar to a diagonal matrix, since both of them have $(k-2)^3$ as their characteristic polynomial, the only diagonal matrix they could be similar to would be $\begin{bmatrix} 2 & 0 & 0 \\ 0 & 2 & 0 \\ 0 & 0 & 2 \end{bmatrix}$, but this is a scalar matrix, and so as we have seen is similar only to itself! Thus neither A nor B can be diagonalized.

Theorem 4.6 *An $n \times n$ matrix A is diagonalizable if and only if there exists a basis of R^n consisting of eigenvectors of A, say $\{\mathbf{v}_1, \mathbf{v}_2, \cdots, \mathbf{v}_n\}$. Let $P = \begin{bmatrix} \mathbf{v}_1 & \mathbf{v}_2 & \cdots & \mathbf{v}_n \end{bmatrix}$, then*

$$P^{-1}AP = \begin{bmatrix} d_1 & 0 & 0 & \cdots & 0 \\ 0 & d_2 & 0 & \cdots & 0 \\ 0 & 0 & d_3 & \ddots & \vdots \\ \vdots & \vdots & \ddots & \ddots & 0 \\ 0 & 0 & \cdots & 0 & d_n \end{bmatrix},$$

where d_i is the eigenvalue of A for the eigenvector $\mathbf{v}_i$.

Proof.   Suppose that there exists such a basis of eigenvectors. Then $A\mathbf{v}_j = d_j\mathbf{v}_j$, so

$$AP = P \begin{bmatrix} d_1 & 0 & 0 & \cdots & 0 \\ 0 & d_2 & 0 & \cdots & 0 \\ 0 & 0 & d_3 & \ddots & \vdots \\ \vdots & \vdots & \ddots & \ddots & 0 \\ 0 & 0 & \cdots & 0 & d_n \end{bmatrix} = PD.$$

Since P is invertible (because its columns are a basis of R^n), we can multiply both sides on the left by P^{-1} obtaining $P^{-1}AP = D$. Suppose conversely that A is diagonalizable. Then $P^{-1}AP = D$. Let $\mathbf{v}_1, \mathbf{v}_2, \cdots, \mathbf{v}_n$ be the columns of P. Since P is invertible, the columns of P are a basis of R^n. Now $AP = PD$, and

$$\begin{aligned} AP &= A \begin{bmatrix} \mathbf{v}_1 & {}_1\mathbf{v}_2 & \cdots & \mathbf{v}_n \end{bmatrix} \\ &= \begin{bmatrix} A\mathbf{v}_1 & A\mathbf{v}_2 & \cdots & A\mathbf{v}_n \end{bmatrix}, \end{aligned}$$

while

$$\begin{aligned} PD &= \begin{bmatrix} \mathbf{v}_1 & {}_1\mathbf{v}_2 & \cdots & \mathbf{v}_n \end{bmatrix} \begin{bmatrix} d_1 & 0 & 0 & \cdots & 0 \\ 0 & d_2 & 0 & \cdots & 0 \\ 0 & 0 & d_3 & \ddots & \vdots \\ \vdots & \vdots & \ddots & \ddots & 0 \\ 0 & 0 & \cdots & 0 & d_n \end{bmatrix} \\ &= \begin{bmatrix} d_1\mathbf{v}_1 & {}_1d_2\mathbf{v}_2 & \cdots & d_n\mathbf{v}_n \end{bmatrix}. \end{aligned}$$

Equating columns of AP and PD gives $A\mathbf{v}_i = d_i\mathbf{v}_i$, which says that $\mathbf{v}_i$ is an eigenvector of A for the eigenvalue d_i. ∎

Definition 4.10 *The algebraic multiplicity of an eigenvalue c is the number of times $k - c$ appears as a factor of the characteristic polynomial $p(k)$. The geometric multiplicity of c is the dimension of the eigenspace of A for the eigenvalue c.*

Remark   *The geometric multiplicity of c equals $n - \mathrm{rank}(A - cI)$.*

Theorem 4.7 *For any* $n \times n$ *matrix* A *and any eigenvalue* c *of* A, *the geometric multiplicity of* c *is less than or equal to the algebraic multiplicity of* c.

Corollary 1 *An eigenvalue of algebraic multiplicity one always has geometric multiplicity equal one.*

Theorem 4.8 *An* $n \times n$ *matrix* A *is diagonalizable if and only if all the eigenvalues of* A *are real and the geometric multiplicity equals the algebraic multiplicity for every eigenvalue of* A.

We postpone the proofs of Theorems 4.7, 4.8, and the Corollary until after we have described a method for determining whether A is diagonalizable, diagonalizing A when possible, and giving some examples.

Procedure 4.1 To determine whether A is diagonalizable and to diagonalize A if possible,

1. Find the characteristic polynomial $p(k)$ of A.

2. Solve $p(k) = 0$ to find the eigenvalues of A.

   (a) If there are no repeated roots and no complex roots, then A is diagonalizable. Moreover if P is the matrix whose columns are the eigenvectors $\mathbf{v}_i$ going with the n distinct eigenvalues d_i then P is invertible and $P^{-1}AP = D$, where the eigenvalues d_i of A appear on the diagonal of D in the same order that the corresponding eigenvectors are listed as columns of P.

   (b) If there are repeated roots, check each repeated root c to see if {geometric multiplicity of c} = {$n - rank(A - cI)$} = {algebraic multiplicity of c}. If equality fails for even one eigenvalue c, the matrix A is not diagonalizable. If equality holds for all c, then find a basis of each eigenspace and take the union of these sets. Use these n vectors as the columns of a matrix P. (That these vectors are linearly independent will be established in the course of proving Theorem 4.8.)

Then P is invertible and $P^{-1}AP = D$, where D is diagonal with the eigenvalues d_i of A on the diagonal in the same order that the corresponding eigenvectors appear as the columns of P.

Example 4.9 Let $A = \begin{bmatrix} 2 & 2 & 1 \\ 2 & -1 & -2 \\ 1 & -2 & 2 \end{bmatrix}$. Determine if A is diagonaliz-

able and if possible find an invertible P such that $P^{-1}AP$ is diagonal. To solve this we consider

$$|A - kI| = \begin{vmatrix} 2-k & 2 & 1 \\ 2 & -1-k & -2 \\ 1 & -2 & 2-k \end{vmatrix}$$

$$= (2-k)(-1-k)(2-k) - 4 - 4 + k + 1 + 4k - 8 + 4k - 8$$
$$= -k^3 + 3k^2 - 4 - 23 + 9k$$
$$= -k^3 + 3k^2 + 9k - 27,$$

using the formula for a 3×3 determinant and collecting terms. Thus

$$p(k) = k^3 - 3k^2 - 9k + 27$$
$$= (k-3)^2(k+3).$$

We see that 3 is a double root and -3 is a single root. We investigate the geometric multiplicity of the double root 3. Since

$$(A - 3I) = \begin{bmatrix} -1 & 2 & 1 \\ 2 & -4 & -2 \\ 1 & -2 & -1 \end{bmatrix}$$

has rank one, the geometric multiplicity for $k = 3$ is two. Therefore A must be diagonalizable. To find the P that diagonalizes A, we need to find a basis of this eigenspace. The Hermite form H of $(A - 3I)$ is

$$H = \begin{bmatrix} 1 & -2 & -1 \\ 0 & 0 & 0 \\ 0 & 0 & 0 \end{bmatrix},$$

and

$$(H - I) = \begin{bmatrix} 0 & -2 & -1 \\ 0 & -1 & 0 \\ 0 & 0 & -1 \end{bmatrix},$$

so $\begin{bmatrix} 2 \\ 1 \\ 0 \end{bmatrix}$ and $\begin{bmatrix} 1 \\ 0 \\ 1 \end{bmatrix}$ form a basis of the eigenspace for $k = 3$. For the third column of P we need an eigenvector for $c = -3$. Now

$$(A + 3I) = \begin{bmatrix} 5 & 2 & 1 \\ 2 & 2 & -2 \\ 1 & -2 & 5 \end{bmatrix}$$

has Hermite form

$$H = \begin{bmatrix} 1 & 0 & 1 \\ 0 & 1 & -2 \\ 0 & 0 & 0 \end{bmatrix},$$

so

$$(H - I) = \begin{bmatrix} 0 & 0 & 1 \\ 0 & 0 & -2 \\ 0 & 0 & -1 \end{bmatrix}$$

and a basis for the null space of $(A + 3I)$ is $\begin{bmatrix} 1 \\ -2 \\ -1 \end{bmatrix}$. Thus

$$P = \begin{bmatrix} 1 & 2 & 1 \\ 0 & 1 & -2 \\ 1 & 0 & -1 \end{bmatrix}$$

has the property that $AP = PD$ or $P^{-1}AP = D$, where

$$D = \begin{bmatrix} 3 & 0 & 0 \\ 0 & 3 & 0 \\ 0 & 0 & -3 \end{bmatrix}.$$

Example 4.10 Let $A = \begin{bmatrix} 1 & 0 & -2 \\ 0 & 1 & 0 \\ 1 & 0 & 4 \end{bmatrix}$. Then

$$|A - kI| = \begin{vmatrix} 1 - k & 0 & -2 \\ 0 & 1 - k & 0 \\ 1 & 0 & 4 - k \end{vmatrix}$$

$$= (1-k) \begin{vmatrix} 1-k & -2 \\ 1 & 4-k \end{vmatrix}$$
$$= (1-k)(k^2 - 5k + 6)$$
$$= (1-k)(k-2)(k-3).$$

Since A has 3 distinct real roots, A must be diagonalizable. To find P we must find an eigenvector for each eigenvalue. For $k = 1$, we wish to solve $(A - I)\mathbf{x} = \mathbf{O}$, or

$$\begin{bmatrix} 0 & 0 & -2 \\ 0 & 0 & 0 \\ 1 & 0 & 3 \end{bmatrix} \begin{bmatrix} x_1 \\ x_2 \\ x_3 \end{bmatrix} = \begin{bmatrix} 0 \\ 0 \\ 0 \end{bmatrix}.$$

Clearly $\begin{bmatrix} 0 \\ 1 \\ 0 \end{bmatrix}$ is a solution.

For $k = 2$, we wish to solve $(A - 2I)\mathbf{x} = \mathbf{O}$, or

$$\begin{bmatrix} -1 & 0 & -2 \\ 0 & -1 & 0 \\ 1 & 0 & 2 \end{bmatrix} \begin{bmatrix} x_1 \\ x_2 \\ x_3 \end{bmatrix} = \begin{bmatrix} 0 \\ 0 \\ 0 \end{bmatrix}.$$

Clearly $\begin{bmatrix} 2 \\ 0 \\ -1 \end{bmatrix}$ is a solution.

Last for $k = 3$ we wish to solve $(A - 3I)\mathbf{x} = \mathbf{O}$, or

$$\begin{bmatrix} -2 & 0 & -2 \\ 0 & -2 & 0 \\ 1 & 0 & 1 \end{bmatrix} \begin{bmatrix} x_1 \\ x_2 \\ x_3 \end{bmatrix} = \begin{bmatrix} 0 \\ 0 \\ 0 \end{bmatrix}.$$

We see that $\begin{bmatrix} 1 \\ 0 \\ -1 \end{bmatrix}$ is a solution. Now if

$$P = \begin{bmatrix} 0 & 2 & 1 \\ 1 & 0 & 0 \\ 0 & -1 & -1 \end{bmatrix},$$

and

$$D = \begin{bmatrix} 1 & 0 & 0 \\ 0 & 2 & 0 \\ 0 & 0 & 3 \end{bmatrix},$$

then $P^{-1}AP = D$.

Example 4.11 *Let* $A = \begin{bmatrix} 2 & 1 & 0 & 0 \\ 0 & 2 & 0 & 0 \\ 1 & 0 & 1 & 1 \\ 2 & -2 & -2 & 4 \end{bmatrix}$. *Then*

$$|A - kI| = \begin{vmatrix} 2-k & 1 & 0 & 0 \\ 0 & 2-k & 0 & 0 \\ 1 & 0 & 1-k & 1 \\ 2 & -2 & -2 & 4-k \end{vmatrix}.$$

Viewing this as a lower triangular matrix of square 2×2 *blocks, the determinant is the product of the determinants of the diagonal blocks, so we get*

$$|A - kI| = \begin{vmatrix} 2-k & 1 \\ 0 & 2-k \end{vmatrix} \cdot \begin{vmatrix} 1-k & 1 \\ -2 & 4-k \end{vmatrix}$$
$$= (k-2)^2(k^2 - 5k + 6)$$
$$= (k-2)^3(k-3).$$

Thus 2 *is an eigenvalue of algebraic multiplicity* 3. *We determine the geometric multiplicity of* 2 *by considering*

$$(A - 2I) = \begin{bmatrix} 0 & 1 & 0 & 0 \\ 0 & 0 & 0 & 0 \\ 1 & 0 & -1 & 1 \\ 2 & -2 & -2 & 2 \end{bmatrix}.$$

This matrix has rank 2, *because the last two rows are linearly independent, while the first row is a linear combination of the last two rows, so the last two rows are a basis of the row space. This means that the null space of* $(A - 2I)$ *has dimension* $4 - 2 = 2$. *Thus for* $k = 2$ *we have geometric multiplicity* 2. *Since this eigenvalue has algebraic multiplicity* 3, A *is not similar to a diagonal matrix.*

In doing these problems it becomes necessary to find roots of the characteristic polynomial. If the degree is two, we have the quadratic formula, so we can always find the roots. For higher degrees things become more difficult. If $p(k)$ is a polynomial of the form

$$p(k) = k^n + b_{n-1}k^{n-1} + \cdots + b_1 k + b_0,$$

where b_i are integers, then there is a theorem that tells us that the only rational roots of $p(k)$ are integer divisors of b_0. If A is a matrix with integer entries, then its characteristic polynomial $p(k)$ will be such a polynomial. There is also a theorem that says that c is a root of $p(k)$ if and only if $k - c$ is a factor of $p(k)$. Thus if n is not too large and if $p(k)$ has integer coefficients and rational roots, then finding them may not be easy, but at least it is a finite process.

Example 4.12 Let $p(k) = k^4 - 8k^3 + 22k^2 - 24k + 9$. Find all rational roots. According to the result mentioned above the only possible rational roots are integer divisors of 9, namely $+3, -3, +1, -1, +9, -9$. One should check 1 first because it is easiest. Substituting $k = 1$ in $p(k)$ gives $p(1) = 1 - 8 + 22 - 24 + 9 = 0$, so 1 is a root. Substituting $k = -1$ will show that -1 is not a root, but substituting $k = 3$ shows that 3 is a root. Thus both $(k - 1)$ and $(k - 3)$ must be factors of $p(k)$. Dividing $p(k)$ by the product $(k - 1)(k - 3) = k^2 - 4k + 3$ gives a quotient of $k^2 - 4k + 3$, showing that $p(k) = (k^2 - 4k + 3)^2 = (k - 1)^2(k - 3)^2$, and the roots are 1 and 3.

Example 4.13 Let $p(k) = k^3 - 6k^2 + 11k - 6$. Find all rational roots. The possible rational roots are the integer divisors of -6, so that would be $+1, -1, +2, -2, +3, -3, +6, -6$. Note that if $k < 0$, $p(k) < 0$, so we don't need to try any of the negative numbers. However substitution shows that $p(1) = 0$, $p(2) = 0$, and $p(3) = 0$, so $p(k) = (k - 1)(k - 2)(k - 3)$, and the roots are 1, 2, and 3.

We now give the proofs of Theorems 4.7 and 4.8.

Proof. (Theorem 4.7) Let A be an $n \times n$ matrix and let c be an eigenvalue of A of geometric multiplicity t. Let $\mathbf{v}_1, \mathbf{v}_2, \cdots, \mathbf{v}_t$ be a basis of the eigenspace of A for the eigenvalue c. Extend the set $\mathbf{v}_1, \mathbf{v}_2, \cdots,$

$\mathbf{v}_t$ to a basis of R^n, $\{\mathbf{v}_1, \mathbf{v}_2, \cdots, \mathbf{v}_n\}$. Then $A\mathbf{v}_j = c\mathbf{v}_j$ for $1 \leq j \leq t$. If $P = \begin{bmatrix} \mathbf{v}_1 & \mathbf{v}_2 & \cdots & \mathbf{v}_n \end{bmatrix}$ is the matrix with column vectors $\mathbf{v}_1, \mathbf{v}_2, \cdots, \mathbf{v}_n$, then P is invertible and

$$
AP = P \begin{bmatrix} c & 0 & \cdots & 0 & & \\ 0 & c & \cdots & \vdots & * & \\ 0 & 0 & \ddots & 0 & & \\ \vdots & \vdots & & c & & \\ \vdots & \vdots & & \vdots & * & \\ 0 & 0 & \cdots & 0 & & \end{bmatrix}
$$

$$
= PB,
$$

where B is a matrix with the first t columns having c on the diagonal and zeros elsewhere. Then $P^{-1}AP = B$, and A and B have the same characteristic polynomial. Thus

$$
p(k) = |kI - B| = (k - c)^t q(k).
$$

This shows that the algebraic multiplicity of c must be t or larger. (It could be larger since c could still be a root of $q(k)$). ∎

Corollary 1 *An eigenvalue of algebraic multiplicity one always has geometric multiplicity equal one.*

Proof.   Let c be an eigenvalue of multiplicity one. Since c is an eigenvalue of A, $[A - cI]\mathbf{x} = \mathbf{O}$ must have at least one nonzero solution, so the geometric multiplicity of c is ≥ 1. But since geometric multiplicity $\leq$ algebraic multiplicity $= 1$, we also have geometric multiplicity of $c \leq 1$. Thus geometric multiplicity of $c = 1$. ∎

Proof.   (Theorem 4.8) Suppose that A is an $n \times n$ matrix, that all the roots of the characteristic polynomial of A are real, and that for each root the geometric multiplicity equals the algebraic multiplicity. Since the algebraic multiplicities must add up to n, in this case the geometric multiplicities also add up to n. This means that if we take a basis for each of the eigenspaces and take the union of these sets we will have n vectors. If we can show that these n vectors are linearly independent, we will have a basis of R^n consisting of eigenvectors of A, and A will be diagonalizable by Theorem 4.6. We now proceed to show that the union of the bases for all the eigenspaces of A is a linearly independent set. We will prove it by contradiction. Suppose that this set is linearly dependent. Then there is a dependence relation on these vectors. Take a dependence relation involving the fewest possible vectors, so that $b_1\mathbf{v}_1 + b_2\mathbf{v}_2 + \cdots + b_s\mathbf{v}_s = \mathbf{O}$, with all $b_i \neq 0$. Now these vectors $\mathbf{v}_i$ cannot all be from the same eigenspace, since within each eigenspace we have a basis, hence a linearly independent set. Without loss of generality, assume that $\mathbf{v}_1$ is from the eigenspace for eigenvalue c, and $\mathbf{v}_2$ is from some other eigenspace, say for eigenvalue d. Multiplying the dependence relation above on the left by A, we get $A(b_1\mathbf{v}_1 + b_2\mathbf{v}_2 + \cdots + b_s\mathbf{v}_s) = b_1c\mathbf{v}_1 + b_2d\mathbf{v}_2 + \cdots + b_sA\mathbf{v}_s = \mathbf{O}$. Multiplying the dependence relation on the left by c gives $cb_1\mathbf{v}_1 + cb_2\mathbf{v}_2 + \cdots + cb_s\mathbf{v}_s = \mathbf{O}$. Subtracting the second of these equations from the first results in the $\mathbf{v}_1$ term canceling, but the $\mathbf{v}_2$ term has coefficient $b_2(d - c)$, which is definitely not zero. Thus we have a dependence relation involving fewer vectors, a contradiction. Hence the union of the bases for the eigenspaces of A is a linearly independent set of n vectors and hence must be a basis of R^n (by Corollary 5 to Theorem 1.10). This completes the proof that if the geometric multiplicities equal the algebraic multiplicities, A is diagonalizable. We now prove the converse. If A is similar to a real diagonal matrix the eigenvalues of A must all be real. We now wish to prove that if A is similar to a diagonal matrix the geometric multiplicities must all equal the algebraic multiplicities. The maximum number of linearly independent eigenvectors we can get for an eigenvalue c is the geometric multiplicity of c. Since every eigenvector is in one of the eigenspaces, the maximum number of linearly independent eigenvectors we can get is the sum of the geometric multiplicities. Thus if the sum of the geometric multiplicities is less than n, we can not get a basis of R^n consisting entirely of eigenvectors of A, and so by Theorem 4.6, A is not diagonalizable. If for even one root c, the geometric multiplicity is less than the algebraic multiplicity, you can never make it up because of Theorem 4.7, so the maximum number of linearly independent eigenvectors of A is less than n and A is not similar to a diagonal matrix. ■

Remark   **(1)** *If all the roots of the characteristic polynomial of A are real and if there are no repeated roots, then combining the Corollary to Theorem 4.7 with Theorem 4.8 shows that A is diagonalizable.*

Remark   **(2)** *The method used in the proof of Theorem 8 can be used to prove that if $\mathbf{v}_1, \mathbf{v}_2, \cdots, \mathbf{v}_k$ are eigenvectors of A each belonging to a different eigenvalue, then $\{\mathbf{v}_1, \mathbf{v}_2, \cdots, \mathbf{v}_k\}$ is a linearly independent set.*

As we have seen in Example 4.4, the roots of the characteristic equation of A can be complex even when A is real. Let us extend our definition of eigenvalue to include all roots of the characteristic polynomial of A. The following theorem narrows the possible locations of all eigenvalues of A in the complex plane without finding them exactly. Sometimes this information is enough.

Theorem 4.9 (Gershgorin) *Let $A = [a_{ij}]$ be a real $n \times n$ matrix. Then all eigenvalues are located inside or on the following disks:*

$$D_j = \{z \, | \, z \in C \text{ and } |z - a_{jj}| \leq \sum_{i=1, i \neq j}^{n} |a_{ji}|.\}$$

In other words the disks in the complex plane with centers at the diagonal elements of A, and having the radius of the disk around the jj element equal to the sum of the absolute values of the other entries in the jth row.

Proof.   Let $\mathbf{v}$ be an eigenvector for the eigenvalue $k = c$. Then $A\mathbf{v} = c\mathbf{v}$. Suppose $\mathbf{v} = \begin{bmatrix} b_1 \\ b_2 \\ \vdots \\ b_n \end{bmatrix}$, where the b_i may be complex numbers. Choose b_s so that $|b_s|$ is maximal in the set $\{|b_1|, |b_2|, \cdots, |b_n|\}$. Since $\mathbf{v} \neq \mathbf{0}$, $|b_s| > 0$. Equating the row s in $A\mathbf{v} = c\mathbf{v}$, we obtain

$$a_{s1}b_1 + a_{s2}b_2 + \cdots + a_{sn}b_n = cb_s,$$

so that

$$(c - a_{ss})b_s = \sum_{i=1, i \neq s}^{n} a_{si}b_i.$$

Taking absolute values of both sides, we get

$$|c - a_{ss}||b_s| = \left| \sum_{i=1,i\neq s}^{n} a_{si}b_i \right| \leq \sum_{i=1,i\neq s}^{n} |a_{si}||b_i|.$$

Dividing by $|b_s| > 0$, the inequality is preserved so that we get

$$|c - a_{ss}| \leq \sum_{i=1,i\neq s}^{n} |a_{si}|.$$

Thus c is in D_s. ∎

Exercises

1. 1. Find all rational roots of the following equations:

   (a) $k^5 - 2k^4 + 3k^3 - 2k^2 + k - 1 = 0$,
   (b) $k^3 + 9k^2 + 27k + 27 = 0$,
   (c) $k^3 - 15k^2 + 57k - 70 = 0$,
   (d) $k^3 + 4k^2 - 16k - 64 = 0$,
   (e) $k^5 - 6k^4 + 3k^3 - 18k^2 + 2k - 12 = 0$.

2. Find the characteristic polynomials and solve for the eigenvalues for the following matrices:

   (a) $\begin{bmatrix} 2 & 5 \\ 4 & 3 \end{bmatrix}$,

   (b) $\begin{bmatrix} 0 & -1 \\ 1 & -1 \end{bmatrix}$,

   (c) $\begin{bmatrix} 3 & 1 \\ -1 & 1 \end{bmatrix}$,

   (d) $\begin{bmatrix} 7 & 4 & -1 \\ 4 & 7 & -1 \\ -4 & -4 & 4 \end{bmatrix}$,

(e) $\begin{bmatrix} 0 & 4 & -1 \\ 4 & 0 & 1 \\ -1 & 1 & 1 \end{bmatrix}$,

(f) $\begin{bmatrix} 2 & 2 & 1 \\ 2 & -1 & -2 \\ 1 & -2 & 2 \end{bmatrix}$,

(g) $\begin{bmatrix} 1 & 1 & -2 \\ 4 & 0 & 4 \\ 1 & -1 & 4 \end{bmatrix}$,

(h) $\begin{bmatrix} 5 & 4 & 4 \\ -7 & -3 & -1 \\ 7 & 4 & 2 \end{bmatrix}$,

(i) $\begin{bmatrix} 2 & 0 & 0 & 0 \\ 1 & 2 & 0 & 0 \\ -2 & 1 & 2 & 0 \\ -5 & -2 & -1 & 3 \end{bmatrix}$,

(j) $\begin{bmatrix} 1 & 1 & 0 & 0 \\ -1 & 2 & 1 & 0 \\ -1 & 0 & 3 & 0 \\ -1 & 0 & 0 & 3 \end{bmatrix}$,

(k) $\begin{bmatrix} 2 & 1 & 0 & 0 \\ 0 & 2 & 0 & 0 \\ 1 & 3 & 1 & 1 \\ 2 & -2 & -2 & 4 \end{bmatrix}$,

(l) $\begin{bmatrix} 2 & 1 & 0 & 0 \\ 1 & 2 & 1 & 0 \\ 0 & 0 & 3 & 0 \\ 0 & 0 & 0 & 1 \end{bmatrix}$.

3. For each matrix in problem 2 determine if the matrix is similar to a diagonal matrix (over the real numbers), and if it is, find a matrix P such that $P^{-1}AP$ is diagonal.

4. Are any of the matrices in problem 2 which are not similar to a diagonal matrix over the real numbers similar to a diagonal

matrix over the complex numbers? If the answer is yes, find a matrix P (entries may be complex) such that $P^{-1}AP$ is diagonal.

5. The leading coefficient of a polynomial $p(k)$ is the coefficient of the highest power of k that has nonzero coefficient. This problem shows that there is nothing special about characteristic polynomials except that the leading coefficient is 1. Any polynomial with leading coefficient 1 is the characteristic polynomial of some matrix. Given a polynomial

$$p(k) = k^n + c_{n-1}k^{n-1} + c_{n-2}k^{n-2} + \cdots + c_1 k + c_0,$$

the following matrix is called the **companion matrix** of p:

$$A = \begin{bmatrix} 0 & 0 & \cdots & 0 & -c_0 \\ 1 & 0 & \cdots & 0 & -c_1 \\ 0 & 1 & \ddots & \vdots & -c_2 \\ \vdots & & \ddots & 0 & \vdots \\ 0 & \cdots & 0 & 1 & -c_{n-1} \end{bmatrix},$$

$a_{21} = a_{32} = \cdots = a_{n\,n-1} = 1$, $a_{1n} = -c_0$, $a_{2n} = -c_1, \cdots,$ $a_{in} = -c_{i-1}, \cdots, a_{nn} = -c_{n-1}$, and all other entries are 0. Show that the characteristic polynomial of A is $p(k)$. (Hint: In $|A - kI|$ add k times row n to row $n-1$, then add k times row $n-1$ to row $n-2$, etc.)

6. Give the companion matrices for the polynomials in problem 1.

7. Show that for a square matrix A, A and A^T have the same characteristic polynomial and hence the same eigenvalues.

While it can be difficult to find the eigenvalues of matrices because we have to find roots of polynomials, it is fairly easy to check whether or not a given number is an eigenvalue of a given matrix. We can use the fact that c is an eigenvalue of an $n \times n$ matrix A if and only if $|A - cI| = 0$, or we can use the fact that c is an eigenvalue of A if and only if $rank(A - cI) < n$, in other words if and only if you get at least one zero row when you row reduce $(A - cI)$.

8 Verify that 1 and 2 are the eigenvalues of $A = \begin{bmatrix} \frac{14}{9} & -\frac{2}{9} & -\frac{4}{9} & 0 \\ -\frac{2}{9} & \frac{17}{9} & -\frac{2}{9} & 0 \\ -\frac{4}{9} & -\frac{2}{9} & \frac{14}{9} & 0 \\ 0 & 0 & 0 & 2 \end{bmatrix}$,

and find the geometric multiplicity of each.

9 Verify that the eigenvalues of $B = \begin{bmatrix} 14 & -2 & -4 & 0 \\ -2 & 17 & -2 & 0 \\ -4 & -2 & 14 & 0 \\ 0 & 0 & 0 & 2 \end{bmatrix}$ are 2, 9,

and 18, and find the geometric multiplicity of each.

10 Apply the theorem due to Gershgorin to find disks in the complex plane within which the eigenvalues of the following matrices will be found:

$$A = \begin{bmatrix} -4 & 1 & 0 \\ 1 & -5 & 1 \\ -1 & 1 & 6 \end{bmatrix}, B = \begin{bmatrix} 5 & 1 & 1 \\ 1 & 6 & -1 \\ 2 & -1 & 7 \end{bmatrix}, C = \begin{bmatrix} -8 & 2 & -2 \\ 0 & 9 & 1 \\ -1 & 1 & -9 \end{bmatrix},$$

$$D = \begin{bmatrix} 7 & 1 & 1 & -1 \\ 0 & 6 & 2 & 1 \\ 1 & 0 & -8 & -1 \\ 0 & 3 & 4 & -9 \end{bmatrix}.$$

Section 4.2 Orthogonal Bases

Definition 4.11 *Two vectors* $\mathbf{u}$ *and* $\mathbf{v}$ *in* R^n *are orthogonal if* $\mathbf{u} \cdot \mathbf{v} = 0$.

Definition 4.12 *A set of nonzero vectors* $\{\mathbf{v}_1, \mathbf{v}_2, \cdots, \mathbf{v}_k\}$ *is called an orthogonal set if the vectors are pairwise orthogonal, i.e. if* $\mathbf{v}_i \cdot \mathbf{v}_j = 0$ *for* $i \neq j$. *A set containing a single nonzero vector is also considered an orthogonal set.*

Lemma 4.2 *An orthogonal set is linearly independent.*

Proof.   Let $\{v_1, v_2, \cdots, v_k\}$ be an orthogonal set. If $a_1v_1 + a_2v_2 + \cdots + a_kv_k = O$, let i be any integer $1 \le i \le k$, then

$$v_i \cdot (a_1v_1 + a_2v_2 + \cdots + a_kv_k) = v_i \cdot O = 0.$$

But $v_i \cdot v_j = 0$ unless $j = i$, so this implies $a_i(v_i \cdot v_i) = 0$. We know that $v_i \cdot v_i \ne 0$ since $v_i \ne O$, thus $a_i = 0$. ∎

Corollary 1 *A set of k orthogonal vectors in a subspace S of dimension k is a basis of S.*

Definition 4.13 *Let S be a subspace of R^n and $B = \{v_1, v_2, \cdots, v_k\}$ a basis of S. If B is an orthogonal set, it is called an **orthogonal basis** of S. If in addition the length of each basis vector is one, i.e. $v_i \cdot v_i = 1$ for every $v_i \in B$, then B is an **orthonormal basis** of S.*

Definition 4.14 *An $n \times n$ matrix P is orthogonal if $P^T P = I$.*

Remark   *Because of Theorem 2.16 this is equivalent to $P^T = P^{-1}$.*

Lemma 4.3 *The following are equivalent:*
   1. *P is orthogonal.*
   2. *The columns of P are an orthonormal basis of R^n.*
   3. *The rows of P are an orthonormal basis of R^n.*

Proof. P is orthogonal if and only if $P^T P = I$. Let $u_1, u_2, \cdots, u_n$ be the columns of P. Then the $[P^T P]_{ij} = u_i \cdot u_j$. Thus $P^T P = I$ if and only if $u_i \cdot u_j = \begin{cases} 1 \text{ if } i = j \\ 0 \text{ if } i \ne j \end{cases}$. Also P is orthogonal if and only if $PP^T = I$. Let $v_1, v_2, \cdots, v_n$ be the rows of P. Then $[PP^T]_{ij} = v_i \cdot v_j$. Thus $PP^T = I$ if and only if $v_i \cdot v_j = \begin{cases} 1 \text{ if } i = j \\ 0 \text{ if } i \ne j \end{cases}$. ∎

Lemma 4.4 *If P and Q are orthogonal, then PQ and P^{-1} are orthogonal.*

Proof. If P and Q are orthogonal, then $(PQ)^{-1} = Q^{-1}P^{-1} = Q^T P^T = (PQ)^T$, so PQ is orthogonal. Since P is orthogonal, $P^{-1} = P^T$, transposing both sides gives $(P^{-1})^T = P = (P^{-1})^{-1}$, thus P^{-1} is orthogonal. ■

In the next section we will want to find an orthonormal basis for each eigenspace of a symmetric matrix A.

Procedure 4.2 How to find an orthonormal basis for an eigenspace of a matrix A. If the eigenspace is of dimension 1, just divide any eigenvector by its length. For eigenspaces of dim > 1, we must first find an orthogonal basis. In this situation we have our eigenspace described as the null space of a matrix $B = (A - cI)$, where c is the eigenvalue. Instead of getting a basis of the null space in the usual way, select just one vector $\mathbf{u}_1$ in the null space of B. Then $\mathbf{u}_1$ is our first basis vector for the eigenspace. We want a second vector still in null space of B, but now also orthogonal to $\mathbf{u}_1$. This means the new vector $\mathbf{u}_2$ must satisfy an additional equation, $\mathbf{u}_1 \cdot \mathbf{u}_2 = 0$. Thus $\mathbf{u}_2$ is in the null space of the matrix $\begin{bmatrix} B \\ \mathbf{u}_1 \end{bmatrix}$, and you can find a vector in this null space and call it $\mathbf{u}_2$. If the dimension of your eigenspace is 2, you are finished. If not you now want to find an eigenvector $\mathbf{u}_3$ orthogonal to both $\mathbf{u}_1$ and $\mathbf{u}_2$, which means you want $\mathbf{u}_3$ in the null space of $\begin{bmatrix} B \\ \mathbf{u}_1 \\ \mathbf{u}_2 \end{bmatrix}$. Since the dimension of the eigenspace is the nullity of B, you will be able to find nonzero solutions until you have a basis of the eigenspace. Some examples illustrate the process.

Example 4.14 If $A = \begin{bmatrix} 1 & -4 & 2 \\ -4 & 1 & -2 \\ 2 & -2 & -2 \end{bmatrix}$, then $k = -3$ is a double root of the characteristic polynomial. We seek an orthogonal basis for the eigenspace E_{-3} of A for $k = -3$. This eigenspace is the null space of the matrix

$$B = (A + 3I) = \begin{bmatrix} 4 & -4 & 2 \\ -4 & 4 & -2 \\ 2 & -2 & 1 \end{bmatrix}.$$

B is row equivalent to $\begin{bmatrix} 2 & -2 & 1 \\ 0 & 0 & 0 \\ 0 & 0 & 0 \end{bmatrix}$, so to be in the null space a vector (a, b, c) need only satisfy $2a - 2b + c = 0$, or $c = 2b - 2a$. Thus the eigenspace consists of vectors of the form

$$(a, b, 2b - 2a).$$

One such vector is $(1, 1, 0)$. To be orthogonal to $(1, 1, 0)$ a vector (a, b, c) need only satisfy $a + b = 0$. If we chose $a = 1$, $b = -1$, we must have $c = -4$. Thus we get $(1, -1, -4)$. We now have an orthogonal basis of E_{-3},

$$\{(1, 1, 0), (1, -1, -4)\}.$$

To get an orthonormal basis, we divide each vector by its length to get

$$\left\{ \left(\frac{1}{\sqrt{2}}, \frac{1}{\sqrt{2}}, 0 \right), \left(\frac{1}{\sqrt{18}}, -\frac{1}{\sqrt{18}}, -\frac{4}{\sqrt{18}} \right) \right\}.$$

This answer is not unique. Another orthonormal basis of E_{-3} is

$$\left\{ \left(\frac{2}{3}, \frac{1}{3}, -\frac{2}{3} \right), \left(\frac{1}{3}, \frac{2}{3}, \frac{2}{3} \right) \right\}.$$

You can get different ones by varying the initial choice of a and b.

Exercises

1. Find an orthonormal basis for the null space of $\begin{bmatrix} 3 & 1 & 3 \\ 6 & 2 & 6 \\ 9 & 3 & 9 \end{bmatrix}$.

2. Find an orthogonal basis of S if $S = \{(a, b, c) \mid c = 4a - 2b\}$.

3. $A = \begin{bmatrix} \frac{14}{9} & -\frac{2}{9} & -\frac{4}{9} & 0 \\ -\frac{2}{9} & \frac{17}{9} & -\frac{2}{9} & 0 \\ -\frac{4}{9} & -\frac{2}{9} & \frac{14}{9} & 0 \\ 0 & 0 & 0 & 2 \end{bmatrix}$ has eigenvalues 1 and 2. Find an orthonormal basis of the eigenspace of A for $k = 2$.

4. Let $B = \begin{bmatrix} 1 & 2 & 3 & -1 \\ 3 & 6 & 9 & -3 \\ 1 & 2 & 3 & -1 \\ -1 & -2 & -3 & 1 \end{bmatrix}$. Find an orthogonal basis for the
null space of B.

5. If P is an orthogonal matrix, show that $|P| = 1$ or-1. (We assume the entries of P are real numbers.)

6. If P is an orthogonal matrix and c a real eigenvalue of P, show that $c = 1$ or -1.

7. If P is an orthogonal matrix, show that the mapping $\mathbf{w} \to P\mathbf{w}$ preserves the dot product, i.e. show that $P\mathbf{w} \cdot P\mathbf{v} = \mathbf{w} \cdot \mathbf{v}$ for all $\mathbf{w}$ and $\mathbf{v}$ in R^n.

8. Show that if $P\mathbf{w} \cdot P\mathbf{v} = \mathbf{w} \cdot \mathbf{v}$ for all $\mathbf{w}$ and $\mathbf{v}$ in R^n, then P is orthogonal.

9. Show that if P is orthogonal, then the mapping $P \to P\mathbf{w}$ preserves length, i.e. $|P\mathbf{w}| = |\mathbf{w}|$, for all $\mathbf{w}$ in R^n.

10. Show that if $|P\mathbf{w}| = |\mathbf{w}|$ for all $\mathbf{w}$ in R^n, then P is orthogonal.

Remark   *A mapping which preserves length is called an isometry. Problems 9 and 10 show that an $n \times n$ matrix P is orthogonal if and only if the mapping $\mathbf{w} \to P\mathbf{w}$ is an isometry.*

Section 4.3 Real Symmetric Matrices

Definition 4.15 *An $n \times n$ matrix A is symmetric if $A^T = A$.*

The question of diagonalization is easy for symmetric matrices. Every symmetric matrix is diagonalizable. You can even do it with an orthogonal P, so that $P^{-1}AP = P^T AP = D$, diagonal. We now proceed to prove this.

Definition 4.16 *If B is a matrix with complex entries, then $\bar{B}$ is the matrix whose entries are the complex conjugates of the entries of B, and $B^* = \bar{B}^T$.*

Lemma 4.5 $\bar{B}\bar{C} = \overline{BC}$, and $(BC)^* = C^*B^*$, and $B^{**} = B$.

Proof. The first equality is follows from the fact that for complex numbers r and s we have $\bar{r}\bar{s} = \overline{rs}$ and $\overline{r+s} = \bar{r} + \bar{s}$. Next we have $(BC)^* = (\overline{BC})^T = (\bar{B}\bar{C})^T = \bar{C}^T\bar{B}^T = C^*B^*$. The last follows since $(B^T)^T = B$ for all matrices, and $\bar{\bar{r}} = r$ for all complex numbers. ∎
  We can regard a vector $\mathbf{v}$ as an $n \times 1$ matrix and $\mathbf{v}^*$ as a $1 \times n$ matrix.

Lemma 4.6 *Let A be any real symmetric $n \times n$ matrix, then all the eigenvalues of A are real.*

Proof. Let c be an eigenvalue of A and $\mathbf{v}$ an eigenvector of A for c. We must initially allow for the possibility that c is a complex number and that $\mathbf{v}$ is an n-tuple with complex entries. Consider the matrix product $\mathbf{v}^*A\mathbf{v}$. Since $\mathbf{v}$ is an eigenvector of A for c, we have $A\mathbf{v} = c\mathbf{v}$, and $\mathbf{v} \neq \mathbf{O}$. Then $\mathbf{v}^*A\mathbf{v} = \mathbf{v}^*(c\mathbf{v}) = c\mathbf{v}^*\mathbf{v}$, so we have

$$\mathbf{v}^*A\mathbf{v} = c\mathbf{v}^*\mathbf{v}, \tag{4.1}$$

and $\mathbf{v}^*\mathbf{v}$ is a positive real 1×1 matrix. In what follows we regard c as a 1×1 matrix. Applying $*$ to both sides of 4.1 and using the previous lemma, we get $\mathbf{v}^*A^*\mathbf{v} = \mathbf{v}^*\mathbf{v}c^* = c^*\mathbf{v}^*\mathbf{v}$, thus

$$\mathbf{v}^*A^*\mathbf{v} = c^*\mathbf{v}^*\mathbf{v}. \tag{4.2}$$

Since A is real and symmetric, $A^* = A$, and substituting this in 4.2 gives

$$\mathbf{v}^*A\mathbf{v} = c^*\mathbf{v}^*\mathbf{v}. \tag{4.3}$$

Comparing 4.1 and 4.3 we see that $c = c^*$, and hence c is real. ∎

Remark *Once c is real, there must be a real eigenvector to go with c, since the system of equations $(A - cI)\mathbf{x} = \mathbf{O}$ has a real coefficient matrix with determinant zero, it must have a real nonzero solution.*

Lemma 4.7 *If A is a real symmetric matrix and if $\mathbf{u}$ and $\mathbf{v}$ are eigenvectors of A for eigenvalues c and d respectively, with $c \neq d$, then $\mathbf{u} \cdot \mathbf{v} = 0$.*

Proof.   First observe that since $A\mathbf{v} = d\mathbf{v}$ we have

$$\mathbf{u}^T A\mathbf{v} = d\mathbf{u}^T\mathbf{v} = d(\mathbf{u} \cdot \mathbf{v}).$$

But since A is symmetric and $A\mathbf{u} = c\mathbf{u}$, we also have

$$\mathbf{u}^T A\mathbf{v} = \mathbf{u}^T A^T\mathbf{v} = (A\mathbf{u})^T\mathbf{v} = (c\mathbf{u})^T\mathbf{v} = c(\mathbf{u}^T\mathbf{v}) = c(\mathbf{u} \cdot \mathbf{v}).$$

Comparing these results we see that

$$d(\mathbf{u} \cdot \mathbf{v}) = c(\mathbf{u} \cdot \mathbf{v}),$$

or

$$(d - c)(\mathbf{u} \cdot \mathbf{v}) = 0.$$

Since $d - c \neq 0$, this implies that $\mathbf{u} \cdot \mathbf{v} = 0$. ∎

Definition 4.17 *If there exists an orthogonal matrix P such that $P^{-1}AP = B$ we say that A is orthogonally similar to B. In this case we also have $P^T AP = B$.*

Remark   *Since the product of orthogonal matrices is orthogonal and the inverse of an orthogonal matrix is orthogonal, one can show (as in Lemma 4.1) that orthogonal similarity is an equivalence relation. In particular A is orthogonally similar to B if and only if B is orthogonally similar to A.*

Lemma 4.8 *If A is orthogonally similar to a symmetric matrix B, then A is symmetric.*

Proof.   Assume that A is orthogonally similar to a symmetric matrix B. There exists a symmetric matrix P such that $A = P^T BP$. Then $A^T = (P^T BP)^T = P^T B^T (P^T)^T = P^T BP = A$, so A is symmetric. ∎

Theorem 4.10 (Principal Axes Theorem) *Every real symmetric matrix A is orthogonally similar to a diagonal matrix D.*

Proof. The proof is by induction on n. For $n = 1$, A itself is diagonal, and since the identity matrix I of any size is orthogonal and $I^{-1}AI = A$, the theorem is true for $n = 1$. Assume that every $(n-1) \times (n-1)$ symmetric matrix is orthogonally similar to a diagonal matrix. Let A by an $n \times n$ symmetric matrix. Let d be an eigenvalue of A and let $\mathbf{v}_1$ be an eigenvector of A for d. Without loss of generality we may assume that the length of $\mathbf{v}_1$ is 1, since we could divide $\mathbf{v}_1$ by its length if necessary and the result would still be an eigenvector for d. Extend $\mathbf{v}_1$ to an orthonormal basis of R^n, $\{\mathbf{v}_1, \mathbf{v}_2, \cdots, \mathbf{v}_k\}$. Let

$$P = \begin{bmatrix} \mathbf{v}_1 & \mathbf{v}_1 & \cdots & \mathbf{v}_n \end{bmatrix},$$

then P is an orthogonal matrix. Let

$$B = P^T A P.$$

Then B is a symmetric matrix and the first column of B is $P^T A \mathbf{v}_1$. Since $\mathbf{v}_1$ is an eigenvector of A for d, $A\mathbf{v}_1 = d\mathbf{v}_1$, thus

$$
\begin{aligned}
P^T A \mathbf{v}_1 &= P^T d\mathbf{v}_1 \\
&= \begin{bmatrix} \mathbf{v}_1 & \mathbf{v}_1 & \cdots & \mathbf{v}_n \end{bmatrix}^T (d\mathbf{v}_1) \\
&= \begin{bmatrix} d(\mathbf{v}_1 \cdot \mathbf{v}_1) \\ d(\mathbf{v}_2 \cdot \mathbf{v}_1) \\ \vdots \\ d(\mathbf{v}_n \cdot \mathbf{v}_1) \end{bmatrix} \\
&= \begin{bmatrix} d \\ 0 \\ \vdots \\ 0 \end{bmatrix}.
\end{aligned}
$$

We have just calculated the first column of B, and since B is symmetric, this also gives us the first row of B. Thus we have

$$B = \begin{bmatrix} d & 0 & \cdots & 0 \\ 0 & & & \\ \vdots & & M & \\ 0 & & & \end{bmatrix},$$

where M is $(n-1) \times (n-1)$ and M is symmetric. By the induction hypothesis there exists an orthogonal matrix Q_1 such that $Q_1^T M Q_1 = D'$, a diagonal matrix. Form

$$Q = \begin{bmatrix} 1 & 0 & \cdots & 0 \\ 0 & & & \\ \vdots & & Q_1 & \\ 0 & & & \end{bmatrix}.$$

One can use block multiplication to check that

$$\begin{aligned} (PQ)^T A (PQ) &= Q^T P^T A P Q \\ &= Q^T B Q \\ &= \begin{bmatrix} d & 0 & \cdots & 0 \\ 0 & & & \\ \vdots & & D' & \\ 0 & & & \end{bmatrix} = D. \end{aligned}$$

Since D' is diagonal, so is D. The product of orthogonal matrices is orthogonal, so PQ is orthogonal and A is orthogonally similar to a diagonal matrix. ∎

Corollary 1 *A real $n \times n$ matrix A is orthogonally similar to a diagonal matrix if and only if A is symmetric.*

Proof.   This combines Lemma 4.8 and Theorem 4.10. ∎

Corollary 2 *If A is symmetric the geometric multiplicity equals the algebraic multiplicity for every eigenvalue of A.*

Proof.   This follows from Lemma 4.6 and Theorems 4.8 and 4.10. ∎

Procedure 4.3 To find an orthogonal P that diagonalizes a symmetric matrix A:

1. Check if A is symmetric.

2. Find all eigenvalues of A.

3. Find an orthogonal basis for each eigenspace of A.

4. Take the union of all the bases from step 3 and divide each vector by its length. Put these vectors as the columns of a matrix P.

Because we know that the geometric multiplicity equals the algebraic multiplicity for each eigenvalue, the union of the bases from step 3 will be a basis of R^n consisting entirely of eigenvectors of A. Because eigenvectors from different eigenspaces are orthogonal, it will also be an orthogonal basis. If all the eigenvalues are distinct and hence all eigenspaces of dimension 1, then we have no extra work in step 3, but if we have a repeated root, and hence an eigenspace of dimension greater than one, we must take care to select an orthogonal basis of that eigenspace. One must not forget to divide each vector by its length in step 4, otherwise the basis won't be orthonormal and so the matrix P will not be orthogonal. We have already shown (Theorem 4.6) that if the columns of P are a basis of R^n consisiting entirely of eigenvectors of A then $P^{-1}AP = D$, a diagonal matrix.

Example 4.15 *Find an orthogonal matrix P that diagonalizes A if*
$A = \begin{bmatrix} 5 & -2 & 2 \\ -2 & 2 & 4 \\ 2 & 4 & 2 \end{bmatrix}$. *Solution: First we find the eigenvalues of A. We have*

$$|A - kI| = \begin{vmatrix} 5 - k & -2 & 2 \\ -2 & 2 - k & 4 \\ 2 & 4 & 2 - k \end{vmatrix}.$$

Since adding the last row to the middle row doesn't change the determinant we have

$$|A - kI| = \begin{vmatrix} 5 - k & -2 & 2 \\ 0 & 6 - k & 6 - k \\ 2 & 4 & 2 - k \end{vmatrix}.$$

Now we can factor $(6 - k)$ out of the second row, getting

$$|A - kI| = (6 - k) \begin{vmatrix} 5 - k & -2 & 2 \\ 0 & 1 & 1 \\ 2 & 4 & 2 - k \end{vmatrix}.$$

Since subtracting the last column from the middle column doesn't change the determinant we have

$$|A - kI| = (6 - k) \begin{vmatrix} 5 - k & -4 & 2 \\ 0 & 0 & 1 \\ 2 & 2 + k & 2 - k \end{vmatrix}.$$

Now expand by minors around the second row, getting

$$\begin{aligned} |A - kI| &= -(6 - k)\left((5 - k)(2 + k) + 8\right) \\ &= -(6 - k)(10 + 3k - k^2 + 8) \\ &= -(k - 6)(k^2 - 3k - 18) \\ &= -(k - 6)(k - 6)(k + 3). \end{aligned}$$

Thus the eigenvalues are 6 and −3. The eigenspace for $k = -3$ is the null space of $(A + 3I)$, and will be of dimension 1. We have

$$(A + 3I) = \begin{bmatrix} 8 & -2 & 2 \\ -2 & 5 & 4 \\ 2 & 4 & 5 \end{bmatrix}.$$

The RREF of this matrix is

$$RREF(A + 3I) = \begin{bmatrix} 1 & 0 & \frac{1}{2} \\ 0 & 1 & 1 \\ 0 & 0 & 0 \end{bmatrix},$$

so a basis of the eigenspace of A for $k = -3$ is $(-\frac{1}{2}, -1, 1)$, or if we prefer, $(1, 2, -2)$.
For $k = 6$, the eigenspace is the nullspace of $(A - 6I)$, and we know it will be of dimension 2. Now

$$(A - 6I) = \begin{bmatrix} -1 & -2 & 2 \\ -2 & -4 & 4 \\ 2 & 4 & -4 \end{bmatrix}.$$

In this case we want an orthogonal basis for the nullspace of this matrix. It is easy to find by inspection a vector in the nullspace of this matrix since we need only satisfy the equation $x + 2y - 2z = 0$. Suppose we pick

$(0, 1, 1)$. *Then for our second vector we want a vector that is orthogonal to $(0, 1, 1)$ and still satisfies the equation $x + 2y - 2z = 0$. Thus we want a vector satisfying*

$$\begin{bmatrix} 1 & 2 & -2 \\ 0 & 1 & 1 \end{bmatrix} \mathbf{x} = \mathbf{O}.$$

$(4, -1, 1)$ *is such a vector. Thus $\{(0, 1, 1), (4, -1, 1)\}$ is an orthogonal basis for the eigenspace of A for $k = 6$. We know from Lemma 4.4 that these vectors will be orthogonal to any eigenvector of A for $k = -3$, thus $\{(1, 2, -2), (0, 1, 1), (4, -1, 1)\}$ is an orthogonal basis of R^3 consisting entirely of eigenvectors of A. We must still divide each vector by its length to get an orthonormal basis,*

$$\left\{ \left(\frac{1}{3}, \frac{2}{3}, -\frac{2}{3}\right), \left(0, \frac{1}{\sqrt{2}}, \frac{1}{\sqrt{2}}\right), \left(\frac{4}{\sqrt{18}}, -\frac{1}{\sqrt{18}}, \frac{1}{\sqrt{18}}\right) \right\}.$$

The orthogonal matrix P which diagonalizes A is

$$P = \begin{bmatrix} \frac{1}{3} & 0 & \frac{4}{\sqrt{18}} \\ \frac{2}{3} & \frac{1}{\sqrt{2}} & -\frac{1}{\sqrt{18}} \\ -\frac{2}{3} & \frac{1}{\sqrt{2}} & \frac{1}{\sqrt{18}} \end{bmatrix}.$$

Application to conic sections

The general equation of a conic section can be written as

$$ax^2 + 2bxy + cy^2 + dx + ey + f = 0. \tag{4.4}$$

This can be rewritten in matrix notation as

$$\begin{bmatrix} x & y \end{bmatrix} \begin{bmatrix} a & b \\ b & c \end{bmatrix} \begin{bmatrix} x \\ y \end{bmatrix} + \begin{bmatrix} d & e \end{bmatrix} \begin{bmatrix} x \\ y \end{bmatrix} + \begin{bmatrix} f \end{bmatrix} = \begin{bmatrix} 0 \end{bmatrix}. \tag{4.5}$$

The matrix $A = \begin{bmatrix} a & b \\ b & c \end{bmatrix}$ is symmetric, so there exists an orthogonal P such that

$$P^T A P = \begin{bmatrix} d_1 & 0 \\ 0 & d_2 \end{bmatrix},$$

where d_1 and d_2 are the eigenvalues of A. Since P is invertible, the columns of P are a basis of R^2. If we define x' and y' to be the coordinates of the point (x, y) with respect to this basis, then

$$\begin{bmatrix} x \\ y \end{bmatrix} = P \begin{bmatrix} x' \\ y' \end{bmatrix}, \text{ and } \begin{bmatrix} x & y \end{bmatrix} = \begin{bmatrix} x' & y' \end{bmatrix} P^T \qquad (4.6)$$

and substituting (4.6) in equation (4.5) gives

$$\begin{bmatrix} x' & y' \end{bmatrix} P^T A P \begin{bmatrix} x' \\ y' \end{bmatrix} + \begin{bmatrix} d & e \end{bmatrix} P \begin{bmatrix} x' \\ y' \end{bmatrix} + \begin{bmatrix} f \end{bmatrix} = \begin{bmatrix} 0 \end{bmatrix}. \qquad (4.7)$$

This can be rewritten as

$$\begin{bmatrix} x' & y' \end{bmatrix} \begin{bmatrix} d_1 & 0 \\ 0 & d_2 \end{bmatrix} \begin{bmatrix} x' \\ y' \end{bmatrix} + \begin{bmatrix} d & e \end{bmatrix} P \begin{bmatrix} x' \\ y' \end{bmatrix} + \begin{bmatrix} f \end{bmatrix} = \begin{bmatrix} 0 \end{bmatrix}, \qquad (4.8)$$

which is

$$d_1(x')^2 + d_2(y')^2 + d'x' + e'y' + f = 0.$$

Thus by a change of basis we have eliminated the xy term. If d_1 and d_2 are not 0, the linear terms can now be eliminated by a translation (complete the square to see where the new origin should be). This will not change the coefficients of the square terms. We will get an equation of the form $d_1(x'')^2 + d_2(y'')^2 = g$, where $x'' = x' - h$ and $y'' = y' - k$. If the conic is not degenerate, it is possible to recognize whether it is an ellipse, a hyperbola or a parabola by whether d_1 and d_2 have the same sign (ellipse), opposite sign (hyperbola) or one of them is zero (parabola). This we can tell by whether the product $d_1 d_2$ is positive, negative or zero. But if this is all we want to know, we don't even have to diagonalize A, since A is similar to $\begin{bmatrix} d_1 & 0 \\ 0 & d_2 \end{bmatrix}$, these two matrices have the same determinant, so $|A| = d_1 d_2$. Thus we can say the conic is an ellipse if $|A| > 0$, a hyperbola if $|A| < 0$, and a parabola if $|A| = 0$. In any of these cases however the conic could be degenerate (two lines, one line a point or empty).

Application to quadric surfaces

An equation of the form

$$g_1 x^2 + g_2 y^2 + g_3 z^2 + 2g_4 xy + 2g_5 xz + 2g_6 yz = g_7 \qquad (4.9)$$

represents a quadric surface in R^3. In matrix form this becomes

$$\begin{bmatrix} x & y & z \end{bmatrix} \begin{bmatrix} g_1 & g_4 & g_5 \\ g_4 & g_2 & g_6 \\ g_5 & g_6 & g_3 \end{bmatrix} \begin{bmatrix} x \\ y \\ z \end{bmatrix} = \begin{bmatrix} g_7 \end{bmatrix},$$

or

$$\begin{bmatrix} x & y & z \end{bmatrix} M \begin{bmatrix} x \\ y \\ z \end{bmatrix} = \begin{bmatrix} g_7 \end{bmatrix}, \qquad (4.10)$$

where $M = \begin{bmatrix} g_1 & g_4 & g_5 \\ g_4 & g_2 & g_6 \\ g_5 & g_6 & g_3 \end{bmatrix}$ is a real symmetric matrix. Let P be

an orthogonal matrix which diagonalizes M, so that $P^T M P = D$, diagonal. Then let x', y', and z' be defined by

$$\begin{bmatrix} x \\ y \\ z \end{bmatrix} = P \begin{bmatrix} x' \\ y' \\ z' \end{bmatrix}. \qquad (4.11)$$

Substituting for $\begin{bmatrix} x \\ y \\ z \end{bmatrix}$ and $\begin{bmatrix} x & y & z \end{bmatrix}$ in equation 4.10 gives

$$[x' \ y' \ z'] P^T M P \begin{bmatrix} x' \\ y' \\ z_1 \end{bmatrix} = [g_7]. \qquad (4.12)$$

Now

$$P^T M P = D = \begin{bmatrix} d_1 & 0 & 0 \\ 0 & d_2 & 0 \\ 0 & 0 & d_3 \end{bmatrix},$$

so equation (4.12) becomes

$$d_1 (x')^2 + d_2 (y')^2 + d_3 (z')^2 = g_7. \qquad (4.13)$$

The d_j are of course the eigenvalues of M. By changing to the x', y', and z' coordinates we have eliminated all the cross terms, as in the case of conic sections. If there were linear terms in 4.9, we could carry them along and then eliminate them by a translation as we did for conic sections. It is the existence of an orthogonal P that diagonalizes M that lets us eliminate the cross terms by changing to basis (axes) given by the columns of P. For this reason Theorem 4.10 is sometimes called the Principal Axes Theorem.

Exercises

1. For each of the following matrices A find an orthogonal matrix P such that $P^{-1}AP = D$, diagonal and verify by showing that $AP = PD$ and that P is invertible.

(a) $\begin{bmatrix} 4 & 1 \\ 1 & 4 \end{bmatrix}$

(b) $\begin{bmatrix} 2 & 2 \\ 2 & 2 \end{bmatrix}$

(c) $\begin{bmatrix} 1 & 1 & -3 \\ 1 & 1 & -3 \\ -3 & -3 & 5 \end{bmatrix}$

(d) $\begin{bmatrix} 0 & 4 & -1 \\ 4 & 0 & 1 \\ -1 & 1 & 1 \end{bmatrix}$

(e) $\begin{bmatrix} 1 & 2 & 1 \\ 2 & 4 & 2 \\ 1 & 2 & 1 \end{bmatrix}$

(f) $\begin{bmatrix} 5 & -2 & 2 \\ -2 & 2 & 4 \\ 2 & 4 & 2 \end{bmatrix}$

(g) $\begin{bmatrix} 3 & 2 & 2 \\ 2 & 2 & 0 \\ 2 & 0 & 4 \end{bmatrix}$

(h) $\begin{bmatrix} 2 & 2 & 1 & 0 \\ 2 & -1 & -2 & 0 \\ 1 & -2 & 2 & 0 \\ 0 & 0 & 0 & 3 \end{bmatrix}$

(i) $\begin{bmatrix} 2 & 1 & 1 & 1 \\ 1 & 2 & 1 & 1 \\ 1 & 1 & 2 & 1 \\ 1 & 1 & 1 & 2 \end{bmatrix}$ (Hint for this last one only: try to find the

eigenvalues without computing the characteristic polynomial. Look for numbers that when subtracted from the diagonal produce a matrix with dependent rows.)

2. Classify each of the following conics as ellipse, hyperbola or parabola, assuming they are not degenerate.

(a) $4x^2 + 4xy + 4y^2 = 18$,

(b) $3x^2 + 10xy + 3y^2 = 8$,

(c) $x^2 + 2xy + y^2 + 8x + y = 0$.

In exercises 3 through 6, answer the following questions for each matrix and justify your answer: a) Is it similar to a diagonal matrix? b) Is it orthogonally similar to a diagonal matrix?

3 $A = \begin{bmatrix} -1 & 2 & 0 \\ 1 & 2 & 1 \\ 0 & 2 & -1 \end{bmatrix}$

4 $B = \begin{bmatrix} 2 & -1 & 0 \\ -1 & 2 & -1 \\ 0 & -1 & 2 \end{bmatrix}$

5 $C = \begin{bmatrix} 2 & 1 & 0 \\ 0 & 2 & 0 \\ 0 & 0 & 3 \end{bmatrix}$

$$6 \quad D = \begin{bmatrix} 2 & 2 & 0 & 0 \\ 2 & 2 & 0 & 0 \\ 0 & 0 & 0 & 1 \\ 0 & 0 & 1 & 0 \end{bmatrix}.$$

Section 4.4 Cayley Hamilton Theorem

Theorem 4.11 *Every $n \times n$ matrix A is similar over the complex numbers to an upper triangular matrix with the eigenvalues of A (possibly complex) on the main diagonal.*

Proof.   The proof is by induction on n. If $n = 1$, A is already in upper triangular form, so the theorem is obvious. Assume that every $(n-1) \times (n-1)$ matrix is similar over the complex numbers to an upper triangular matrix. Let A be an $n \times n$ matrix. Let $p(k)$ be the characteristic polynomial of A. By the fundamental theorem of algebra $p(k)$ has a root in the complex numbers. Let c be a root of $p(k)$. Then c is an eigenvalue of A. Let v_1 be an eigenvector of A for c. Then $Av_1 = cv_1$. Extend v_1 to a basis $\{v_1, v_2, \cdots, v_n\}$ of R^n, and let $P = \begin{bmatrix} v_1 & v_2 & \cdots & v_n \end{bmatrix}$. P is invertible by Theorem 2.18. The first column of AP equals Av_1 which equals cv_1, so we have

$$AP = P \begin{bmatrix} c & * & \cdots & * \\ 0 & & & \\ \vdots & & M & \\ 0 & & & \end{bmatrix}.$$

Thus

$$P^{-1}AP = \begin{bmatrix} c & * & \cdots & * \\ 0 & & & \\ \vdots & & M & \\ 0 & & & \end{bmatrix}.$$

By the induction hypothesis there exists an invertible $(n-1) \times (n-1)$ matrix Q_1 such that $Q_1^{-1} M Q_1$ is upper triangular. Form

$$Q = \begin{bmatrix} 1 & 0 & \cdots & 0 \\ 0 & & & \\ \vdots & & Q_1 & \\ 0 & & & \end{bmatrix},$$

then one can check by block multiplication that

$$Q^{-1} = \begin{bmatrix} 1 & 0 & \cdots & 0 \\ 0 & & & \\ \vdots & & Q_1^{-1} & \\ 0 & & & \end{bmatrix},$$

and

$$\begin{bmatrix} 1 & 0 & \cdots & 0 \\ 0 & & & \\ \vdots & & Q_1^{-1} & \\ 0 & & & \end{bmatrix} \begin{bmatrix} c & * & \cdots & * \\ 0 & & & \\ \vdots & & M & \\ 0 & & & \end{bmatrix} \begin{bmatrix} 1 & 0 & \cdots & 0 \\ 0 & & & \\ \vdots & & Q_1 & \\ 0 & & & \end{bmatrix}$$

equals

$$\begin{bmatrix} c & * & \cdots & * \\ 0 & & & \\ \vdots & & Q_1^{-1} M Q_1 & \\ 0 & & & \end{bmatrix},$$

which is upper triangular since $Q_1^{-1} M Q_1$ is upper triangular. Thus we have

$$Q^{-1} P^{-1} A P Q = \begin{bmatrix} c & * & \cdots & * \\ 0 & & & \\ \vdots & & Q_1^{-1} M Q_1 & \\ 0 & & & \end{bmatrix},$$

so A is similar to an upper triangular matrix. This completes the induction. ∎

Corollary 1 *If A is an $n \times n$ matrix then*

$$|A| = \prod (eigenvalues \ of \ A)$$

and

$$tr(A) = \sum (eigenvalues \ of \ A).$$

In both cases each eigenvalue is repeated j times if it has algebraic multiplicity j.

Proof. Since the result is obvious for upper triangular matrices, the general case follows from the preceding theorem, and the result that similar matrices have the same eigenvalues, trace, and determinant (Theorem 4.4). ∎

Corollary 2 *Let A be an $n \times n$ matrix, then c is an eigenvalue of A^j if and only if $c = d^j$ for some eigenvalue d of A.*

Proof. This result is obvious for a triangular matrix and the general case then follow from the Theorems 4.11 and 4.4. ∎

Theorem 4.12 (Cayley-Hamilton) *If $p(k)$ is the characteristic polynomial of an $n \times n$ matrix A, say $p(k) = k^n + b_{n-1}k^{n-1} + \cdots + b_1 k + b_0$, then $p(A) = A^n + b_{n-1}IA^{n-1} + \cdots + b_1 IA + b_0 I = [0]$, where the $b_j I$ are scalar matrices.*

Proof. For every polynomial $g(k)$, clearly $g(P^{-1}AP) = P^{-1}g(A)P$. Thus if A and B are similar, they satisfy (make equal to the zero matrix) the same polynomials. As a result of Theorems 4.11 and 4.4, it suffices to prove the Cayley-Hamilton Theorem for upper triangular matrices. We assume A is an $n \times n$ upper triangular matrix. The proof is by induction on n. If $n = 1$, $A = [c]$, and $A - cI = [0]$, so the theorem is true for $n = 1$. Assume that if A is an upper triangular matrix of size $(n-1) \times (n-1)$ then A satisfies its characteristic polynomial. Now Let A be $n \times n$ and upper triangular. Let $d_1, d_2, \cdots, d_n$ be the eigenvalues of A, and thus also the elements on the main diagonal of A, listed in order as they appear, so that

$$
A = \begin{bmatrix} d_1 & * & \cdots & * \\ 0 & d_2 & \ddots & \vdots \\ \vdots & \ddots & \ddots & * \\ 0 & \cdots & 0 & d_n \end{bmatrix}.
$$

Notice that the first column of $A - d_1I$ is the zero vector. Also note that $(A-d_1I)(A-d_2I) = (A-d_2I)(A-d_1I)$ because the only matrices involved are A and I, which commute. Let $M = (A - d_1I)(A - d_2I) = (A - d_2I)(A - d_1I)$. We now prove that the first two columns of M are zero. The first column of $(A - d_1I)$ has already been shown to be zero, and so the first column of $(A - d_2I)(A - d_1I)$ must be zero. But then

$$
\begin{aligned}
M &= (A - d_1I)(A - d_2I) \\
&= \begin{bmatrix} 0 & * & \cdots & * \\ 0 & d_2 - d_1 & \ddots & \vdots \\ \vdots & \ddots & \ddots & * \\ 0 & \cdots & 0 & d_n - d_1 \end{bmatrix} \begin{bmatrix} d_1 - d_2 & * & \cdots & * \\ 0 & 0 & \ddots & \vdots \\ \vdots & \ddots & \ddots & * \\ 0 & \cdots & 0 & d_n - d_1 \end{bmatrix}
\end{aligned}
$$

clearly has second column zero also. Assume inductively that the first $t - 1$ columns of $(A - d_1I)(A - d_2I) \cdots (A - d_{t-1}I)$ are zero. This is passed on to

$$
\begin{aligned}
N &= (A - d_1I)(A - d_2I) \cdots (A - d_tI) \\
&= (A - d_tI)(A - d_1I)(A - d_2I) \cdots (A - d_{t-1}I).
\end{aligned}
$$

Thus

$$
N = \begin{bmatrix}
\overset{first\ t-1\ cols\ 0}{} & & & & \\
0 & \cdots & 0 & * & \cdots & * \\
0 & \cdots & 0 & * & \cdots & * \\
0 & \cdots & 0 & * & \cdots & * \\
\vdots & & \vdots & \vdots & & \vdots \\
0 & \cdots & 0 & * & \cdots & * \\
0 & \cdots & 0 & * & \cdots & *
\end{bmatrix}
\begin{bmatrix}
\overset{0\ in\ position\ tt}{} & & & & \\
d_1 - d_t & & & & \\
0 & \ddots & & & \\
0 & & 0 & & * \\
\vdots & & & \ddots & \\
0 & & & & \\
0 & \cdots & \cdots & 0 & d_n - d_t
\end{bmatrix}.
$$

When one computes column t of this product, one sees that it is zero. Thus
$$(A - d_1 I)(A - d_2 I) \cdots (A - d_n I) = [0].$$
But $p(A) = (A - d_1 I)(A - d_2 I) \cdots (A - d_n I)$, so $p(A) = [0]$. ∎

Exercises

1. For each of the matrices in section 4.1, problem 2, verify the Cayley-Hamilton Theorem.

2. The minimal polynomial is the (unique) monic polynomial of least degree satisfied by an $n \times n$ matrix A. Give an example of a matrix for which the characteristic polynomial is not the minimal polynomial. (A polynomial of degree n is called monic if the coefficient if x^n is 1.)

3. Give an example of a matrix A whose characteristic and minimal polynomial are the same.

4. Use the Cayley-Hamilton Theorem to show that any $n \times n$ matrix A with $A^k = [0]$ for some positive integer k, must satisfy $A^n = [0]$.

5. Verify that the matrix $A = \begin{bmatrix} 1 & 2 \\ 2 & 3 \end{bmatrix}$ satisfies its characteristic equation. Show how this equation, when multiplied by A^{-1} yields a formula for A^{-1} as a linear combination of I and A.

Section 4.5 Applications of Diagonalization

Powers of a square matrix A

If A is a matrix which can be diagonalized then we can derive a formula for A^j in terms of d_i^j, where d_i are the eigenvalues of A as follows: If A can be diagonalized, there exists an invertible matrix P such that

$$P^{-1}AP = D = \begin{bmatrix} d_1 & 0 & \cdots & 0 \\ 0 & d_2 & & \vdots \\ \vdots & & \ddots & 0 \\ 0 & \cdots & 0 & d_n \end{bmatrix}.$$

Taking j^{th} powers of both sides we see that

$$(P^{-1}AP)^j = D^j = \begin{bmatrix} d_1^j & 0 & \cdots & 0 \\ 0 & d_2^j & & \vdots \\ \vdots & & \ddots & 0 \\ 0 & \cdots & 0 & d_n^j \end{bmatrix}.$$

Since $(P^{-1}AP)^j = P^{-1}A^jP$, we have $P^{-1}A^jP = D^j$, and we can solve for A^j, obtaining

$$A^j = PD^jP^{-1}. \tag{4.14}$$

Example 4.16 *Let* $A = \begin{bmatrix} 2 & 3 \\ 3 & 2 \end{bmatrix}$. *Then*

$$\begin{aligned} |A - kI| &= \begin{vmatrix} 2-k & 3 \\ 3 & 2-k \end{vmatrix} \\ &= k^2 - 4k - 5 \\ &= (k-5)(k+1). \end{aligned}$$

Thus the eigenvalues of A are 5 and -1. To find the eivenvectors of A for 5, we solve $(A - 5I)\mathbf{x} = \mathbf{O}$, or

$$\begin{bmatrix} -3 & 3 \\ 3 & -3 \end{bmatrix} \mathbf{x} = \begin{bmatrix} 0 \\ 0 \end{bmatrix},$$

and $\mathbf{v} = \begin{bmatrix} 1 \\ 1 \end{bmatrix}$ *is a solution. To find the eigenvectors of A for -1, we solve $(A + I)\mathbf{x} = \mathbf{O}$, or*

$$\begin{bmatrix} 3 & 3 \\ 3 & 3 \end{bmatrix} \mathbf{x} = \begin{bmatrix} 0 \\ 0 \end{bmatrix},$$

and $\mathbf{u} = \begin{bmatrix} -1 \\ 1 \end{bmatrix}$ *is a solution. Let*

$$P = \begin{bmatrix} 1 & -1 \\ 1 & 1 \end{bmatrix},$$

then

$$P^{-1} = \frac{1}{2}\begin{bmatrix} 1 & 1 \\ -1 & 1 \end{bmatrix} = \begin{bmatrix} \frac{1}{2} & \frac{1}{2} \\ -\frac{1}{2} & \frac{1}{2} \end{bmatrix},$$

and

$$P^{-1}AP = \begin{bmatrix} 5 & 0 \\ 0 & -1 \end{bmatrix}.$$

Now

$$D^j = \begin{bmatrix} 5^j & 0 \\ 0 & (-1)^j \end{bmatrix},$$

so

$$A^j = \begin{bmatrix} 1 & -1 \\ 1 & 1 \end{bmatrix}\begin{bmatrix} 5^j & 0 \\ 0 & (-1)^j \end{bmatrix}\begin{bmatrix} \frac{1}{2} & \frac{1}{2} \\ -\frac{1}{2} & \frac{1}{2} \end{bmatrix},$$

whence

$$A^j = \begin{bmatrix} 5^j & (-1)^{j+1} \\ 5^j & (-1)^j \end{bmatrix}\begin{bmatrix} \frac{1}{2} & \frac{1}{2} \\ -\frac{1}{2} & \frac{1}{2} \end{bmatrix},$$

which gives

$$A^j = \frac{1}{2}\begin{bmatrix} 5^j - (-1)^{j+1} & 5^j + (-1)^{j+1} \\ 5^j - (-1)^j & 5^j + (-1)^j \end{bmatrix}.$$

Example 4.17 *In a certain city there are two major modes of transportation, private and public. Each year 20% switch from public to private and 10% switch from private to public. Assuming no change in the population what will be the distribution after j years? As $j \to \infty$? Solution: Let x_0 be the number of people who use private transport, and y_0 the number who use public transport at the start, year 0. Let $A = \begin{bmatrix} .9 & .2 \\ .1 & .8 \end{bmatrix}$, then*

$$A\begin{bmatrix} x_0 \\ y_0 \end{bmatrix} = \begin{bmatrix} .9 & .2 \\ .1 & .8 \end{bmatrix}\begin{bmatrix} x_0 \\ y_0 \end{bmatrix} = \begin{bmatrix} .9x_0 + .2y_0 \\ .1x_0 + .8y_0 \end{bmatrix} = \begin{bmatrix} x_1 \\ y_1 \end{bmatrix},$$

where x_1 and y_1 are the numbers using private and public respectively one year later. Let x_j and y_j be the numbers using private and public respectively after j years. Then we have $\begin{bmatrix} x_j \\ y_j \end{bmatrix} = A^j \begin{bmatrix} x_0 \\ y_0 \end{bmatrix}$. *We now find a formula for A^j as in the previous example. First we find the eigenvalues of A.*

$$|A - kI| = \begin{vmatrix} .9 - k & 2 \\ .1 & .8 - k \end{vmatrix}$$
$$= k^2 - 1.7k + .7$$
$$= (k - 1)(k - .7),$$

so the eigenvalues are 1 and .7. Thus

$$D = \begin{bmatrix} 1 & 0 \\ 0 & .7 \end{bmatrix},$$

and

$$D^j = \begin{bmatrix} 1 & 0 \\ 0 & (.7)^j \end{bmatrix}.$$

Then find the eigenvectors. For $k = 1$, $(A - 1I)\,\mathbf{x} = \mathbf{O}$ is

$$\begin{bmatrix} -.1 & .2 \\ .1 & -.2 \end{bmatrix} \mathbf{x} = \begin{bmatrix} 0 \\ 0 \end{bmatrix},$$

and has solution $\begin{bmatrix} 2 \\ 1 \end{bmatrix}$. *For $k = .7$, $(A - .7I)\mathbf{x} = \mathbf{O}$ is*

$$\begin{bmatrix} .2 & .2 \\ .1 & .1 \end{bmatrix} \mathbf{x} = \begin{bmatrix} 0 \\ 0 \end{bmatrix},$$

and has solution $\begin{bmatrix} 1 \\ -1 \end{bmatrix}$, *so*

$$P = \begin{bmatrix} 2 & 1 \\ 1 & -1 \end{bmatrix}$$

and

$$P^{-1} = \begin{bmatrix} \frac{1}{3} & \frac{1}{3} \\ \frac{1}{3} & -\frac{2}{3} \end{bmatrix}.$$

As before, $A^j = PD^jP^{-1}$, thus

$$\begin{bmatrix} x_j \\ y_j \end{bmatrix} = A^j \begin{bmatrix} x_0 \\ y_0 \end{bmatrix}$$

$$= \begin{bmatrix} 2 & 1 \\ 1 & -1 \end{bmatrix} \begin{bmatrix} 1 & 0 \\ 0 & (.7)^j \end{bmatrix} \begin{bmatrix} \frac{1}{3} & \frac{1}{3} \\ \frac{1}{3} & -\frac{2}{3} \end{bmatrix} \begin{bmatrix} x_0 \\ y_0 \end{bmatrix}.$$

Computing this product we get

$$\begin{bmatrix} x_j \\ y_j \end{bmatrix} = \frac{1}{3} \begin{bmatrix} 2 + (.7)^j & 2 - 2(.7)^j \\ 1 - (.7)^j & 1 + 2(.7)^j \end{bmatrix} \begin{bmatrix} x_0 \\ y_0 \end{bmatrix}.$$

Now as $j \to \infty$,

$$\begin{bmatrix} x_j \\ y_j \end{bmatrix} \to \begin{bmatrix} \frac{2}{3} & \frac{2}{3} \\ \frac{1}{3} & \frac{1}{3} \end{bmatrix} \begin{bmatrix} x_0 \\ y_0 \end{bmatrix}.$$

Thus eventually $(\frac{2}{3})(x_0+y_0)$ will use private transportation and $(\frac{1}{3})(x_0+y_0)$ will use public transport. Note that the eventual distribution is $\frac{2}{3}$ and $\frac{1}{3}$ respectively of the total population and does not depend on the initial numbers, x_0, y_0.

Nilpotent Matrices

Definition 4.18 *An $n \times n$ matrix A is said to be nilpotent if for some positive integer j, $A^j = [0]$. The least positive integer for which this is true is called the degree of nilpotence.*

Lemma 4.9 *If A and B are similar, then $A^j = [0]$ if and only if $B^j = [0]$, so they are either both nilpotent of the same index of nilpotency, or both not nilpotent.*

Proof.   If A and B are similar, then there is an invertible P such that $B = P^{-1}AP$. Then $B^j = P^{-1}A^jP$, so B^j and A^j are similar for every positive integer j. But the zero matrix is similar only to itself, whence $A^j = [0]$ if and only if $B^j = [0]$. ∎

Theorem 4.13 *The largest degree of nilpotence of any nilpotent $n \times n$ matrix A is n.*

Proof.   Suppose A is nilpotent, $A^j = [0]$ for some positive integer j. By Theorem 4.11, A is similar to an upper triangular matrix

$$B = \begin{bmatrix} d_1 & & & * \\ 0 & d_2 & & \\ \vdots & & \ddots & \\ 0 & \cdots & 0 & d_n \end{bmatrix},$$

then

$$B^j = \begin{bmatrix} d_1^j & & & * \\ 0 & d_2^j & & \\ \vdots & & \ddots & \\ 0 & \cdots & 0 & d_n^j \end{bmatrix},$$

and since $A^j = [0]$, $B^j = [0]$. Thus we must have $d_i^j = 0$ for all integers i from 1 to n, and hence $d_i = 0$ for all integers i from 1 to n. Such a matrix is known as strictly upper triangular. The matrix E_{ij} has been defined earlier as the $n \times n$ matrix with a one in the ij position and zeros elsewhere. Then

$$B = \sum_{i<j} b_{ij} E_{ij}.$$

Let us define the spread of E_{ij} for matrices E_{ij} with $j > i$ as $j - i$. Now $E_{ij} E_{kl} = \begin{cases} 0 \text{ if } j \neq k \\ E_{il} \text{ if } j = l \end{cases}$, so if E_{ij} and E_{kl} are two matrices with $j > i$ and $k > l$ then either $E_{ij} E_{kl} = [0]$ or (if $j = k$) $E_{ij} E_{kl} = E_{il}$ with $i > k = j > l$, which has spread $l - i$. Now when $j = k$ we have $l - i = (l - k) + (j - i)$, thus either the product is the zero matrix or the spread of the product is the sum of the spreads of the factors. Since the largest spread possible for a strictly upper triangular E_{ij} is $n-1$, it follows that a product of n such matrices must be $[0]$. Since the strictly upper triangular matrix B can be written as $B = \sum_{i<j} b_{ij} E_{ij}$, each term in the expansion of $B^n = \left(\sum_{i<j} b_{ij} E_{ij} \right)^n$ contains a product of n upper triangular matrices of the type E_{ij} with $j > i$, and hence must be $[0]$. Since A is similar to B, $A^n = [0]$ also, hence the index of nilpotency of A is $\leq n$. Since $(E_{12} + E_{23} + \cdots + E_{n-1,n})^{n-1} = E_{1n}$, there does exist an $n \times n$ matrix whose degree of nilpotence is n. ∎

Idempotent matrices

Definition 4.19 *A matrix A is said to be idempotent if $A^2 = A$.*

We have already proved that a matrix in Hermite form is idempotent. The matrix $A = \begin{bmatrix} 1 & 0 \\ 0 & 0 \end{bmatrix}$, is a matrix in Hermite form and is idempotent. Two additional examples of matrices which are idempotent are $B = \begin{bmatrix} 25 & -20 \\ 30 & -24 \end{bmatrix}$, and $C = \begin{bmatrix} -26 & -18 & -27 \\ 21 & 15 & 21 \\ 12 & 8 & 13 \end{bmatrix}$. Verification is left to the reader.

Theorem 4.14 *If A is an $n \times n$ idempotent matrix of rank r then A is similar to a matrix of the form $\begin{bmatrix} [0] & [0] \\ [0] & I \end{bmatrix}$, where I is the $r \times r$ identity matrix and the rest of the matrix elements are zero.*

Proof.   Let A be an $n \times n$ idempotent matrix of rank r and let c be an eigenvalue of A and $\mathbf{v}$ an eigenvector of A for c. Then $A\mathbf{v} = c\mathbf{v}$. Thus $A^2\mathbf{v} = Ac\mathbf{v} = cA\mathbf{v} = c^2\mathbf{v}$, but on the other hand $A^2\mathbf{v} = A\mathbf{v} = c\mathbf{v}$, so we must have $(c^2 - c)\mathbf{v} = \mathbf{O}$, and since $\mathbf{v} \neq \mathbf{O}$ this implies $c^2 = c$. Hence $c = 0$ or 1. Thus an idempotent matrix can have no eigenvalues other than 0 and 1. If 0 is not an eigenvalue of A, let S be the empty set. If 0 is an eigenvalue of A, then $n - r$ must be the dimension of the eigenspace of A for $c = 0$, because the eigenspace of A for $c = 0$ is the null space of A, which has dimension $n - r$. Let $S = \{\mathbf{u}_1, \mathbf{u}_2, \cdots, \mathbf{u}_{n-r}\}$ be a basis of this eigenspace. Hence $A\mathbf{u}_i = \mathbf{O}$ for $i = 1, 2, \cdots, n - r$. Extend S to a basis of R^n, say $\{\mathbf{u}_1, \mathbf{u}_2, \cdots, \mathbf{u}_{n-r}, \mathbf{v}_{n-r+1}, \cdots, \mathbf{v}_n\}$. Let $\mathbf{w}_j = A\mathbf{v}_j$. Then $A\mathbf{w}_j = A(A\mathbf{v}_j) = A^2\mathbf{v}_j = A\mathbf{v}_j = \mathbf{w}_j$, so $\mathbf{w}_j$ is in the eigenspace of A for the eigenvalue 1. If $\mathbf{v}$ is an arbitrary vector in R^n, we can express $\mathbf{v}$ as a linear combination of $\mathbf{u}_1, \mathbf{u}_2, \cdots, \mathbf{u}_{n-r}, \mathbf{v}_{n-r+1}, \cdots, \mathbf{v}_n$, and since $A(\mathbf{u}_i) = \mathbf{O}$, we see that

$$Av = \sum_{i=n-r}^{n} c_i\mathbf{w}_i,$$

so that Av is in the eigenspace of A for $c = 1$. Now we claim that the set $\{\mathbf{u}_1, \mathbf{u}_2, \cdots, \mathbf{u}_{n-r}, \mathbf{w}_{n-r+1}, \cdots, \mathbf{w}_n\}$ is a basis for R^n. We need only verify that it is a linearly independent set. Suppose

$$k_1 \mathbf{u}_1 + \cdots + k_{n-r}\mathbf{u}_{n-r} + k_{n-r+1}\mathbf{w}_{n-r+1} + \cdots + k_n \mathbf{w}_n = \mathbf{O}. \quad (4.15)$$

Multiplying both sides of this equation on the left by A gives

$$k_{n-r+1}\mathbf{w}_{n-r+1} + \cdots + k_n \mathbf{w}_n = \mathbf{O},$$

from which it follows that

$$k_{n-r+1}A\mathbf{v}_{n-r+1} + \cdots + k_n A\mathbf{v}_n = \mathbf{O},$$

and thus

$$A(k_{n-r+1}\mathbf{v}_{n-r+1} + \cdots + k_n \mathbf{v}_n) = \mathbf{O}.$$

It follows that $k_{n-r+1}\mathbf{v}_{n-r+1} + \cdots + k_n \mathbf{v}_n$ is in the eigenspace of A for $c = 0$. But then

$$k_{n-r+1}\mathbf{v}_{n-r+1} + \cdots + k_n \mathbf{v}_n = s_1 \mathbf{u}_1 + s_2 \mathbf{u}_2 + \cdots + s_{n-r}\mathbf{u}_{n-r}.$$

Since $\{\mathbf{u}_1, \mathbf{u}_2, \cdots, \mathbf{u}_{n-r}, \mathbf{v}_{n-r+1}, \cdots, \mathbf{v}_n\}$ is a basis of R^n it follows that all coefficients are zero in the above dependence relation. Thus $k_i = 0$ for $i = n-r+1, \cdots, n$. Substituting this in 4.15 gives $k_1 \mathbf{u}_1 + \cdots + k_{n-r}\mathbf{u}_{n-r} = \mathbf{O}$, and from this we conclude that $k_i = 0$ for $i = 1, 2, \cdots, n - r$. Thus $\{\mathbf{u}_1, \mathbf{u}_2, \cdots, \mathbf{u}_{n-r}, \mathbf{w}_{n-r+1}, \cdots, \mathbf{w}_n\}$ is a basis for R^n. Let

$$P = \begin{bmatrix} \mathbf{u}_1 & \mathbf{u}_2 & \cdots & \mathbf{u}_{n-r} & \mathbf{w}_{n-r+1} & \cdots & \mathbf{w}_n \end{bmatrix}.$$

Then P is an invertible matrix and

$$AP = P \begin{bmatrix} [0] & [0] \\ [0] & I_{r \times r} \end{bmatrix}.$$

This last equation follows because for the first $n - r$ columns of P we have $A\mathbf{u}_i = \mathbf{O}$, while for the remaining r columns of P we have $A\mathbf{w}_i = \mathbf{w}_i$. Thus

$$P^{-1}AP = \begin{bmatrix} [0] & [0] \\ [0] & I_{r \times r} \end{bmatrix}.$$

■

Remark   *If we had used* $Q = \begin{bmatrix} \mathbf{w}_1 & \mathbf{w}_2 & \cdots & \mathbf{w}_r & \mathbf{u}_1 & \cdots & \mathbf{u}_{n-r} \end{bmatrix}$ *we would get* $Q^{-1}AQ = \begin{bmatrix} I_{r \times r} & [0] \\ [0] & [0] \end{bmatrix}.$

Corollary 1 *If A is an idempotent matrix then* $tr(A) = rank(A)$.

Proof. If D is of the form $\begin{bmatrix} [0] & [0] \\ [0] & I_{r \times r} \end{bmatrix}$ then clearly $tr(D) =$ $rank(D) = r$. If A is an idempotent matrix, then A is similar to a matrix of this form. If A is similar to D then $tr(A) = tr(D) =$ $rank(D) = rank(A)$. ∎

Systems of first order linear differential equations

Consider the following system of first order linear differential equations

$$
\begin{aligned}
x_1'(t) &= a_{11}x_1(t) + a_{12}x_2(t) + \cdots + a_{1n}x_n(t) \\
x_2'(t) &= a_{21}x_1(t) + a_{22}x_2(t) + \cdots + a_{2n}x_n(t) \\
&\;\;\vdots \\
x_n'(t) &= a_{n1}x_1(t) + a_{n2}x_2(t) + \cdots + a_{nn}x_n(t),
\end{aligned}
$$

which can be written in matrix form as

$$
\begin{bmatrix} x_1'(t) \\ \vdots \\ x_n'(t) \end{bmatrix} = A \begin{bmatrix} x_1(t) \\ \vdots \\ x_n(t) \end{bmatrix}, \tag{4.16}
$$

or even more simply as

$$
\mathbf{x}' = A\mathbf{x},
$$

where $\mathbf{x} = \mathbf{x}(t)$ is the vector whose components are $x_i(t)$.

For $n = 1$ the system becomes $x_1'(t) = a_{11}x_1(t)$, so we drop the subscripts and just write $x'(t) = ax(t)$. In this case the variables separate if we rewrite the equation as

$$
\frac{x'(t)}{x(t)} = a,
$$

we can integrate both sides with respect to t and get $\ln(x) = at + b$, or $x(t) = e^{at+b} = e^{at}e^b = e^b e^{at}$. Substituting $t = 0$ we get $x(0) = e^b$. Thus we have that a solution is $x(t) = x(0)e^{at}$.

If the coefficient matrix A of our system is diagonal, then the system is just n separate equations of this type

$$
\begin{aligned}
x_1'(t) &= a_{11}x_1(t) \\
x_2'(t) &= a_{22}x_2(t) \\
&\vdots \\
x_n'(t) &= a_{nn}x_n(t),
\end{aligned}
$$

so a solution would be

$$
\begin{aligned}
x_1(t) &= k_1 e^{a_{11}t} \\
x_2(t) &= k_2 e^{a_{22}t} \\
&\vdots \\
x_n(t) &= k_n e^{a_{nn}t},
\end{aligned}
$$

where the k_i are constants, in this case $k_i = x_i(0)$.

Now if A is similar to a diagonal matrix D, we can reduce the general case to the diagonal case by a change of variable. Let P be an invertible matrix such that

$$
P^{-1}AP = D = \begin{bmatrix} d_1 & 0 & \cdots & 0 \\ 0 & d_2 & & \vdots \\ \vdots & & \ddots & 0 \\ 0 & \cdots & 0 & d_n \end{bmatrix},
$$

where the d_i are the eigenvalues of A. Then make the substitution

$$
\mathbf{x}(t) = P\mathbf{u}(t),
$$

or

$$
\begin{bmatrix} x_1(t) \\ \vdots \\ x_n(t) \end{bmatrix} = P \begin{bmatrix} u_1(t) \\ \vdots \\ u_n(t) \end{bmatrix}. \tag{4.17}
$$

Differentiating both sides of 4.17 gives

$$
\begin{bmatrix} x_1'(t) \\ \vdots \\ x_n'(t) \end{bmatrix} = P \begin{bmatrix} u_1'(t) \\ \vdots \\ u_n'(t) \end{bmatrix}. \tag{4.18}
$$

Substituting 4.17 and 4.18 in 4.16 gives

$$P \begin{bmatrix} u_1'(t) \\ \vdots \\ u_n'(t) \end{bmatrix} = AP \begin{bmatrix} u_1(t) \\ \vdots \\ u_n(t) \end{bmatrix}, \qquad (4.19)$$

or

$$\begin{bmatrix} u_1'(t) \\ \vdots \\ u_n'(t) \end{bmatrix} = P^{-1}AP \begin{bmatrix} u_1(t) \\ \vdots \\ u_n(t) \end{bmatrix} = D \begin{bmatrix} u_1(t) \\ \vdots \\ u_n(t) \end{bmatrix}. \qquad (4.20)$$

This is now the diagonal case

$$\begin{bmatrix} u_1'(t) \\ \vdots \\ u_n'(t) \end{bmatrix} = D \begin{bmatrix} u_1(t) \\ \vdots \\ u_n(t) \end{bmatrix}$$

and we know the solution is

$$\begin{bmatrix} u_1(t) \\ \vdots \\ u_n(t) \end{bmatrix} = \begin{bmatrix} k_1 e^{d_1 t} \\ \vdots \\ k_n e^{d_n t} \end{bmatrix}.$$

We can now find $\mathbf{x}(t)$ because $\mathbf{x}(t) = P\mathbf{u}(t)$, so that

$$\begin{bmatrix} x_1(t) \\ \vdots \\ x_n(t) \end{bmatrix} = P \begin{bmatrix} k_1 e^{d_1 t} \\ \vdots \\ k_n e^{d_n t} \end{bmatrix}.$$

The k_i are constants that we can find if we are given some initial conditions. To use this method it is essential that A be similar to a diagonal matrix and that you be able to find the eigenvalues of A (the d_i) and a matrix P that diagonalizes A.

Procedure 4.4 To solve $\mathbf{x}' = A\mathbf{x}$, assuming that A is diagonalizable.

1. Find an invertible matrix P such that $P^{-1}AP = D$.

2. The solution is then given by $\mathbf{x} = P \begin{bmatrix} k_1 e^{d_1 t} \\ \vdots \\ k_n e^{d_n t} \end{bmatrix}$.

3. The d_i are the eigenvalues of A, in the same order as the eigenvectors occur as columns of P.

4. The k_i are constants to be determined by substituting $t = 0$ and using the given initial conditions, (or some other value of t for which values of x_i are given).

Example 4.18 *Two species live in the same habitat and compete for food. By observation $x_1(0) = 600$ and $x_2(0) = 200$, time $t = 0$ refers to the present. Future population prediction is based on the differential equations*

$$\begin{array}{rcl} x_1'(t) & = & -2x_1(t) + 4x_2(t) \\ x_2'(t) & = & x_1(t) - 2x_2(t) \end{array}$$

where t is measured in years. Find $x_1(t)$ and $x_2(t)$. Make some inferences. Solution:

$$\begin{bmatrix} x_1' \\ x_2' \end{bmatrix} = A \begin{bmatrix} x_1 \\ x_2 \end{bmatrix},$$

where

$$A = \begin{bmatrix} -2 & 4 \\ 1 & -2 \end{bmatrix}.$$

The characteristic polynomial of A is $p(k) = k^2 + 4k$. The eigenvectors of A are 0 and -4. Now $(A + 4I) = \begin{bmatrix} 2 & 4 \\ 1 & 2 \end{bmatrix}$, so $\begin{bmatrix} 2 \\ -1 \end{bmatrix}$ is an eigenvector for $c = -4$. We have $(A - 0I) = A = \begin{bmatrix} -2 & 4 \\ 1 & -2 \end{bmatrix}$, so $\begin{bmatrix} 2 \\ 1 \end{bmatrix}$ is an eigenvector for $c = 0$. If we let $P = \begin{bmatrix} 2 & 2 \\ -1 & 1 \end{bmatrix}$, then $P^{-1}AP = D = \begin{bmatrix} -4 & 0 \\ 0 & 0 \end{bmatrix}$, so a solution is given by

$$\begin{bmatrix} x_1 \\ x_2 \end{bmatrix} = \begin{bmatrix} 2 & 2 \\ -1 & 1 \end{bmatrix} \begin{bmatrix} k_1 e^{-4t} \\ k_2 e^{0t} \end{bmatrix}$$

$$= \begin{bmatrix} 2 & 2 \\ -1 & 1 \end{bmatrix} \begin{bmatrix} k_1 e^{-4t} \\ k_2 \end{bmatrix}$$

$$= \begin{bmatrix} 2k_1 e^{-4t} + 2k_2 \\ -k_1 e^{-4t} + k_2 \end{bmatrix}.$$

Substituting the initial conditions $x_1(0) = 600$ *and* $x_2(0) = 200$, *we get* $2k_1 + 2k_2 = 600 - k_1 + k_2 = 200$, *which has solution* $k_1 = 50$, $k_2 = 250$. *Thus the model predicts that after* t *years* $x_1(t) = 100e^{-4t} + 500$ *and* $x_2(t) = -50e^{-4t} + 250$. *In the long run (as* $t \to \infty$*)* $x_1(t) \to 500$ *and* $x_2(t) \to 250$. *The populations stabilizes and both species survive.*

Exercises

1. Use diagonalization of the matrix $A = \begin{bmatrix} 1 & 1 & 2 \\ 0 & 1 & 0 \\ 0 & 1 & 3 \end{bmatrix}$ to derive a
   formula for all powers of A, A^k, where k is a positive integer.

2. In a certain location the probability is $\frac{1}{2}$ that following a rainy
   day it will rain. Following a dry day, the probability is $\frac{2}{3}$ that the
   next day will be dry. In the long run, find the probability that a
   given day will be rainy.

3. Show that the 3×3 matrix $\begin{bmatrix} a & * & * \\ 0 & b & * \\ 0 & 0 & c \end{bmatrix}$ is nilpotent if and only
   if $a = b = c = 0$.

4. Show that a square matrix A is nilpotent if and only if it is similar
   to a strictly upper triangular matrix.

5. Verify that the matrices A, B, C, defined at the beginning of the
   section on idempotent matrices are idempotent. Also verify that
   $rank = trace$ for these matrices.

6. Prove that the square matrix $\begin{bmatrix} 0 & & & * \\ & 0 & & \\ & & \ddots & \\ * & & & 0 \end{bmatrix}$ (zeros on the diagonal, and unspecified elsewhere) is idempotent if and only if it

is the zero matrix.

7. Solve the system of differential equations given by:

$$\begin{aligned} x_1' &= x_2 \\ x_2' &= x_3 \\ x_3' &= 8x_1 - 14x_2 + 7x_3, \end{aligned}$$

with initial conditions $x_1(0) = 3$, $x_2(0) = 7$, and $x_3(0) = 21$.

8. Solve the system of differential equations given by:

$$\begin{aligned} x_1' &= -x_1 + 2x_2 \\ x_2' &= x_1 + 2x_2 + x_3 \\ x_3' &= 2x_2 - x_3 \end{aligned}$$

Chapter 5

Subspaces, Orthogonality, and Projections

Section 5.1 Finding Vectors in $S \sqcap T$

In Section 1.8 we defined $S \sqcap T$ and $S + T$ for subspaces S and T of a vector space V and proved that when S and T are finite dimensional then $\dim(S + T) = \dim(S) + \dim(T) - \dim(S \sqcap T)$. If we are in R^n and if S and T are given we can use this theorem and our knowledge of what the RREF of a matrix tells us about the linear dependence relations among the columns of that matrix to find vectors in $S \sqcap T$. Indeed we can find a basis of $S \sqcap T$.

Example 5.1 *Let* $S = sp\{(4, 1, -2), (3, 2, -1)\}$, *and* $T = sp\{(1, -1, 2), (1, 4, -3)\}$. *Find* $\dim(S)$, $\dim(T)$, $\dim(S + T)$, $\dim(S \sqcap T)$ *and a basis for* $S \sqcap T$.
Solution: It is obvious that both S and T are of dimension 2. Now we consider the matrix

$$A = \begin{bmatrix} 4 & 3 & 1 & 1 \\ 1 & 2 & -1 & 4 \\ -2 & -1 & 2 & -3 \end{bmatrix},$$

whose columns are the basis vectors of S and T. Then

$$RREF(A) = \begin{bmatrix} 1 & 0 & 0 & -\frac{2}{3} \\ 0 & 1 & 0 & \frac{5}{3} \\ 0 & 0 & 1 & -\frac{4}{3} \end{bmatrix}.$$

Since $(S + T)$ is the column space of this matrix, we see that the dimension of $(S + T)$ is 3. Since $\dim(S) + \dim(T) = 4$, we conclude that $\dim(S \sqcap T) = 1$. Now if we call the columns of the matrix A s_1, s_2, t_1, t_2, we can see from $RREF(A)$ that

$$t_2 = (-2/3)s_1 + (5/3)s_2 - (4/3)t_1.$$

Putting all the t_i on the left (these are in T and the s_i are in S), this becomes

$$(4/3)t_1 + t_2 = (-2/3)s_1 + (5/3)s_2.$$

Since one side of this equation is a vector in T and the other side is a vector in S, we have found a vector in $S \sqcap T$. We can use either side of the equation to compute the vector we want. So we have

$$(-2/3)s_1 + (5/3)s_2 = (-2/3)\begin{bmatrix} 4 \\ 1 \\ -2 \end{bmatrix} + (5/3)\begin{bmatrix} 3 \\ 2 \\ -1 \end{bmatrix} = \begin{bmatrix} \frac{7}{3} \\ \frac{8}{3} \\ -\frac{1}{3} \end{bmatrix}$$

$\in S \sqcap T$. *Since $\dim(S \sqcap T) = 1$, this vector is a basis for $S \sqcap T$. We can check by computing the other side of the equation,*

$$(4/3)t_1 + t_2 = (4/3)\begin{bmatrix} 1 \\ -1 \\ 2 \end{bmatrix} + \begin{bmatrix} 1 \\ 4 \\ -3 \end{bmatrix} = \begin{bmatrix} \frac{7}{3} \\ \frac{8}{3} \\ -\frac{1}{3} \end{bmatrix},$$

which checks. Thus $S \sqcap T = sp\left\{\left(\frac{7}{3}, \frac{8}{3}, -\frac{1}{3}\right)\right\} = sp\{(7, 8, -1)\}$. Geometrically S and T are planes through the origin and $S \sqcap T$ is their line of intersection. An alternate method of finding $S \sqcap T$ would be to find the equation of each of the planes and solve these equations simultaneously to get the line of intersection.

The method just used is entirely general, and will work whatever the dimensions of S and T as long as they are subspaces of R^n and you have a basis for each. The following is an example in R^4.

Example 5.2 *Let* $S = sp\{(1,1,2,1),(1.-1,0,2),(1,2,0,1)\}$ *and* $T = sp\{(2,1,0,1),(-1,0,-2,1),(1,1,1,0)\}$. *Find a basis for* $S \sqcap T$. *Solution: S has dimension 3 and T has dimension 3, which can be seen by inspection, or by finding the rank of the matrix with the spanning vectors as rows (or columns). We form the matrix A with columns the given basis vectors of S followed by the given basis vectors of T,*

$$A = \begin{bmatrix} 1 & 1 & 1 & 2 & -1 & 1 \\ 1 & -1 & 2 & 1 & 0 & 1 \\ 2 & 0 & 0 & 0 & -2 & 1 \\ 1 & 2 & 1 & 1 & 1 & 0 \end{bmatrix}.$$

Then

$$RREF(A) = \begin{bmatrix} 1 & 0 & 0 & 0 & -1 & \frac{1}{2} \\ 0 & 1 & 0 & 0 & \frac{5}{6} & -\frac{5}{12} \\ 0 & 0 & 1 & 0 & \frac{3}{2} & -\frac{1}{4} \\ 0 & 0 & 0 & 1 & -\frac{7}{6} & \frac{7}{12} \end{bmatrix}.$$

We see that $\dim(S+T) = 4$, *so since* $\dim(S) + \dim(T) = 6$, *we must have* $\dim(S \sqcap T) = 2$. *If we call the columns of A* $s_1, s_2, s_3, t_1, t_2, t_3$, *the last two columns of $RREF(A)$ give us the following equations:*

$$t_3 = (1/2)s_1 - (5/12)s_2 - (1/4)s_3 + (7/12)t_1$$

and

$$t_2 = -s_1 + (5/6)s_2 + (3/2)s_3 - (7/6)t_1.$$

Putting all the vectors in T on the left hand side of each equation we see that this will give us $t_3 - \frac{7}{12}t_1$ *equals a vector in S and* $t_2 + \frac{7}{6}t_1$ *equals a vector in S. Thus these two vectors are in* $S \sqcap T$. *The first is*

$$(1,1,1,0) - \frac{7}{12}(2,1,0,1) = \left(-\frac{2}{12}, \frac{5}{12}, 1, -\frac{7}{12}\right),$$

and the second is

$$(-1,0,-2,1) + \frac{7}{6}(2,1,0,1) = \left(\frac{8}{6}, \frac{7}{6}, -2, \frac{13}{6}\right).$$

These two vectors are linearly independent and so form a basis for $S \sqcap T$. . We could clear fractions and write

$$S \sqcap T = sp\{(-2,5,12,-7),(8,7,-12,13)\}.$$

Sometimes we can see what's happening without doing a row reduction.

Example 5.3 *Let* $S = sp\{(3, 1, -1), (-1, 2, 1)\}$, *and*
$T = sp\{(1, 0, 0), (0, 1, 0)\}$. *Find* $\dim(S)$, $\dim(T)$, $\dim(S + T)$, $\dim(S \sqcap T)$ *and a basis for* $S \sqcap T$.
Solution: This example is easy enough to do by inspection if you notice a few things. S and T are both of dimension 2. Note that since $T = sp\{(1, 0, 0), (0, 1, 0)\}$, the vectors in T are of the form $(a, b, 0)$. The first two can be any numbers, but the last entry must be 0. All we need to do to find a vector in $S \sqcap T$ is to find a vector in S with the last component 0. We don't have to look far, because $S = sp\{(3, 1, -1), (-1, 2, 1)\}$, and the sum of these two vectors has the last component 0. Thus $(2, 3, 0) \in S \sqcap T$. Since the two basis vectors of T are linearly independent and $(3, 1, -1)$ is not in T, these three vectors are linearly independent, hence $\dim(S + T)$ is at least 3, but we are in R^3, so $\dim(S + T)$ is at most 3. Thus $\dim(S \sqcap T)$ must be 1, and $(2, 3, 0)$ is a basis for $S \sqcap T$.

Example 5.4 *Let* $S = \{(a, b, c)|a - b = 0\} = \{(a, a, c)\}$ *and* $T = \{(a, b, c)|a + 2b - c = 0\} = \{(a, b, a + 2b)\}$. *Find* $\dim(S)$, $\dim(T)$, $\dim(S + T)$, *and* $\dim(S \sqcap T)$, *and find a basis for* $S \sqcap T$. *Solution: Here the subspaces are described by homogeneous equations, or equivalently by giving the general form of vectors in the subspace, and not by a basis or spanning set as in the previous examples. We could find a basis for S and a basis for T and proceed as in the previous examples, but that is unnecessary. It is easy to see that S and T both have dimension 2, since in their general form, both have two components that can be chosen freely and the third then determined. To be in S, a triple must be of the form (a, a, c). If it is also to be in T, it must additionally satisfy "third component is first component plus twice second component", so it must be of the form $(a, a, 3a)$. Thus in $S \sqcap T$ we have only one free choice, so $\dim(S \sqcap T) = 1$, and a basis for $S \sqcap T$ is $(1, 1, 3)$. Again S and T are planes through the origin and $S \sqcap T$ is their line of intersection. Since $\dim(S) + \dim(T) = 4$ and $\dim(S \sqcap T) = 1$, we must have $\dim(S + T) = 3$, so we don't have to worry about a basis for $S + T$, since it is all of R^3. In this case the planes S and T were given by their equations instead of by bases, and the intersection is simply a matter of solving both equations simultaneously.*

Procedure 5.1 To find a basis for $S \sqcap T$ when spanning sets of S and T are given:

1. Refine the spanning set of S to a basis of S by putting the spanning vectors as columns of a matrix, finding the RREF, and keeping the leading columns.

2. Refine the spanning set of T to a basis of T in the same way.

3. Put the basis of S followed by the basis of T as the columns of a matrix A and find $RREF(A)$.

4. The leading columns of A are a basis for $S + T$, and each non-leading column of A can be expressed as a linear combination of preceding leading columns. Write out each of these equations (see Corollary 2 to Lemma 2.14 and the examples following it.)

5. In each equation from step 4, put all the vectors from T on one side, and all the vectors from S on the other. Each equation gives a vector in $S \sqcap T$ that can be computed from either side of the equation. These will be a basis of $S \sqcap T$.

Remark   *If you start out with spanning sets of S and T that are not linearly independent as the columns of A, this method will still find* $\dim(S+T)$ *and vectors in* $S \sqcap T$*, but now you will have extra dependence relations among the columns.*

Exercises
In the following problems find $\dim(S)$, $\dim(T)$, $\dim(S+T)$, $\dim(S \sqcap T)$, and a basis for each of these spaces.

1. Let $S = sp\{(2, 1, -1), (-1, 3, 0)\}$ and
   $T = sp\{(-1, 0, 1), (2, 2, -1)\}$.

2. Let $S = \{(x, y, x - y)\}$ and $T = \{(x, -x, y)\}$.

3. Let $S = sp\{(0, 0, 1, 0), (1, 1, 0, 0)\}$ and
   $T = sp\{(1, 1, 0, 0,), (0, 0, 0, 1)\}$.

4. Let $S = sp\{(2,1,2,1),(1,2,1,1)\}$ and
   $T = sp\{(7,5,7,4),(1,0,2,0)\}$.

5. Let $S = sp\{(1,1,2,1),(1,-1,0,2),(1,2,0,1)\}$ and
   $T = sp\{(2,0,2,3),(2,3,2,2)\}$.

Section 5.2 Orthogonal Complement and Projection

Recall that two vectors in R^n are orthogonal if $\mathbf{u} \cdot \mathbf{v} = 0$ (Definition 4.11).

Definition 5.1 *If S is any set of vectors in R^n, the orthogonal complement of S, written $S^\perp$, is the set of vectors orthogonal to every vector in S.*

Lemma 5.1 *For any nonempty set of vectors S in R^n, $S^\perp$ is a subspace.*

Proof. $S^\perp$ is not empty, since $\mathbf{O} \cdot \mathbf{y} = 0$ for any vector $\mathbf{y}$, thus $\mathbf{O} \in S^\perp$. Let $\mathbf{u_1}$ and $\mathbf{u_2}$ be in $S^\perp$, then for every $\mathbf{y} \in S$ we have $\mathbf{u_1} \cdot \mathbf{y} = 0$, and $\mathbf{u_2} \cdot \mathbf{y} = 0$. Then $(\mathbf{u_1} + \mathbf{u_2}) \cdot \mathbf{y} = \mathbf{u_1} \cdot \mathbf{y} + \mathbf{u_2} \cdot \mathbf{y} = 0$, so $\mathbf{u_1} + \mathbf{u_2} \in S^\perp$. Also $(r\mathbf{u_1}) \cdot \mathbf{y} = r(\mathbf{u_1} \cdot \mathbf{y}) = 0$ for any real number r, thus $r\mathbf{u_1} \in S^\perp$. ∎

Lemma 5.2 *For any (real) matrix A, $nullsp(A) = (rowsp(A))^\perp$.*

Proof.   Let $\mathbf{u}_i$ be the row vectors of A. It's clear from the definition of matrix multiplication that $A\mathbf{x} = \mathbf{O}$ is equivalent to $\mathbf{u}_i \cdot \mathbf{x} = 0$. Thus $\mathbf{x} \in nullsp(A)$ if and only if $\mathbf{u}_i \cdot \mathbf{x} = 0$ for every row vector $\mathbf{u}_i$ of A. Thus if $\mathbf{x} \in nullsp(A)$, all the rows of A are in $\mathbf{x}^\perp$. By the previous lemma $\mathbf{x}^\perp$ is a subspace of R^n, hence $rowsp(A) \subset \mathbf{x}^\perp$ and so $\mathbf{x} \in (rowsp(A))^\perp$. We have shown that $nullsp(A) \subset (rowsp(A))^\perp$. Now assume that $\mathbf{v} \in (rowsp(A))^\perp$. Then $\mathbf{v}$ is orthogonal to all the rows of A, hence $\mathbf{v} \in nullsp(A)$. ∎

Lemma 5.3 *If S is any subspace of R^n, $S \sqcap S^\perp = \mathbf{O}$.*

Proof.   If $\mathbf{x} \in S \sqcap S^\perp$, then $\mathbf{x} \cdot \mathbf{x} = 0$, but this is only true if $\mathbf{x} = \mathbf{O}$. ∎

Theorem 5.1 *If S is a subspace of R^n, then $R^n = S \oplus S^\perp$.*

Proof.   Let A be a matrix with rows a basis of S. Then $rowsp(A) = S$, A is $m \times n$, and $nullsp(A) = S^\perp$. Thus $\dim(S) + \dim(S^\perp) = n$ by Theorem 2.15. We also have $S \sqcap S^\perp = \mathbf{O}$, so $\dim(S + S^\perp) = n$. Thus $S + S^\perp = R^n$, and since $S \sqcap S^\perp = \mathbf{O}$, the sum is direct. ∎

Corollary 1 *If S is a subspace of R^n, every vector $\mathbf{u}$ in R^n can be written uniquely as a sum of a vector in S and a vector in $S^\perp$.*

Proof.   This is a direct result of Theorem 1.13 and the preceding theorem. ∎

Definition 5.2 *If S is a subspace of R^n and if $\mathbf{u} = \mathbf{w} + \mathbf{v}$ is the unique expression of $\mathbf{u}$ as a sum of a vector $\mathbf{w} \in S$ and a vector $\mathbf{v} \in S^\perp$, then $\mathbf{w}$ is the projection of $\mathbf{u}$ on S, written $\mathbf{w} = proj_S \mathbf{u}$.*

Theorem 5.2 *If S is a subspace of R^n, and if $\{\mathbf{u}_1, \mathbf{u}_2, \cdots, \mathbf{u}_k\}$ is a basis of S and $\{\mathbf{v}_1, \mathbf{v}_2, \cdots, \mathbf{v}_{n-k}\}$ a basis of $S^\perp$, then $B = \{\mathbf{u}_1, \mathbf{u}_2, \cdots, \mathbf{u}_k, \mathbf{v}_1, \mathbf{v}_2, \cdots \mathbf{v}_{n-k}\}$ is a basis of R^n.*

Proof.   From the preceding theorem, $R^n = S + S^\perp$, thus B is a spanning set of R^n. Since we also know that the number of vectors in B is n, B must be a basis of R^n. (This also follows from the corollary to Theorem 2.15). ∎

As described in Procedure 2.5, this gives us a very good way to extend a linearly independent set in R^n to a basis of R^n. This method is usually easier than that outlined in Procedure 1.2 and has the added advantage that the new basis vectors are all orthogonal to the original vectors.

Procedure 5.2 (Same as Procedure 2.5) If $\{\mathbf{u}_1, \mathbf{u}_2, \cdots, \mathbf{u}_k\}$ is a linearly independent set in R^n then to extend to a basis of R^n :

1. Let A be the matrix with rows $\mathbf{u}_1, \mathbf{u}_2, \cdots, \mathbf{u}_k$.

2. Find a basis of $nullsp(A)$, call them $\{v_1, v_2, \cdots, v_{n-k}\}$.

3. Then $B = \{u_1, u_2, \cdots, u_k, v_1, v_2, \cdots, v_{n-k}\}$ is a basis of R^n.

Theorem 5.2 also gives a way to find the projection of a vector u on a subspace S. We outline this method.

Procedure 5.3 To find the projection of u on a subspace S :

1. Find a basis $\{u_1, u_2, \cdots, u_k\}$ of S and extend to a basis of R^n with a basis $\{v_1, v_2, \cdots, v_{n-k}\}$ of $S^\perp$ as in Procedure 5.2.

2. Express u with respect to this basis, $u = a_1 u_1 + \cdots + a_k u_k + b_1 v_1 + \cdots + b_{n-k} v_{n-k}$.

3. Then $proj_S u = a_1 u_1 + \cdots + a_k u_k$.

Remark   *There are several other ways of finding the projection of a vector on a subspace. We will discuss some of them in later sections.*

Example 5.5 *Let $S = sp\{(1,1,-1,1),(1,0,1,1)\}$ and $u = (3,3,0,2)$. Find $proj_S u$. Solution: First let*

$$A = \begin{bmatrix} 1 & 1 & -1 & 1 \\ 1 & 0 & 1 & 1 \end{bmatrix},$$

then $S = rowsp(A)$ and $S^\perp = nullsp(A)$. We find a basis of $nullsp(A)$.

$$RREF(A) = \begin{bmatrix} 1 & 0 & 1 & 1 \\ 0 & 1 & -2 & 0 \end{bmatrix},$$

adding two zero rows we get

$$Hermite(A) = \begin{bmatrix} 1 & 0 & 1 & 1 \\ 0 & 1 & -2 & 0 \\ 0 & 0 & 0 & 0 \\ 0 & 0 & 0 & 0 \end{bmatrix}$$

and

$$I - Hermite(A) = \begin{bmatrix} 0 & 0 & -1 & -1 \\ 0 & 0 & 2 & 0 \\ 0 & 0 & 1 & 0 \\ 0 & 0 & 0 & 1 \end{bmatrix}.$$

The nonzero columns are a basis of $nullsp(A) = S^\perp$, *so a basis for* $S^\perp$
is $\{(-1, 2, 1, 0), (-1, 0, 0, 1)\}$. *Forming the union of this basis of* $S^\perp$
and our original basis of S, *we see that a basis of* R^n *is*

$$\{(1, 1, -1, 1), (1, 0, 1, 1), (-1, 2, 1, 0), (-1, 0, 0, 1)\}.$$

Let B *be the matrix with these basis vectors as columns. To find the
coordinates of* $u = (3, 3, 0, 2)$ *with respect to this basis we must solve*
$Bx = u$. *The augmented matrix of this system is*

$$\begin{bmatrix} B & u \end{bmatrix} = \begin{bmatrix} 1 & 1 & -1 & -1 & 3 \\ 1 & 0 & 2 & 0 & 3 \\ -1 & 1 & 1 & 0 & 0 \\ 1 & 1 & 0 & 1 & 2 \end{bmatrix},$$

which row reduces to

$$\begin{bmatrix} 1 & 0 & 0 & 0 & \frac{19}{11} \\ 0 & 1 & 0 & 0 & \frac{12}{11} \\ 0 & 0 & 1 & 0 & \frac{7}{11} \\ 0 & 0 & 0 & 1 & -\frac{9}{11} \end{bmatrix}.$$

Thus

$$u = \begin{bmatrix} 3 \\ 3 \\ 0 \\ 2 \end{bmatrix} = \frac{19}{11} \begin{bmatrix} 1 \\ 1 \\ -1 \\ 1 \end{bmatrix} + \frac{12}{11} \begin{bmatrix} 1 \\ 0 \\ 1 \\ 1 \end{bmatrix} + \frac{7}{11} \begin{bmatrix} -1 \\ 2 \\ 1 \\ 0 \end{bmatrix} - \frac{9}{11} \begin{bmatrix} -1 \\ 0 \\ 0 \\ 1 \end{bmatrix}.$$

and

$$proj_S u = \frac{19}{11} \begin{bmatrix} 1 \\ 1 \\ -1 \\ 1 \end{bmatrix} + \frac{12}{11} \begin{bmatrix} 1 \\ 0 \\ 1 \\ 1 \end{bmatrix} = \begin{bmatrix} \frac{31}{11} \\ \frac{19}{11} \\ -\frac{7}{11} \\ \frac{31}{11} \end{bmatrix}.$$

Remark   *Here we cannot clear fractions by multiplying by 11.* $Proj_S u$
is a unique vector.

Definition 5.3 *If* u *and* v *are vectors in* R^n *with* $v \neq O$, *we define*
$proj_v u$ *to be the projection of* u *on the subspace* $sp\{v\}$.

Lemma 5.4

$$Proj_v u = \left(\frac{u \cdot v}{v \cdot v} \right) v.$$

Proof.   Let $S = sp\{\mathbf{v}\}$ and let $\mathbf{w} = proj_S\mathbf{u}$. Then $\mathbf{u} = \mathbf{w} + \mathbf{t}$, where
$\mathbf{t} \in (sp\{\mathbf{v}\})^{\perp}$ and $\mathbf{w} \in sp\{\mathbf{v}\}$, so $\mathbf{w} = k\mathbf{v}$, for some scalar k. Then
$\mathbf{t} = \mathbf{u} - k\mathbf{v}$ and $\mathbf{t}$ is orthogonal to $\mathbf{v}$, so that $(\mathbf{u} - k\mathbf{v}) \cdot \mathbf{v} = 0$. This
gives $\mathbf{u} \cdot \mathbf{v} - k\mathbf{v} \cdot \mathbf{v} = 0$. since $\mathbf{v} \cdot \mathbf{v} \neq 0$ we can solve for k, and

$$k = \left(\frac{\mathbf{u} \cdot \mathbf{v}}{\mathbf{v} \cdot \mathbf{v}}\right).$$

■

Thus projecting on a single vector, or equivalently on a one dimensional space is quite easy. In R^3 this can be used to good advantage,
because since $\dim(S)$ and $\dim(S^{\perp})$ must add up to 3, one of them must
be one (if they are proper subspaces). Since $\mathbf{u} = proj_S\mathbf{u} + proj_{S^{\perp}}\mathbf{u}$, if
we find either one of the projections we can find the other.

Lemma 5.5 *If S is a subspace of R^n, then $(S^{\perp})^{\perp} = S$.*

Proof.   Clearly $S \subset (S^{\perp})^{\perp}$. We also know that $R^n = S^{\perp} \oplus (S^{\perp})^{\perp}$,
using Theorem 5.1. Since we also have $R^n = S \oplus S^{\perp}$, we have $\dim(S) =$
$\dim(S^{\perp})^{\perp}$ Thus $S = (S^{\perp})^{\perp}$, using Corollary 7 to the Replacement
Theorem. ■

Corollary 1 *If the rows of B are a basis of $S^{\perp}$ then $nullsp(B) = S$.*

Proof. $Nullsp(B) = (rowsp(B))^{\perp} = (S^{\perp})^{\perp} = S$. ■
Since for any subspace of R^n we can find a basis of $S^{\perp}$, this means
that every subspace of R^n can be described as the solution space of a
system of linear homogeneous equations $B\mathbf{x} = \mathbf{O}$, or equivalently every
subspace of R^n is the null space of some matrix B. The basis vectors
of $S^{\perp}$ are the coefficients of the equations describing S. A special case
of this: the coefficients of the equation of a plane through the origin in
R^3 are the components of a vector orthogonal (normal, perpendicular)
to the plane.

Procedure 5.4 To find a system of equations describing a subspace
$S = sp\{\mathbf{u}_1, \mathbf{u}_2, \cdots, \mathbf{u}_k\}$.

1. Let A be the matrix with rows $\mathbf{u}_1, \mathbf{u}_2, \cdots, \mathbf{u}_k$.

2. Find a basis of $nullsp(A)$.

3. Let B be the matrix with rows the basis of $nullsp(A)$.

4. Then S is $nullsp(B)$.

Example 5.6 *If* $S = sp\{(1,0,2),(1,1,1)\}$ *then* S *is a plane in* R^3.
Find the equation of the plane. Solution: Let

$$A = \begin{bmatrix} 1 & 0 & 2 \\ 1 & 1 & 1 \end{bmatrix},$$

we find a basis of $S^\perp$ *by finding a basis of* $nullsp(A)$.

$$RREF(A) = \begin{bmatrix} 1 & 0 & 2 \\ 0 & 1 & -1 \end{bmatrix},$$

so a basis of $nullsp(A)$ *is* $(-2,1,1)$. *Thus the equation of the plane is*

$$-2x + y + z = 0.$$

Example 5.7 *For the subspace* $S = sp\{(1,1,-1,1),(1,0,1,1)\}$ *of* R^4,
find a system of linear homogeneous equations that characterizes S.
Solution: In Example 5.5 we found a basis for $S^\perp$, *namely* $(-1,2,1,0)$
and $(-1,0,0,1)$, *thus* S *is the solution space of*

$$\begin{bmatrix} -1 & 2 & 1 & 0 \\ -1 & 0 & 0 & 1 \end{bmatrix} \begin{bmatrix} w \\ x \\ y \\ z \end{bmatrix} = \begin{bmatrix} 0 \\ 0 \\ 0 \\ 0 \end{bmatrix}.$$

Example 5.8 *For the plane* $S = sp\{(1,0,2),(1,1,1)\}$ *find the* $proj_S u$
if $u = (3,1,2)$. *Solution: In Example 5.6 we found that a basis for* $S^\perp$
is $(-2,1,1)$. *Now*

$$proj_{(-2,1,1)}(3,1,2) = \frac{(3,1,2)\cdot(-2,1,1)}{(-2,1,1)\cdot(-2,1,1)}(-2,1,1),$$

$$= -\frac{3}{6}(-2,1,1)$$

$$= \left(1,-\frac{1}{2},-\frac{1}{2}\right),$$

and so

$$proj_S\mathbf{u} = (3,1,2) - \left(1, -\frac{1}{2}, -\frac{1}{2}\right) = \left(2, \frac{3}{2}, \frac{5}{2}\right).$$

Remark   *The projection of $\mathbf{u}$ on a subspace S of dimension > 1 is not in general equal to the sum of the projections of $\mathbf{u}$ on the basis vectors of S. You can verify this in the preceding examples. However if the basis vectors of S are pairwise orthogonal, $\mathbf{v}_i \cdot \mathbf{v}_j = 0$ for $i \neq j$, the projection of $\mathbf{u}$ on S is the sum of the projections of $\mathbf{u}$ on the basis vectors of S. We prove this in the next section.*

Theorem 5.3 (Generalized Pythagorean Theorem) *If $\mathbf{u}$ and $\mathbf{v}$ are orthogonal, then*

$$|\mathbf{u} + \mathbf{v}|^2 = |\mathbf{u}|^2 + |\mathbf{v}|^2.$$

Proof.   We have $|\mathbf{u} + \mathbf{v}|^2 = (\mathbf{u} + \mathbf{v}) \cdot (\mathbf{u} + \mathbf{v}) = \mathbf{u} \cdot \mathbf{u} + 2(\mathbf{u} \cdot \mathbf{v}) + \mathbf{v} \cdot \mathbf{v}$. But $\mathbf{u} \cdot \mathbf{v} = 0$, $\mathbf{u} \cdot \mathbf{u} = |\mathbf{u}|^2$ and $\mathbf{v} \cdot \mathbf{v} = |\mathbf{v}|^2$. ■

Theorem 5.4 (Best Approximation Theorem) *$Proj_S\mathbf{u}$ is the closest vector in S to $\mathbf{u}$ in the sense that*

$$|\mathbf{u} - proj_S\mathbf{u}| \leq |\mathbf{u} - s|, \text{ for all } s \in S.$$

Proof.   Let $\mathbf{s}$ be a vector in S. Then

$$|\mathbf{u} - \mathbf{s}| = |(\mathbf{u} - proj_S\mathbf{u}) + (proj_S\mathbf{u} - \mathbf{s})|.$$

Now $(\mathbf{u} - proj_S\mathbf{u}) \in S^{\perp}$, while $(proj_S\mathbf{u} - \mathbf{s}) \in S$. Let $\mathbf{x} = (\mathbf{u} - proj_S\mathbf{u})$ and $\mathbf{y} = (proj_S\mathbf{u} - \mathbf{s})$. Then $\mathbf{x}$ and $\mathbf{y}$ are orthogonal and

$$|\mathbf{u} - \mathbf{s}| = |\mathbf{x} + \mathbf{y}|.$$

Using the previous theorem, we have

$$|\mathbf{u} - \mathbf{s}|^2 = |\mathbf{x}|^2 + |\mathbf{y}|^2.$$

This is minimized if we chose $\mathbf{s}$ so that $|\mathbf{y}| = 0$, and this happens if and only if $\mathbf{s} = proj_S\mathbf{u}$. ■

Remark *If we let* $\hat{\mathbf{u}} = proj_S \mathbf{u}$, *then then Theorem 5.4 says that* $|\mathbf{u} - \hat{\mathbf{u}}| \leq |\mathbf{u} - \mathbf{s}|$ *for all* $\mathbf{s} \in S$. *Using the definition of length this implies*

$$\sum_{i=1}^{n}(u_i - \hat{u}_i)^2 \leq \sum_{i=1}^{n}(u_i - s_i)^2, \text{ for all } \mathbf{s} \in S.$$

For this reason $\hat{\mathbf{u}}$ *is called the least squares approximation of* **u** *in* S.

Remark *In this section we have been working in* R^n *and assuming that all matrices have real entries. For* C^n, *n-tuples of complex numbers, the definition of orthogonal must be modified. For complex n-tuples we define* **x** *orthogonal to* **y** *if* $\bar{\mathbf{x}} \cdot \mathbf{y} = 0$, *so that we can retain the property that only the zero vector is orthogonal to itself. With this modification, things go through as before, except that we have* $(rowsp(A))^{\perp} = nullsp(\bar{A})$ *for matrices with complex entries.*

Exercises

1. Extend the set $\{(1, 2, 1, 1), (1, 0, 1, 0)\}$ to a basis of R^4.

2. Find the $proj_S \mathbf{u}$ if $S = sp\{(1, 2, 1, 1), (1, 0, 1, 0)\}$, and $\mathbf{u} = (-1, 3, 0, -1)$.

3. For the same subspace S as in problem 2, find a matrix B such that S is the null space of B.

4. Find $S^{\perp}$ if $S = \{(1, 1, 3), (2, -1, 2)\}$.

5. Find $proj_S \mathbf{u}$ if S is as in problem 4 and $\mathbf{u} = (1, 2, 1)$.

6. What is the equation of the plane S in question 4.

7. Find an equation describing the subspace

$$S = sp\{(1, 1, 0, 2), (1, 2, 1, 1), (0, 1, 0, 1)\}.$$

8. If $S = sp\{(3, 1, 2), (-1, 2, 2)\}$ and $\mathbf{u} = (3, 3, 3)$, find $proj_S \mathbf{u}$.

9. Let W be the subspace of R^4 spanned by $(3, 2, 5, 3)$, $(0, 1, 1, 0)$, and $(1, 1, 2, 1)$. Find a system of homogeneous equations that characterize the subspace W.

10. Find necessary and sufficient conditions on a, b, c, and d so that the following system has a solution. (Hint: Compare this with problem 9. Recall the system has a solution if and only if the vector (a, b, c, d) is in the column space of the coefficient matrix)

$$\begin{array}{rcl} 3x_1 + 0x_2 + 1x_3 & = & a \\ 2x_1 + 1x_2 + 1x_3 & = & b \\ 5x_1 + 1x_2 + 2x_3 & = & c \\ 3x_1 + 0x_2 + 1x_3 & = & d \end{array} .$$

Section 5.3 More on Orthogonal Bases

Recall that a set of vectors is orthogonal if they are all nonzero and pairwise orthogonal (Definition 4.12) We have seen that an orthogonal set is linearly independent (Lemma 4.2). A basis of S that is also an orthogonal set is an orthogonal basis of S (Definition 4.13), and an orthogonal basis whose vectors all have length one is an orthonormal basis. The following theorem gives a nice property of orthogonal bases.

Theorem 5.5 *If* $B = \{v_1, v_2, \cdots, v_n\}$ *is an orthogonal basis of* R^n, *and if* $u = a_1v_1 + a_2v_2 + \cdots + a_nv_n$, *then*

$$a_k = \frac{u \cdot v_k}{v_k \cdot v_k}.$$

If B *is an orthonormal basis*

$$a_k = u \cdot v_k.$$

Remark   *This means that the coordinates of a vector* u *with respect to an orthogonal basis are easy to find. They can be computed just by dot products instead of having to solve a system of equations.*

Proof.   If

$$u = a_1v_1 + a_2v_2 + \cdots + a_nv_n,$$

then

$$\begin{aligned} \mathbf{u} \cdot \mathbf{v}_k &= (a_1\mathbf{v}_1 + a_2\mathbf{v}_2 + \cdots + a_n\mathbf{v}_n) \cdot \mathbf{v}_k \\ &= a_k(\mathbf{v}_k \cdot \mathbf{v}_k), \end{aligned}$$

the rest of the terms being 0. Solving for a_k gives the desired result. ■

The natural basis vectors are an orthonormal basis of R^n. In Section 4.2 we found orthonormal bases for eigenspaces. In the proof of Theorem 5.6 below, we use this method to prove that every nonzero subspace of R^n has an orthogonal basis. An eigenspace appears as the null space of a matrix B, however our method will work in general, since every subspace is the null space of some matrix (Corollary to Lemma 5.5). We will also show that any orthogonal basis of S can be extended to an orthogonal basis of R^n. Once you have an orthogonal basis it is easy to turn it into an orthonormal basis. Just divide every vector by its length. This process is called normalizing and it is best to get an orthogonal basis first, and only normalize at the end, since this is likely to introduce irrational numbers such as $\sqrt{2}$, $\sqrt{3}$, etc.

Theorem 5.6 *Every nonzero subspace S or R^n has an orthogonal basis.*

Proof. The proof is constructive and describes a process that will result in an orthogonal basis of S. Find a basis of $S^\perp$ and let B be the matrix with the basis of $S^\perp$ as rows. Then $S = nullsp(B)$. Let $k = \dim(S)$. We may assume $0 < k < n$, since S is nonzero, and if $S = R^n$ the natural basis vectors are an orthogonal basis and we are done. Then $\dim(S^\perp) = n - k$, so B is a matrix with $n - k$ linearly independent rows. Select any nonzero vector $\mathbf{u}_1 \in S$ and form the matrix B_1 by adding $\mathbf{u}_1$ to B as an extra row. Now select a nonzero vector $\mathbf{u}_2 \in nullsp(B_1)$. Notice that $B_1\mathbf{x} = \mathbf{O}$ represents the system of equations $B\mathbf{x} = \mathbf{O}$ with one more equation added, that new equation being $\mathbf{u}_1 \cdot \mathbf{u}_2 = 0$. Thus $\mathbf{u}_2$ is in S and $\mathbf{u}_2$ is orthogonal to $\mathbf{u}_1$. Now form the matrix B_2 by adding $\mathbf{u}_2$ to B_1 as an extra row. Select a nonzero vector $\mathbf{u}_3 \in nullsp(B_2)$. Then $\mathbf{u}_3$ is in S and $\mathbf{u}_3$ is orthogonal to both $\mathbf{u}_1$ and $\mathbf{u}_2$. Continue in this way until $nullsp(B_j) = \mathbf{O}$. The process must come to an end with $j \leq k$ because the vectors $\{\mathbf{u}_i\}$ are an orthogonal set and hence linearly independent. To see that the process does not end before we have k vectors we consider the rank of B_j. B_j has $n - k + j$ rows. The first $n - k$ rows are a linearly independent set in $S^\perp$. The next j rows are a linearly independent set in S. Since $S \sqcap S^\perp = \mathbf{O}$ the $n - k + j$ rows of B_j are linearly independent (see proof of Theorem 1.12) and thus B_j has rank $n - k + j$. As long as $rank(B_j) < n$, $B_j\mathbf{x} = \mathbf{O}$ has a nonzero solution, so the process won't stop until $j = k$. At this point we have $\{\mathbf{u}_1, \mathbf{u}_2, \cdots \mathbf{u}_k\}$ is an orthogonal basis of S. ∎

Remark   *If we start with an orthogonal basis of $S^\perp$ the vectors* $\mathbf{u}_1, \mathbf{u}_2, \cdots \mathbf{u}_k$ *extend this to an orthogonal basis of R^n. Any orthogonal set can be used as the basis of $S^\perp$ and thus extended to an orthogonal basis of R^n. So we have proved:*

Corollary 1 *Any orthogonal set in R^n can be extended to an orthogonal basis of R^n.*

Corollary 2 *If $\{\mathbf{v}_1, \mathbf{v}_2, \cdots, \mathbf{v}_k\}$ is an orthogonal basis of a subspace S of R^n, then*

$$proj_S\mathbf{u} = \left(\frac{\mathbf{u} \cdot \mathbf{v}_1}{\mathbf{v}_1 \cdot \mathbf{v}_1}\right)\mathbf{v}_1 + \left(\frac{\mathbf{u} \cdot \mathbf{v}_2}{\mathbf{v}_2 \cdot \mathbf{v}_2}\right)\mathbf{v}_2 + \cdots + \left(\frac{\mathbf{u} \cdot \mathbf{v}_k}{\mathbf{v}_k \cdot \mathbf{v}_k}\right)\mathbf{v}_k$$

$$= \sum_{i=1}^{k} proj_{\mathbf{v}_i}\mathbf{u}.$$

Proof.   Extend $\{\mathbf{v}_1, \mathbf{v}_2, \cdots, \mathbf{v}_k\}$ to an orthogonal basis $\{\mathbf{v}_1, \mathbf{v}_2, \cdots, \mathbf{v}_k, \mathbf{v}_{k+1}, \cdots, \mathbf{v}_n\}$ of R^n. Then if

$$\mathbf{u} = a_1\mathbf{v}_1 + \cdots + a_k\mathbf{v}_k + a_{k+1}\mathbf{v}_{k+1} + \cdots + a_n\mathbf{v}_n,$$

we have

$$proj_S\mathbf{u} = a_1\mathbf{v}_1 + \cdots + a_k\mathbf{v}_k.$$

Because the basis is orthogonal, Theorem 5.5 tells us that

$$a_i = \frac{\mathbf{u} \cdot \mathbf{v}_i}{\mathbf{v}_i \cdot \mathbf{v}_i}.$$

∎

Procedure 5.5 To find an orthogonal basis for a subspace S:

1. Find a basis of $S^{\perp}$ and put as rows of a matrix B.

2. Adjoin a nonzero vector in S as an extra row to B, forming B_1.

3. Find a nonzero vector in $nullsp(B_1)$ and adjoin as extra row to B_1 forming B_2.

4. Repeat step 3 using B_2 and keep repeating until B_k has rank n.

5. The nonzero vectors you adjoined to B form an orthogonal basis of S.

Procedure 5.6 To extend an orthogonal set to a basis of R^n.

1. Put the orthogonal set as the rows of a matrix B.

2. Find a nonzero vector in $nullsp(B)$ and adjoin as extra row to B, forming B_1.

3. Find a nonzero vector in $nullsp(B_1)$ and adjoin as extra row to B_1, forming B_2.

4. Repeat step 3 using B_2, and continue until B_k has rank n.

5. The rows of B_k are an orthogonal basis of R^n.

Example 5.9 Let $S = sp\{(1,1,1,0),(2,1,0,1),(-1,0,-2,1)\}$. *Find an orthogonal basis of S. We use Procedure 2.10. First we need to find a basis of $S^{\perp}$. Let*

$$A = \begin{bmatrix} 1 & 1 & 1 & 0 \\ 2 & 1 & 0 & 1 \\ -1 & 0 & -2 & 1 \end{bmatrix},$$

then $S^{\perp} = nullsp(A)$.

$$RREF(A) = \begin{bmatrix} 1 & 0 & 0 & \frac{1}{3} \\ 0 & 1 & 0 & \frac{1}{3} \\ 0 & 0 & 1 & -\frac{2}{3} \end{bmatrix},$$

so a basis for $S^\perp$ is $\left(\frac{1}{3}, \frac{1}{3}, -\frac{2}{3}, -1\right)$, or we can use $(1, 1, -2, -3)$. Now we let

$$B = \begin{bmatrix} 1 & 1 & -2 & -3 \end{bmatrix},$$

and choose $\mathbf{u}_1 = (1, 1, 1, 0) \in S = nullsp(B)$. *Then*

$$B_1 = \begin{bmatrix} 1 & 1 & -2 & -3 \\ 1 & 1 & 1 & 0 \end{bmatrix}.$$

An obvious solution of $B_1\mathbf{x} = \mathbf{O}$ is $(1, -1, 0, 0)$, so we form

$$B_2 = \begin{bmatrix} 1 & 1 & -2 & -3 \\ 1 & 1 & 1 & 0 \\ 1 & -1 & 0 & 0 \end{bmatrix}.$$

Not seeing an obvious solution to $B_2\mathbf{x} = \mathbf{O}$, we find

$$RREF(B_2) = \begin{bmatrix} 1 & 0 & 0 & -\frac{1}{2} \\ 0 & 1 & 0 & -\frac{1}{2} \\ 0 & 0 & 1 & 1 \end{bmatrix},$$

and see that $\left(\frac{1}{2}, \frac{1}{2}, -1, 1\right)$ is a solution, but we may chose $(1, 1, -2, 2)$, which is also in $nullsp(B_2)$. Since we know $\dim(S) = 3$, we are done, and $\{(1, 1, 1, 0), (1, -1, 0, 0), (1, 1, -2, 2)\}$ is an orthogonal basis of S.

Example 5.10 *Extend the set $\{(1, 2, 3, 1)\}$ to an orthogonal basis of R^4. Solution: we seek a vector orthogonal to $(1, 2, 3, 1)$, equivalently a vector in the null space of*

$$B = \begin{bmatrix} 1 & 2 & 3 & 1 \end{bmatrix}.$$

It is easy to see that $(-1, 0, 0, 1)$ is such a vector. Now we seek a third vector orthogonal to both of these, equivalently in the null space of

$$B_1 = \begin{bmatrix} 1 & 2 & 3 & 1 \\ -1 & 0 & 0 & 1 \end{bmatrix}.$$

Again we can see by inspection that $(0, 3, -2, 0)$ is such a vector. Now to get a vector orthogonal to all three is not so easy to solve by inspection. We want a vector in the null space of

$$B_2 = \begin{bmatrix} 1 & 2 & 3 & 1 \\ -1 & 0 & 0 & 1 \\ 0 & 3 & -2 & 0 \end{bmatrix}.$$

Row reduction gives

$$RREF(B_2) = \begin{bmatrix} 1 & 0 & 0 & -1 \\ 0 & 1 & 0 & \frac{4}{13} \\ 0 & 0 & 1 & \frac{6}{13} \end{bmatrix},$$

so a vector in the null space is $\left(1, -\frac{4}{13}, -\frac{6}{13}, 1\right)$, but we prefer to use $(13, -4, -6, 13)$. One can check the dot products and see that this is orthogonal to the three rows of B_2. Thus the required orthogonal basis of R^4 including the vector $(1, 2, 3, 1)$ is

$$\{(1, 2, 3, 1), (-1, 0, 0, 1), (0, 3, -2, 0), (13, -4, -6, 13)\}.$$

Example 5.11 *Find the projection of $\mathbf{u} = (3, 3, 0, 2)$ on*

$$S = sp\{(1, 1, -1, 1), (1, 0, 1, 1)\}$$

by finding an orthogonal basis of S and adding the projections of $\mathbf{u}$ on the orthogonal basis vectors. Solution: This is the same S and $\mathbf{u}$ as in Example 5.5 of the previous section. In that example we found that a basis of $S^\perp$ is $\{(-1, 2, 1, 0), (-1, 0, 0, 1)\}$. Thus S is the null space of

$$B = \begin{bmatrix} -1 & 2 & 1 & 0 \\ -1 & 0 & 0 & 1 \end{bmatrix}.$$

For $\mathbf{u}_1$ we choose $(1, 0, 1, 1)$ in the null space of B, and then form

$$B_1 = \begin{bmatrix} -1 & 2 & 1 & 0 \\ -1 & 0 & 0 & 1 \\ 1 & 0 & 1 & 1 \end{bmatrix}.$$

A vector in $nullsp(B_1)$ will be in S and orthogonal to $(1, 0, 1, 1)$, as desired. We use row reduction on B_1 to find

$$RREF(B_1) = \begin{bmatrix} 1 & 0 & 0 & -1 \\ 0 & 1 & 0 & -\frac{3}{2} \\ 0 & 0 & 1 & 2 \end{bmatrix},$$

so a solution is $(1, 3/2, -2, 1)$ or we can chose $(2, 3, -4, 2)$. Thus $\mathbf{v}_1 = (1, 0, 1, 1)$, and $\mathbf{v}_2 = (2, 3, -4, 2)$ is an orthogonal basis of S, and so Corollary 2 to Theorem 5.6 implies

$$proj_S\mathbf{u} = \left(\frac{\mathbf{u} \cdot \mathbf{v}_1}{\mathbf{v}_1 \cdot \mathbf{v}_1}\right)\mathbf{v}_1 + \left(\frac{\mathbf{u} \cdot \mathbf{v}_2}{\mathbf{v}_2 \cdot \mathbf{v}_2}\right)\mathbf{v}_2.$$

Calculating this gives

$$projs\mathbf{u} = \left(\frac{5}{3}\right)(1,0,1,1) + \left(\frac{19}{33}\right)(2,3,-4,2) = \left(\frac{31}{11}, \frac{19}{11}, -\frac{7}{11}, \frac{31}{11}\right),$$

which agrees with the calculation in Example 5.5.

Remark   *The Gram-Schmidt Process for finding an orthogonal basis of a subspace S given a basis of S is outlined in the next chapter.*

Exercises

Since a vector space has infinitely many orthogonal bases, the answers to the following problems are not unique.

1. Extend the orthogonal set $\{(1,1,0,0),(0,0,2,1)\}$ to an orthogonal basis of R^4.

2. Extend the orthogonal set $\{(1,2,-1,1),(2,1,2,-2)\}$ to an orthogonal basis of R^4.

3. Find an orthogonal basis for S if

$$S = sp\{(1,1,0,2),(1,2,1,1),(0,1,0,1)\}.$$

4. Extend $(1,-1,3,1)$ to a basis of R^4.

5. Find an orthogonal basis of S if $S = sp\{(1,1,3),(2,1,0)\}$.

6. Find an orthogonal basis of S if $S = \{(a,b,c)|c = 2a - 3b\}$.

Section 5.4 Least Squares Solutions

We know that $A\mathbf{x} = \mathbf{b}$ has a solution if and only if $\mathbf{b} \in colsp(A)$. Assume now that $A\mathbf{x} = \mathbf{b}$ has no solution, or equivalently that $\mathbf{b} \notin colsp(A)$. Let $S = colsp(A)$, and let $\hat{\mathbf{b}} = projs\mathbf{b}$. Then $A\mathbf{x} = \hat{\mathbf{b}}$ has a solution and $\hat{\mathbf{b}}$ is the closest vector to $\mathbf{b}$ for which this is true (Theorem 5.4). We know that $\hat{\mathbf{b}} \in S = colsp(A)$, so $\hat{\mathbf{b}} = A\hat{\mathbf{x}}$ for some vector $\hat{\mathbf{x}}$.

We also know that $(\mathbf{b} - \hat{\mathbf{b}}) \in S^\perp$. Now $S = colsp(A) = rowsp(A^T)$, so $S^\perp = nullsp(A^T)$. Thus we must have $A^T(\hat{\mathbf{b}} - \mathbf{b}) = \mathbf{O}$. Substituting $\hat{\mathbf{b}} = A\hat{\mathbf{x}}$, this gives $A^T(A\hat{\mathbf{x}} - \mathbf{b}) = \mathbf{O}$, or $(A^T A)\hat{\mathbf{x}} = A^T\mathbf{b}$. Thus $\hat{\mathbf{x}}$ is a solution of the equation

$$(A^T A)\mathbf{x} = A^T\mathbf{b}. \tag{5.1}$$

Equation 5.1 is called the normal equation associated with $A\mathbf{x} = \mathbf{b}$. The calculation above shows that $\hat{\mathbf{x}}$ is a solution of 5.1 if and only if $A\hat{\mathbf{x}} = \hat{\mathbf{b}}$. A solution to $A\mathbf{x} = \hat{\mathbf{b}}$ always exists, and is unique if and only if the columns of A are linearly independent.

Solving the normal equation associated with $A\mathbf{x} = \mathbf{b}$ gives us another way of finding $proj_S\mathbf{b}$, where $S = colsp(A)$.

Procedure 5.7 To find $proj_S\mathbf{b}$:

1. Let A be the matrix whose columns are a basis of S.

2. Solve $A^T A\mathbf{x} = A^T\mathbf{b}$. Let $\hat{\mathbf{x}}$ denote a solution.

3. Multiply $\hat{\mathbf{x}}$ on the left by A. $Proj_S\mathbf{b} = A\hat{\mathbf{x}}$.

The projection computed in the next example has already been done by two different methods in Examples 5.5 and 5.11.

Example 5.12 *For $S = sp\{(1, 1, -1, 1), (1, 0, 1, 1)\}$ and $\mathbf{b} = (3, 3, 0, 2)$, find $proj_S\mathbf{b}$. Solution: let*

$$A = \begin{bmatrix} 1 & 1 \\ 1 & 0 \\ -1 & 1 \\ 1 & 1 \end{bmatrix}.$$

Then $S = colsp(A)$. We will solve $A^T A\mathbf{x} = A^T\mathbf{b}$.

$$A^T A = \begin{bmatrix} 1 & 1 & -1 & 1 \\ 1 & 0 & 1 & 1 \end{bmatrix} \begin{bmatrix} 1 & 1 \\ 1 & 0 \\ -1 & 1 \\ 1 & 1 \end{bmatrix} = \begin{bmatrix} 4 & 1 \\ 1 & 3 \end{bmatrix},$$

and

$$A^T\mathbf{b} = \begin{bmatrix} 1 & 1 & -1 & 0 \\ 1 & 0 & 1 & 1 \end{bmatrix} \begin{bmatrix} 3 \\ 3 \\ 0 \\ 2 \end{bmatrix} = \begin{bmatrix} 8 \\ 5 \end{bmatrix},$$

so we must solve

$$\begin{bmatrix} 4 & 1 \\ 1 & 3 \end{bmatrix} \begin{bmatrix} x_1 \\ x_2 \end{bmatrix} = \begin{bmatrix} 8 \\ 5 \end{bmatrix}.$$

Now

$$\begin{bmatrix} 4 & 1 & 8 \\ 1 & 3 & 5 \end{bmatrix} \sim \begin{bmatrix} 1 & 0 & \frac{19}{11} \\ 0 & 1 & \frac{12}{11} \end{bmatrix}.$$

Thus $\hat{\mathbf{x}} = \begin{bmatrix} \frac{19}{11} \\ \frac{12}{11} \end{bmatrix}$, *and* $\hat{\mathbf{b}} = A\hat{\mathbf{x}} = \begin{bmatrix} 1 & 1 \\ 1 & 0 \\ -1 & 1 \\ 1 & 1 \end{bmatrix} \begin{bmatrix} \frac{19}{11} \\ \frac{12}{11} \end{bmatrix} = \begin{bmatrix} \frac{31}{11} \\ \frac{19}{11} \\ -\frac{7}{11} \\ \frac{31}{11} \end{bmatrix}.$ *This*

checks with what we got in Examples 5.5 and 5.11

Definition 5.4 *A solution* $\hat{\mathbf{x}}$ *of 5.1, the normal equation associated with* $A\mathbf{x} = \mathbf{b}$, *is called a least squares solution of* $A\mathbf{x} = \mathbf{b}$.

Remark   *The least squares solution of* $A\mathbf{x} = \mathbf{b}$ *is not a solution of* $A\mathbf{x} = \mathbf{b}$, *but rather a solution of* $A\mathbf{x} = \hat{\mathbf{b}}$, *where* $\hat{\mathbf{b}} = proj_{colsp(A)}\mathbf{b}$, *the least squares approximation of* $\mathbf{b}$ *in* $colsp(A)$, *(Theorem 5.4).*

We turn for a moment to another problem. Suppose that

$$(x_1, y_1), (x_2, y_2), \cdots (x_{n+1}, y_{n+1})$$

are $n+1$ points in R^2, and assume that all the x_i are distinct. Consider the matrix equation

$$\begin{bmatrix} 1 & x_1 & x_1^2 & x_1^3 & \cdots & x_1^n \\ 1 & x_2 & x_2^2 & x_2^3 & \cdots & x_2^n \\ \vdots & \vdots & \vdots & \vdots & & \vdots \\ 1 & x_{n+1} & x_{n+1}^2 & x_{n+1}^3 & \cdots & x_{n+1}^n \end{bmatrix} \begin{bmatrix} a_0 \\ a_1 \\ \vdots \\ a_n \end{bmatrix} = \begin{bmatrix} y_1 \\ y_2 \\ \vdots \\ y_{n+1} \end{bmatrix}. \tag{5.2}$$

This is equivalent to the $n + 1$ equations $P(x_i) = y_i$, where P is the polynomial

$P(x) = a_0 + a_1 x + a_2 x^2 + a_3 x^3 + \cdots + a_n x^n$. The x_i and y_i are given. If we can solve for the a_i we will have the coefficients of a polynomial going through the given points.

Theorem 5.7 *There is a unique polynomial of degree n or less passing through any $n + 1$ points with distinct x coordinates.*

Proof.   The coefficient matrix in 5.2 is a Vandermonde matrix, and has nonzero determinant when the x_i are all distinct (Vandermonde Determinants, Section 3.2), and therefore equation 5.2 has a unique solution. ∎

Example 5.13 *Find a polynomial of degree 3 or less through the points* $(1, 2), (-1, 3), (4, 1)$, *and* $(2, 0)$. *We must solve*

$$\begin{bmatrix} 1 & 1 & 1 & 1 \\ 1 & -1 & 1 & -1 \\ 1 & 4 & 16 & 64 \\ 1 & 2 & 4 & 8 \end{bmatrix} \begin{bmatrix} a_0 \\ a_1 \\ a_2 \\ a_3 \end{bmatrix} = \begin{bmatrix} 2 \\ 3 \\ 1 \\ 0 \end{bmatrix}.$$

$$\begin{bmatrix} 1 & 1 & 1 & 1 & 2 \\ 1 & -1 & 1 & -1 & 3 \\ 1 & 4 & 16 & 64 & 1 \\ 1 & 2 & 4 & 8 & 0 \end{bmatrix} \sim \begin{bmatrix} 1 & 0 & 0 & 0 & \frac{53}{15} \\ 0 & 1 & 0 & 0 & -\frac{23}{30} \\ 0 & 0 & 1 & 0 & -\frac{31}{30} \\ 0 & 0 & 0 & 1 & \frac{4}{15} \end{bmatrix},$$

so the solution is

$$P(x) = \frac{52}{15} - \frac{23}{30}x - \frac{31}{30}x^2 + \frac{4}{15}x^3.$$

We said a unique solution existed, we didn't say it would be pretty!

If you want a polynomial of degree n or less to go through some given points, $(x_1, y_1), (x_2, y_2), \cdots, (x_k, y_k)$ but you have more than $n+1$ points, the coefficient matrix in equation 5.2 will have more rows than columns, *i.e.* more equations than unknowns, so there may not be a solution. In this case we can find a least squares solution. This gives a polynomial P that is the "least squares fit" to the given points. This polynomial will satisfy

$$\sum_{i=1}^{k} (P(x_i) - y_i)^2$$

is minimal for all polynomials of degree n.

Example 5.14 *Find a least squares fit polynomial of degree 2 or less through the four points* $(1,1)$, $(2,0)$, $(-1,3)$, *and* $(0,0)$. *Solution: if we were allowed a polynomial of degree 3 or less, there would be a unique solution, but in this case the matrix equation we are trying to solve is*

$$\begin{bmatrix} 1 & 1 & 1 \\ 1 & 2 & 4 \\ 1 & -1 & 1 \\ 1 & 0 & 0 \end{bmatrix} \begin{bmatrix} a_0 \\ a_1 \\ a_2 \end{bmatrix} = \begin{bmatrix} 1 \\ 0 \\ 3 \\ 0 \end{bmatrix}. \tag{5.3}$$

Let $M = \begin{bmatrix} 1 & 1 & 1 \\ 1 & 2 & 4 \\ 1 & -1 & 1 \\ 1 & 0 & 0 \end{bmatrix}$, $\mathbf{v} = \begin{bmatrix} a_0 \\ a_1 \\ a_2 \end{bmatrix}$, *and* $\mathbf{y} = \begin{bmatrix} 1 \\ 0 \\ 3 \\ 0 \end{bmatrix}$. *So 5.3 becomes*

$$M\mathbf{v} = \mathbf{y}.$$

This system has no solution, as can be seen by row reduction of the augmented matrix

$$\begin{bmatrix} M & \mathbf{y} \end{bmatrix} = \begin{bmatrix} 1 & 1 & 1 & 1 \\ 1 & 2 & 4 & 0 \\ 1 & -1 & 1 & 3 \\ 1 & 0 & 0 & 0 \end{bmatrix} \sim \begin{bmatrix} 1 & 0 & 0 & 0 \\ 0 & 1 & 0 & 0 \\ 0 & 0 & 1 & 0 \\ 0 & 0 & 0 & 1 \end{bmatrix}.$$

Thus we solve the normal equation

$$M^T M \mathbf{v} = M^T \mathbf{y},$$

which is

$$\begin{bmatrix} 4 & 2 & 6 \\ 2 & 6 & 8 \\ 6 & 8 & 18 \end{bmatrix} \begin{bmatrix} a_0 \\ a_1 \\ a_2 \end{bmatrix} = \begin{bmatrix} 4 \\ -2 \\ 4 \end{bmatrix}.$$

This has solution

$$\begin{bmatrix} a_0 \\ a_1 \\ a_2 \end{bmatrix} = \begin{bmatrix} \frac{9}{10} \\ -\frac{13}{10} \\ \frac{1}{2} \end{bmatrix}.$$

Thus the polynomial

$$P(x) = \frac{1}{2}x^2 - \frac{13}{10}x + \frac{9}{10}$$

is the least squares fit quadratic to the given points.

Example 5.15 *Find the least squares fit line to the data points* $(-1, -2)$, $(0, 1)$, $(1, 3)$, $(2, 6)$. *Solution: if we were allowed to use a polynomial of degree 3 or less, we could fit these points exactly. The equation of a line is a polynomial of degree 1, and the matrix equation we want to solve is*

$$\begin{bmatrix} 1 & -1 \\ 1 & 0 \\ 1 & 1 \\ 1 & 2 \end{bmatrix} \begin{bmatrix} a_o \\ a_1 \end{bmatrix} = \begin{bmatrix} -2 \\ 1 \\ 3 \\ 6 \end{bmatrix}.$$

The associated normal equation is

$$\begin{bmatrix} 1 & 1 & 1 & 1 \\ -1 & 0 & 1 & 2 \end{bmatrix} \begin{bmatrix} 1 & -1 \\ 1 & 0 \\ 1 & 1 \\ 1 & 2 \end{bmatrix} \begin{bmatrix} a_o \\ a_1 \end{bmatrix} = \begin{bmatrix} 1 & 1 & 1 & 1 \\ -1 & 0 & 1 & 2 \end{bmatrix} \begin{bmatrix} -2 \\ 1 \\ 3 \\ 6 \end{bmatrix},$$

or

$$\begin{bmatrix} 4 & 2 \\ 2 & 6 \end{bmatrix} \begin{bmatrix} a_o \\ a_1 \end{bmatrix} = \begin{bmatrix} 8 \\ 17 \end{bmatrix},$$

which has solution

$$\begin{bmatrix} a_o \\ a_1 \end{bmatrix} = \begin{bmatrix} \frac{7}{10} \\ \frac{13}{5} \end{bmatrix}.$$

The least squares fit line is thus

$$y = \frac{13}{5}x + \frac{7}{10}.$$

Exercises

1. Use the method described in Procedure 5.7 to find the $proj_S \mathbf{b}$ where $S = \{(2, 1, 0, 1), (1, 1, 2, 1)\}$ and $\mathbf{b} = (3, 0, 3, 2)$.

2. Find a polynomial of degree 3 or less through the points $(3, 2)$, $(7, 30)$, $(2, 0)$, $(-4, 30)$.

3. For the data points $(-1, -2)$, $(0, 1)$, $(1, 3)$, $(2, 6)$ of Example 3.15,

   (a) Show that the least squares fit polynomial of degree 2 or less is still the same line we got in Example 3.15. This means you can't do any better with a quadratic than with a line.

(b) Find the cubic that goes through the given points.

4. Show that there is a parabola passing through the points $(0, 1)$, $(1, 3)$, $(2, 6)$, $(3, 10)$ and find its equation. (Graph of $y = a_2 x^2 + a_1 x + a_0$ is a parabola.)

5. Find the least squares fit line through the data points in problem 4.

6. For a real matrix A, prove that $rank(A^T A) = rank(A)$. As a corollary deduce that if the columns of A are linearly independent, $A^T A$ is invertible.

Chapter 6

How To ...

When you have the RREF of A you can readily obtain the following information:

1. Basis of $rowsp(A)$

2. Basis of $colsp(A)$

3. Basis of $nullsp(A)$

4. How to write each column of A as a linear combination of preceding columns of A if possible

5. Basis of $(rowsp(A))^{\perp}$

6. The equations that characterize $rowsp(A)$

7. The rank of A.

In what follows we explain how to get each of the above from the RREF of A, and how to solve many other problems. Most of the procedures use the RREF at some time. We assume all vectors and subspaces are in R^n.

1. How to find a basis of the row space of a matrix A:

   (a) Find $RREF(A)$.

   (b) The nonzero rows of $RREF(A)$ are a basis of the row space.

2. How to find a basis of the column space of a matrix A:

    (a) Find $RREF(A)$.

    (b) Note which columns of $RREF(A)$ have leading ones.

    (c) In the matrix A, select the corresponding columns, *i.e.* if columns 1,2, and 4 of $RREF(A)$ have leading ones, select columns 1,2, and 4 of A. (These are called the leading columns of A.)

    (d) These columns (from the original matrix A) are a basis of $colsp(A)$.

3. How to find a basis of the null space of a matrix A:

    (a) Find $RREF(A)$.

    (b) Add or delete zero rows until the matrix is square.

    (c) Rearrange the rows so that the leading ones come on the main diagonal. Call this matrix H.

    (d) Subtract the identity matrix of the same size from H.

    (e) The nonzero columns of $H - I$ are a basis of the null space of A.
    . See Ex. 2.37.

4. How to write each column of A as a linear combination of preceding columns of A if possible:

    (a) Find $RREF(A)$.

    (b) The leading columns of A cannot be written as linear combinations of preceding columns of A.

    (c) A nonleading column v_i of $RREF(A)$ contains the coefficients that allow you to express the corresponding column u_i of A as a linear combination of preceding columns of A. See Ex. 2.31 and Ex. 2.32.

5. How to find a basis of $S^\perp$ given a basis of S :

    (a) Put the basis of S as rows of a matrix A. ($rowsp(A) = S$)

(b) Find a basis of $nullsp(A)$.

(c) This is a basis of $S^\perp$, since $nullsp(A) = (rowsp(A))^\perp$.

6. How to find a system of equations that characterizes the row space of A:

   (a) Find a basis of the $nullsp(A)$.

   (b) Take the dot product of each of the null space basis vectors with $x = (x_1, x_2, \cdots, x_n)$ and set these dot products equal to zero.

   (c) The homogeneous linear equations you get in step 2 characterize the row space of A. (The basis vectors of $nullsp(A)$ are the coefficients.)

7. How to find the rank of A:

   (a) Find $RREF(A)$

   (b) Count the number of leading ones.

   (c) This is $rank(A)$.

8. How to tell if two sets of vectors in R^n span the same subspace of R^n (equivalently how to tell if two matrices have the same row space):

   (a) Put one set of vectors as the rows of a matrix A.

   (b) Put the other set of vectors as rows of a matrix B.

   (c) Find $RREF(A)$ and $RREF(B)$.

   (d) The two sets span the same subspace if and only if {nonzero rows of $RREF(A)$} = {nonzero rows of $RREF(B)$}.

9. How to tell if a given set of k vectors (n-tuples) is a basis of R^n :

   (a) If $k \neq n$, the set of vectors is not a basis of R^n.

   (b) If $k = n$, put the vectors as rows (or columns) of a matrix A.

(c) Find $|A|$, the determinant of A.

(d) The set of vectors is a basis of R^n if and only if $|A| \neq 0$. (Equivalently, the set of vectors is a basis if and only if RREF$(A) = I$.)

10. How to tell if a given set of k vectors (n-tuples) is linearly independent:

(a) If $k > n$ the set is linearly dependent.

(b) If $k \leq n$ put the vectors as columns of a matrix A.

(c) Find $RREF(A)$.

(d) The set of vectors is linearly independent if and only if every column of $RREF(A)$ has a leading one.(Also if $k = n$ the vectors are linearly independent if and only if they are a basis for R^n so you can use the determinant as in #9.)

11. How to tell if a given set of vectors (n-tuples) span R^n :

(a) If $k < n$, they do not span.

(b) If $k \geq n$, put the vectors as columns of a matrix A.

(c) Find $RREF(A)$.

(d) The vectors span R^n if and only if every row of $RREF(A)$ has a leading one. (Also if $k = n$ the vectors span R^n if and only if they are a basis of R^n, so you can use the determinant as in #9.)

12. How to find a matrix B such that S is the null space of B, given a basis or a spanning set of S :

(a) Put the basis (or spanning) vectors of S as rows of A.

(b) Find a basis of $nullsp(A)$. (This is a basis of $S^{\perp}$.)

(c) Put these basis vectors (of $S^{\perp}$) as rows of a matrix B.

(d) S is the null space of B.

13. How to extend a linearly independent set of vectors in R^n to a basis of R^n :

(a) Put the linearly independent vectors as rows of a matrix A.

(b) Find basis of $nullsp(A)$.

(c) The original vectors and the basis vectors found in step 2 together form a basis of R^n.

14. How to extract a basis of a subspace S from a spanning set of S :

(a) Put the spanning set as columns of a matrix A.

(b) Find $RREF(A)$.

(c) The leading columns of A are a basis of S.

If you just want to find out if A is invertible, but don't need to find the inverse the following will do.

15. How to tell if A is invertible:

(a) A must be square, only square matrices are invertible.

(b) A is invertible if and only if determinant of A is not zero. (Also A is invertible if and only if $RREF(A) = I$.)

16. How to tell if A is invertible and find A^{-1} if it exists:

(a) Check that A is square. Only square matrices are invertible.

(b) Form $\begin{bmatrix} A & I \end{bmatrix}$, where I is the identity matrix of same size as A.

(c) Perform elementary row operations on $\begin{bmatrix} A & I \end{bmatrix}$ until A is in RREF.

(d) If $RREF(A) = I$, then the same row operations that changed A into the identity matrix will have changed I into A^{-1}.

(e) If $RREF(A) \neq I$, then A is not invertible.

17. How to find A^{-1}, another method:

(a) Compute all cofactors of A.

(b) Let $C = matrix$ with $c_{ij} = cofactor$ of the ij^{th} position of A.

(c) $Adj(A) = C^T$.

(d) Compute one of the diagonal elements of A times $adj(A)$. This will be $|A|$, the determinant of A.

(e) If $|A| \neq 0$, then $A^{-1} = \frac{1}{|A|}adj(A)$, if $|A| = 0$, A is not invertible.

18. How to solve $Ax = b$: If A is invertible, $A^{-1}b$ gives a unique solution. The following procedure will work whether A is invertible or not.

(a) Form the augmented matrix $\begin{bmatrix} A & b \end{bmatrix}$.

(b) Find RREF of $\begin{bmatrix} A & b \end{bmatrix}$.

(c) If there is a leading one in the last column, $Ax = b$ has no solution.

(d) If there is no leading one in the last column, assign a letter to each variable corresponding to a nonleading column. These are the free variables and can be assigned any values.

(e) Write out the equations corresponding the the rows of $RREF\left(\begin{bmatrix} A & b \end{bmatrix}\right)$.

(f) Each of the variables corresponding to a leading column occurs in one and only one of these equations. Solve for each of them in terms of the free variables.

19. How to express a vector v as a linear combination of vectors $u_1, u_2, \cdots, u_k$:

(a) Put the vectors $u_1, u_2, \cdots, u_k$ as columns of a matrix A.

(b) Solve $Ax = v$.

(c) If no solution exists, v is not expressible as a linear combination of $u_1, u_2, \cdots, u_k$. If one or more solutions exist, any solution vector can be used as coefficients to express v as a linear combination of $u_1, u_2, \cdots, u_k$.

20. How to find the coordinate vector $[\mathbf{v}]_B$ (coefficients) of a vector $\mathbf{v}$ with respect to a basis $\{\mathbf{u}_1, \mathbf{u}_2, \cdots, \mathbf{u}_k\}$.

   (a) Put the vectors $\mathbf{u}_1, \mathbf{u}_2, \cdots, \mathbf{u}_k$ as the columns of a matrix B.

   (b) Solve $B\mathbf{x} = \mathbf{v}$, the solution vector is $[\mathbf{v}]_B$.

   (c) In this case we know that B is invertible, so $[\mathbf{v}]_B = B^{-1}\mathbf{v}$. (We could also solve $B\mathbf{x} = \mathbf{v}$ by row reducing $\begin{bmatrix} B & \mathbf{v} \end{bmatrix}$ as in #17, and we know the solution is unique, so there will be no free variables.)

21. How to tell if solution to $A\mathbf{x} = \mathbf{b}$ exists and is unique:

   (a) If A is square with $|A| \neq 0$, then a unique solution exists for every $\mathbf{b}$.

   (b) If A is not square, or if $|A| = 0$, a solution exists if and only if $RREF\left(\begin{bmatrix} A & \mathbf{b} \end{bmatrix}\right)$ has no leading one in the last column (or equivalently if $rank(A) = rank\left(\begin{bmatrix} A & \mathbf{b} \end{bmatrix}\right)$.)

   (c) A solution to $A\mathbf{x} = \mathbf{b}$ is unique when it exists if $RREF(A)$ has a leading one in every column.

22. How to find a basis of $S+T$ and $S \sqcap T$ given bases of the subspaces S and T :

   (a) Put the basis of S first and then the basis of T as columns of a matrix A.

   (b) Find $RREF(A)$.

   (c) The leading columns of A are a basis of $S + T$.

   (d) Each nonleading column of the $RREF(A)$ gives a linear dependence relation on the columns of A. Write these out.

   (e) In each equation from (d), put all the vectors from S on one side, and all the vectors from T on the other.

   (f) Each equation gives a vector in $S \sqcap T$. These are a basis for $S \sqcap T$ (assuming you started with bases of S and T.)

23. How to find a "least squares" solution to $A\mathbf{x} = \mathbf{b}$ when $A\mathbf{x} = \mathbf{b}$ has no solution:

    (a) Solve $A^T A\mathbf{x} = A^T\mathbf{b}$.

    (b) A solution $\hat{\mathbf{x}}$ to this equation is a "least squares" solution to $A\mathbf{x} = \mathbf{b}$.

24. How to find $proj_S\mathbf{v}$, the projection of a vector $\mathbf{v}$ on a subspace S:

    (a) Find a basis of S and put them as the columns of a matrix A.

    (b) Solve $A^T A\mathbf{x} = A^T\mathbf{v}$.

    (c) Multiply the solution $\hat{\mathbf{x}}$ by the matrix A. $Proj_S\mathbf{v} = A\hat{\mathbf{x}}$.

25. How to find a polynomial of degree n or less through $n+1$ given points $(x_1, y_1), (x_2, y_2), \cdots, (x_{n+1}, y_{n+1})$ with distinct x values:

    (a) Solve

$$
\begin{bmatrix}
1 & x_1 & x_1^2 & x_1^3 & \cdots & x_1^n \\
1 & x_2 & x_2^2 & x_2^3 & \cdots & x_2^n \\
\vdots & \vdots & \vdots & \vdots & & \vdots \\
1 & x_{n+1} & x_{n+1}^2 & x_{n+1}^3 & \cdots & x_{n+1}^n
\end{bmatrix}
\begin{bmatrix}
a_0 \\
a_1 \\
\vdots \\
a_n
\end{bmatrix}
=
\begin{bmatrix}
y_1 \\
y_2 \\
\vdots \\
y_{n+1}
\end{bmatrix}
$$

for
$$
\begin{bmatrix}
a_0 \\
a_1 \\
\vdots \\
a_n
\end{bmatrix}.
$$

    (b) Then $P(x) = a_0 + a_1 x + a_2 x^2 + \cdots + a_n x^n$ is the desired polynomial. A unique solution always exists because the Vandermonde matrix has nonzero determinant. (See Ex. 5.13.)

26. How to find the least squares fit polynomial of degree n or less to a set of $k \geq n+1$ points:

(a) Find the least squares solution to

$$\begin{bmatrix} 1 & x_1 & x_1^2 & x_1^3 & \cdots & x_1^n \\ 1 & x_2 & x_2^2 & x_2^3 & \cdots & x_2^n \\ \vdots & \vdots & \vdots & \vdots & & \vdots \\ 1 & x_{n+1} & x_{n+1}^2 & x_{n+1}^3 & \cdots & x_{n+1}^n \\ \vdots & \vdots & \vdots & \vdots & & \vdots \\ 1 & x_k & x_k^2 & x_k^3 & \cdots & x_k^n \end{bmatrix} \begin{bmatrix} a_0 \\ a_1 \\ \vdots \\ a_n \end{bmatrix} = \begin{bmatrix} y_1 \\ y_2 \\ \vdots \\ y_{n+1} \\ \vdots \\ y_k \end{bmatrix}$$

(b) If $(a_0, a_1, \cdots, a_n)$ is the least squares solution to this matrix equation, then the polynomial $P(x) = a_0 + a_1x + a_2x^2 + \cdots + a_nx^n$ is the least squares fit polynomial of degree n or less to the given points.

A special case is finding the least squares fit line to a given set of data points, (x_i, y_i). In this case the equation above becomes

$$\begin{bmatrix} 1 & x_1 \\ 1 & x_2 \\ \vdots & \vdots \\ 1 & x_k \end{bmatrix} \begin{bmatrix} a_0 \\ a_1 \end{bmatrix} = \begin{bmatrix} y_1 \\ y_2 \\ \vdots \\ y_k \end{bmatrix}.$$

Find the least squares solution for (a_o, a_1), and then the line $a_0 + a_1x = y$ is the least squares fit line to the given data. (See Ex. 5.14, Ex.5.15.)

27 How to find the eigenvalues of a matrix A:

(a) Subtract k from the diagonal elements of A, forming $A - kI$.

(b) Solve the polynomial equation $|A - kI| = 0$ for k.

(c) The roots of this equation are the eigenvalues of A. (See Ex. 4.1 through 4.6.)

28. How to find the eigenvectors of a matrix A for an eigenvalue c:

(a) Subtract the given eigenvalue c from the diagonal elements of A.

(b) The eigenvectors of A for c are the nonzero solutions of $(A - cI)\mathbf{x} = \mathbf{O}$. (See Ex. 4.1 through 4.6.)

29. How to find the eigenspace of a matrix A for an eigenvalue c:

   (a) The eigenspace of A for c is the null space of $(A - cI)$.

30. How to tell if a matrix A is similar to a diagonal matrix:

   (a) If A is symmetric, it is similar to a diagonal matrix.

   (b) If A is not symmetric, find all eigenvalues of A. If A is $n \times n$ and has n distinct real eigenvalues, it is similar to a diagonal matrix.

   (c) If A has complex eigenvalues, it is not similar to a diagonal matrix (using only matrices with real entries.)

   (d) If A has all real eigenvalues, but some are repeated roots of the characteristic equation $|A - kI| = 0$, then one must find the eigenspace for each repeated root.

   (e) A is similar to a diagonal matrix if and only if all the eigenvalues are real and the (algebraic) multiplicity of each repeated root equals the dimension of the corresponding eigenspace (geometric multiplicity of the root).

31. How to find P such that $P^{-1}AP$ is diagonal, when possible:

   (a) Find the eigenvalues of A.

   (b) Find a basis for each eigenspace of A.

   (c) Put the vectors found in (b) as the columns of a matrix P.

   (d) If P is not square because there are too few linearly independent eigenvectors found in step (b), then A is not similar to a diagonal matrix.

   (e) If P is square, $P^{-1}AP$ will be diagonal with the eigenvalues of A on the diagonal.
   (See Ex. 4.9 through 4.11.)

32.   **(Gram-Schmidt)** How to find an orthogonal basis of a subspace S given a basis $\mathbf{u}_1, \mathbf{u}_2, \cdots, \mathbf{u}_n$ of S:

(a) Let

$$\mathbf{v}_1 = \mathbf{u}_1.$$

(b) Let

$$\mathbf{v}_2 = \mathbf{u}_2 - \left(\frac{\mathbf{u}_2 \cdot \mathbf{v}_1}{\mathbf{v}_1 \cdot \mathbf{v}_1}\right)\mathbf{v}_1.$$

(c) Let

$$\mathbf{v}_3 = \mathbf{u}_3 - \left(\frac{\mathbf{u}_3 \cdot \mathbf{v}_1}{\mathbf{v}_1 \cdot \mathbf{v}_1}\right)\mathbf{v}_1 - \left(\frac{\mathbf{u}_3 \cdot \mathbf{v}_2}{\mathbf{v}_2 \cdot \mathbf{v}_2}\right)\mathbf{v}_2.$$

(d) Let

$$\mathbf{v}_4 = \mathbf{u}_4 - \left(\frac{\mathbf{u}_4 \cdot \mathbf{v}_1}{\mathbf{v}_1 \cdot \mathbf{v}_1}\right)\mathbf{v}_1 - \left(\frac{\mathbf{u}_4 \cdot \mathbf{v}_2}{\mathbf{v}_2 \cdot \mathbf{v}_2}\right)\mathbf{v}_2 - \left(\frac{\mathbf{u}_4 \cdot \mathbf{v}_3}{\mathbf{v}_3 \cdot \mathbf{v}_3}\right)\mathbf{v}_3.$$

(e) Continue to

$$\mathbf{v}_n = \mathbf{u}_n - \left(\frac{\mathbf{u}_n \cdot \mathbf{v}_1}{\mathbf{v}_1 \cdot \mathbf{v}_1}\right)\mathbf{v}_1 - \left(\frac{\mathbf{u}_n \cdot \mathbf{v}_2}{\mathbf{v}_2 \cdot \mathbf{v}_2}\right)\mathbf{v}_2 \cdots - \left(\frac{\mathbf{u}_n \cdot \mathbf{v}_{n-1}}{\mathbf{v}_{n-1} \cdot \mathbf{v}_{n-1}}\right)\mathbf{v}_{n-1}.$$

(f) Now $\{\mathbf{v}_1, \mathbf{v}_2, \cdots, \mathbf{v}_n\}$ is an orthogonal basis of S.
The process given above amounts to subtracting from each $\mathbf{u}_i$ its projection on the preceding $\mathbf{v}_i$, starting off with $\mathbf{v}_1 = \mathbf{u}_1$. For subspaces of dimension 2 or 3 this method isn't too bad. If you want an orthonormal basis, just divide each $\mathbf{v}_i$ by its length.

If S is defined by a system of linear homogeneous equations (*i.e.* as the null space of a matrix B), the following may be a more efficient way of getting an orthogonal basis of S. Instead of first finding a basis of nullspace B in the usual way and then applying the Gram-Schmidt method, with a little more effort we can pick an orthogonal basis.

33. How to find an orthogonal basis of a subspace S given S as the null space of a matrix B.

(a) Pick one nonzero vector $\mathbf{u}_1$ in $S = nullsp(B)$.

(b) Adjoin $\mathbf{u}_1$ as an additional row to B.

(c) Find a nonzero vector $\mathbf{u}_2$ in the null space of $B_1 = \begin{bmatrix} B \\ \mathbf{u}_1 \end{bmatrix}$.

(d) Adjoin $\mathbf{u}_2$ as an additional row to B_1.

(e) Find a nonzero vector $\mathbf{u}_3$ in the null space of $B_2 = \begin{bmatrix} B \\ \mathbf{u}_1 \\ \mathbf{u}_2 \end{bmatrix}$.

(f) Continue until there are no nonzero vectors in the null space of B_k. At this point $\{\mathbf{u}_1, \mathbf{u}_2, \cdots, \mathbf{u}_k\}$ is an orthogonal basis of S.

This method is particularly useful for finding orthogonal bases of eigenspaces, since we encounter the eigenspace as the null space of $B = (A - cI)$.

(See Procedure 4.2 and Ex.4.14.)

If you are given S by being given a basis or spanning set of S and you want to use this method instead of Gram-Schmidt to find an orthogonal basis of S, you can do it but it takes one additional step. You must first find a matrix B such that S is the null space of B. We can do this by the method of #12 above.

34. How to find find an orthogonal matrix P such that $P^{-1}AP$ is diagonal, where A is symmetric:

(a) Find the eigenvalues of A.

(b) Find an orthogonal basis for each eigenspace of A.

(c) Normalize (divide each vector by its length).

(d) Put these vectors as the columns of P.

35. How to extend an orthogonal set of vectors to an orthogonal basis of R^n.

(a) Put the orthogonal set as the rows of a matrix B.

(b) Follow the method of #33. The vectors $\mathbf{u}_1, \mathbf{u}_2, \cdots, \mathbf{u}_k$ will extend the original set to an orthogonal basis of R^n.

Remark (**Warning**) *Many of the procedures will give wrong answers if rows are used instead of columns or vice-versa. A general rule is that when you want information about the vectors themselves, put them as columns. When you want information about the subspace they span, put them as rows.*

Now that we have the $RREF(A)$, the determinant, and some theorems, we can do some of the examples from Chapter 1 much more easily.

Example 6.1 (1.12) *Is the triple* $(2, -1, 3)$ *a linear combination of the triples* $(1, 0, 2)$ *and* $(1, 3, -1)$ *? To answer this we look at the matrix*

$$\begin{bmatrix} 1 & 1 & 2 \\ 0 & 3 & 1 \\ 2 & -1 & 3 \end{bmatrix}.$$

The RREF of this matrix is

$$\begin{bmatrix} 1 & 0 & 0 \\ 0 & 1 & 0 \\ 0 & 0 & 1 \end{bmatrix}.$$

The leading one in the last column tells us there is no solution. In fact we can see that these three column vectors are linearly independent and a basis for R^3 *because of the leading one in every column.*

Example 6.2 (1.15) *Do the triples* $(1, 0, 2)$, $(2, -1, 3)$, $(1, 3, -1)$ *span* R^3 *? The answer is yes because we already concluded in the example above that they are a basis of* R^3. *If we want to express a triple* (a, b, c) *in terms of this basis we must solve the system*

$$x_1 \begin{bmatrix} 1 \\ 0 \\ 2 \end{bmatrix} + x_2 \begin{bmatrix} 2 \\ -1 \\ 3 \end{bmatrix} + x_3 \begin{bmatrix} 1 \\ 3 \\ -1 \end{bmatrix} = \begin{bmatrix} a \\ b \\ c \end{bmatrix}.$$

or

$$\begin{bmatrix} 1 & 2 & 1 \\ 0 & -1 & 3 \\ 2 & 3 & -1 \end{bmatrix} \begin{bmatrix} x_1 \\ x_2 \\ x_3 \end{bmatrix} = \begin{bmatrix} a \\ b \\ c \end{bmatrix}.$$

We find the inverse of this matrix is

$$\begin{bmatrix} -\frac{4}{3} & \frac{5}{6} & \frac{7}{6} \\ 1 & -\frac{1}{2} & -\frac{1}{2} \\ \frac{1}{3} & \frac{1}{6} & -\frac{1}{6} \end{bmatrix},$$

so

$$\begin{bmatrix} x_1 \\ x_2 \\ x_3 \end{bmatrix} = \begin{bmatrix} -\frac{4}{3} & \frac{5}{6} & \frac{7}{6} \\ 1 & -\frac{1}{2} & -\frac{1}{2} \\ \frac{1}{3} & \frac{1}{6} & -\frac{1}{6} \end{bmatrix} \begin{bmatrix} a \\ b \\ c \end{bmatrix}.$$

Example 6.3 (1.16) *Show that the triples* $(1, 4, -3)$, $(2, 0, -1)$, *and* $(3, 4, -4)$ *do not span* R^3. *Put these triples as the columns of a matrix*

$$A = \begin{bmatrix} 1 & 2 & 3 \\ 4 & 0 & 4 \\ -3 & -1 & -4 \end{bmatrix}.$$

We compute

$$|A| = 0,$$

thus $rank(A) < 3$ *so the columns don't span* R^3. *(Dimension of* $colsp(A) < 3$.)

Example 6.4 (1.21) *Show that the triples* $(1, 1, 1)$, $(2, 1, 0)$, $(3, 1, 1)$ *are linearly independent. We compute*

$$\begin{vmatrix} 1 & 2 & 3 \\ 1 & 1 & 1 \\ 1 & 0 & 1 \end{vmatrix} = -2.$$

Since this determinant is not zero, the vectors are linearly independent, (and thus also a basis for R^3.)

Example 6.5 (1.22) *Show that the triples* $(1, 2, 1)$, $(2, 4, 1)$, *and* $(3, 6, 0)$ *form a linearly dependent set. Let*

$$A = \begin{bmatrix} 1 & 2 & 3 \\ 2 & 4 & 6 \\ 1 & 1 & 0 \end{bmatrix}.$$

Clearly

$$|A| = 0,$$

since the second row is twice the first. Thus the triples are linearly dependent. The relationship between the columns is not as obvious as that between the rows, but when we find $RREF(A) = \begin{bmatrix} 1 & 0 & -3 \\ 0 & 1 & 3 \\ 0 & 0 & 0 \end{bmatrix}$,

we see that the third column of A is -3 *times the first column* $+3$ *times the second column.*

We now do some further examples.

Example 6.6 *Are the vectors* $(1,1,2,0)$, $(1,3,1,1)$, *and* $(1,5,6,1)$ *linearly independent?*

Using #10 above, we let

$$A = \begin{bmatrix} 1 & 1 & 1 \\ 1 & 3 & 5 \\ 2 & 1 & 6 \\ 0 & 1 & 1 \end{bmatrix}.$$

$$RREF(A) = \begin{bmatrix} 1 & 0 & 0 \\ 0 & 1 & 0 \\ 0 & 0 & 1 \\ 0 & 0 & 0 \end{bmatrix},$$

so they are linearly independent.

Example 6.7 *Are the vectors* $(1,1,3,2)$, $(1,3,2,1)$, *and* $(1,-3,5,4)$ *linearly independent? Using #10 we let*

$$A = \begin{bmatrix} 1 & 1 & 1 \\ 1 & 3 & -3 \\ 3 & 2 & 5 \\ 2 & 1 & 4 \end{bmatrix}.$$

We find

$$RREF(A) = \begin{bmatrix} 1 & 0 & 3 \\ 0 & 1 & -2 \\ 0 & 0 & 0 \\ 0 & 0 & 0 \end{bmatrix}.$$

The vectors are not linearly independent, in fact the third is 3 times the first minus twice the second, using #4.

Example 6.8 *Let* $S = colsp(A)$ *where* A *is the matrix from the previous example, find* $S^\perp$. *We know that a basis for* S *is* $(1, 1, 3, 2)$ *and* $(1, 3, 2, 1)$. *Thus using #5, we put these basis vectors as the rows of a matrix and find the null space. Let*

$$C = \begin{bmatrix} 1 & 1 & 3 & 2 \\ 1 & 3 & 2 & 1 \end{bmatrix}.$$

Then $S^\perp = nullsp(C)$. *We use #3 to find a basis of* $nullsp(C)$.

$$RREF(C) = \begin{bmatrix} 1 & 0 & \frac{7}{2} & \frac{5}{2} \\ 0 & 1 & -\frac{1}{2} & -\frac{1}{2} \end{bmatrix}.$$

Adding two zero rows we get

$$H = \begin{bmatrix} 1 & 0 & \frac{7}{2} & \frac{5}{2} \\ 0 & 1 & -\frac{1}{2} & -\frac{1}{2} \\ 0 & 0 & 0 & 0 \\ 0 & 0 & 0 & 0 \end{bmatrix},$$

and

$$H - I = \begin{bmatrix} 0 & 0 & \frac{7}{2} & \frac{5}{2} \\ 0 & 0 & -\frac{1}{2} & -\frac{1}{2} \\ 0 & 0 & -1 & 0 \\ 0 & 0 & 0 & -1 \end{bmatrix},$$

thus $\left(\frac{7}{2}, -\frac{1}{2}, -1, 0\right)$ and $\left(\frac{5}{2}, -\frac{1}{2}, 0, -1\right)$ form a basis of $S^\perp$. We can multiply each by two and still have a basis so $(7, -1, -2, 0)$ and $(5, -1, 0, -2)$ form a basis of $S^\perp$.

Example 6.9 *Characterize the subspace* S *of the previous example by a system of linear homogeneous equations. Using #12, and what we did in the previous example, the equations are* $7x_1 - x_2 - 2x_3 = 0$, *and* $5x_1 - x_2 - 2x_4 = 0$.

Example 6.10 *Find an orthogonal basis of S, where S is the same space as in the previous two examples.* **Using #32 (Gram-Schmidt).** *We have a basis for S, namely* $(1, 1, 3, 2)$ *and* $(1, 3, 2, 1)$. *We let*

$$\mathbf{v}_1 = (1, 1, 3, 2),$$

and

$$\mathbf{v}_2 = (1, 3, 2, 1) - \left(\frac{(1, 3, 2, 1) \cdot (1, 1, 3, 2)}{(1, 1, 3, 2) \cdot (1, 1, 3, 2)} \right) (1, 1, 3, 2)$$

$$= (1, 3, 2, 1) - \frac{12}{15}(1, 1, 3, 2)$$

$$= \left(\frac{3}{15}, \frac{33}{15}, -\frac{6}{15} - \frac{9}{15} \right).$$

We can use a multiple of this if we like, since it will still be in S and still be orthogonal to $\mathbf{v}_1$. *Thus we get an orthogonal basis of S is*

$$\{(1, 1, 3, 2), (1, 11, -2, -3)\}.$$

Using #33. *We already know that S is the null space of the matrix* $B = \begin{bmatrix} 7 & -1 & -2 & 0 \\ 5 & -1 & 0 & -2 \end{bmatrix}$ *from Example 6.4.*

$$S = nullsp B = nullsp \begin{bmatrix} 7 & -1 & -2 & 0 \\ 5 & -1 & 0 & -2 \end{bmatrix}.$$

Now we choose any vector in S, for example $(1, 1, 3, 2)$ *as our* $\mathbf{v}_1$. *We then seek a vector which is both in* $nullsp(B)$ *and orthogonal to* $(1, 1, 3, 2)$, *this is the same as being in the null space of the matrix*

$$\begin{bmatrix} B \\ \mathbf{v}_1 \end{bmatrix} = \begin{bmatrix} 7 & -1 & -2 & 0 \\ 5 & -1 & 0 & -2 \\ 1 & 1 & 3 & 2 \end{bmatrix}.$$

Now this matrix row reduces to

$$\begin{bmatrix} 1 & 0 & 0 & \frac{1}{3} \\ 0 & 1 & 0 & \frac{11}{3} \\ 0 & 0 & 1 & -\frac{2}{3} \end{bmatrix},$$

so a basis for the null space is $\left(\frac{1}{3}, \frac{11}{3}, -\frac{2}{3}, -1\right)$, *or we can use a multiple* $(1, 11, -2, -3)$. *Thus* $\{(1, 1, 3, 2), (1, 11, -2, -3)\}$ *is an orthogonal basis of* S. *Even though we got the same answer twice, this basis is not unique. For example in either or both methods we could have started by keeping the other basis vector* $(1, 3, 2, 1)$ *as* $\mathbf{v}_1$, *and we would have gotten a different basis.*

Chapter 7

Geometry

Section 7.1 Vectors and Lines

We assume you are familiar with the Cartesian coordinate system in a plane which represents each point by an ordered pair of real numbers (Figure 7.1a), and the Cartesian coordinate system for 3-space which represents each point by an ordered triple of real numbers (Figure 7.1b).

If you have studied physics, you have probably heard a vector defined as something that has magnitude and direction, and seen a vector represented by a directed line segment or arrow (Figure 7.2).

The length of the segment is the magnitude of the vector and the direction of the segment is the direction of the vector. If two directed line segments (arrows) have the same length and direction, they represent the same vector. If P and Q are points, we denote by $\overrightarrow{PQ}$ the directed line segment that starts at P and ends at Q. Let $\mathbf{v}$ be a vector represented by $\overrightarrow{PQ}$ and $\mathbf{u}$ be a vector represented by $\overrightarrow{RS}$. Then $\mathbf{v} = \mathbf{u}$ if and only if $\overrightarrow{PQ}$ and $\overrightarrow{RS}$ have the same length and direction (Figure 7.3).

We note that for any vector $\mathbf{v}$ and any point P there is a representative of $\mathbf{v}$ starting at P. We add vectors as follows: To find $\mathbf{u} + \mathbf{v}$, take a representative $\overrightarrow{PQ}$ of $\mathbf{u}$ and a representative $\overrightarrow{QR}$ of $\mathbf{v}$. Then $\overrightarrow{PR}$ is a representative of $\mathbf{u} + \mathbf{v}$ (Figure 7.4).

A scalar is defined as something that has only magnitude, and these are represented by numbers. We can multiply a vector $\mathbf{u}$ by a scalar k

247

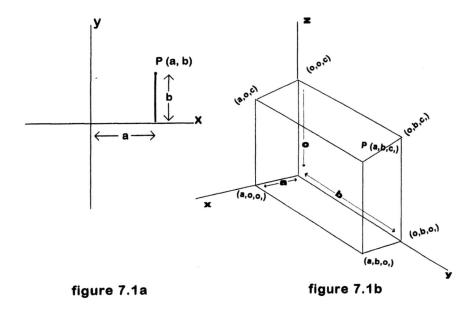

figure 7.1a   **figure 7.1b**

by taking a representative $\overrightarrow{PQ}$ of **u** and forming the arrow in the same direction (if $r > 0$, opposite direction if $r < 0$) and of length k times the length of $\overrightarrow{PQ}$ (Figure 7.5).

Let **v** be a vector represented by the arrow $\overrightarrow{PQ}$. If $P = (p_1, p_2)$ and $Q = (q_1, q_2)$, then The components of **v** are $q_1 - p_1$ and $q_2 - p_2$, (Figure 7.6).

We notice that if $\overrightarrow{PQ}$ and $\overrightarrow{RS}$ are two representatives of **v**, then the components of **v** as computed from $\overrightarrow{PQ}$ and the components of **v** as computed from $\overrightarrow{RS}$ are the same (Figure 7.7).

This is because $\overrightarrow{PQ}$ and $\overrightarrow{RS}$ have the same length and direction, so the triangles in Figure 7.7 are congruent. Thus if $\overrightarrow{PQ}$ is a representative of **v**, we can associate with **v** a unique ordered pair, namely the components of **v**, $(q_1 - p_1, q_2 - p_2)$. We also see that the components of **u** + **v** are the components of **u** plus the components of **v** (Figure 7.8), since $r_1 - p_1 = (r_1 - q_1) + (q_1 - p_1)$ and $r_2 - p_2 = (r_2 - q_2) + (q_2 - p_2)$.

The components of $k\mathbf{u}$ are k times the components of **u**, since the triangles in Figure 7.9 are similar.

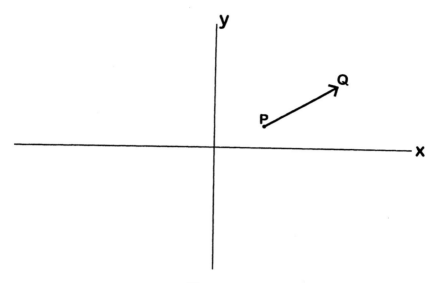

Figure 7.2

We have seen that in the plane each vector can be represented by many directed line segments or by a unique ordered pair of real numbers called the components of v. Two vectors can be added by adding their components and a vector can be multiplied by a scalar by multiplying its components by that scalar. Thus in a plane an ordered pair can represent either a vector or a point. If the vector v is represented by an arrow $\overrightarrow{OP}$ starting at the origin, the same pair represents both the vector v and the point P (Figure 7.10). In the same way an ordered triple can represent either the vector represented by $\overrightarrow{OP}$ or the point P (Figure 7.11).

We can exploit the fact that a pair or triple can represent either a point or a vector to solve problems of the following kind:

Example 7.1 *Find the coordinates of a point $\frac{3}{4}$ of the way from the point $P = (1,1,1)$ to the point $Q = (3,2,4)$ and on the line joining them.*

Solution: Let $R = (x,y,z)$ be the point we are looking for. Then looking

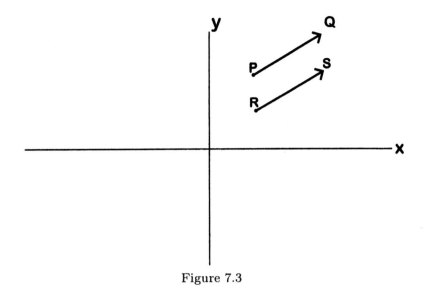

Figure 7.3

at Figure 7.12 we get the vector equation

$$\overrightarrow{OR} = \overrightarrow{OP} + \frac{3}{4}\overrightarrow{PQ}.$$

Translating to components gives

$$(x, y, z) = (1, 1, 1) + \frac{3}{4}(3 - 1, 2 - 1, 4 - 1) = \left(\frac{5}{2}, \frac{7}{4}, \frac{13}{4}\right).$$

There is nothing special about the fraction $\frac{3}{4}$. This same method would have worked for any number t. The point R on the line joining $P = (p_1, p_2, p_3)$ and $Q = (q_1, q_2, q_3)$ and t of the way from P to Q is given by the vector equation

$$\overrightarrow{OR} = \overrightarrow{OP} + t\overrightarrow{PQ},$$

or in component terms,

$$(x, y, z) = (p_1, p_2, p_3) + t(q_1 - p_1, q_2 - p_2, q_3 - p_3). \qquad (7.1)$$

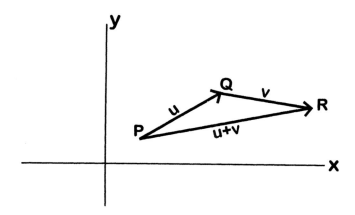

Figure 7.4

Now what if t is greater than 1? Suppose instead of going $\frac{1}{3}$ or $\frac{3}{4}$ of the way from P to Q we go $\frac{5}{2}$ of the way from P to Q along the line joining P and Q. Referring to Figure 7.12 again, we would find ourselves still on the line joining P and Q, but now we would be beyond Q. Similarly if t is negative, this would mean we go out to the point P and walk along the line joining P and Q but in the direction away from Q. We would still be on the same line, but now we would be on the other side of P from Q. Thus 7.1 describes a line through the points P and Q. Each value of t produces a point (x, y, z) on the line, and each point on the line will occur for some value of t. Equating components, 7.1 is equivalent to

$$
\begin{aligned}
x &= p_1 + t(q_1 - p_1) \\
y &= p_2 + t(q_2 - p_2) \\
z &= p_3 + t(q_3 - p_3)
\end{aligned}
\qquad (7.2)
$$

which are **parametric equations describing the line through the points P and Q.**

Note that the constant terms here are the coordinates of a point on the line, and the coefficients of the parameter t are the components

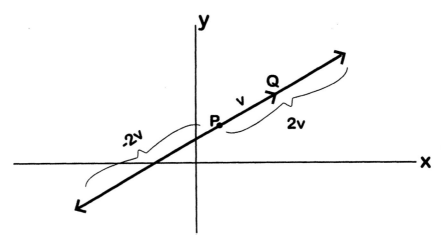

Figure 7.5

of a vector in the direction of the line. These **coefficients of t are called direction numbers of the line.** If we are given a point $P = (x_0, y_0, z_0)$ and a vector $\mathbf{v}$ represented by (a, b, c), we can write the equation of the line through P in the direction of $\mathbf{v}$ as

$$(x, y, z) = (x_0, y_0, z_0) + t(a, b, c), \qquad (7.3)$$

or equating components as

$$
\begin{aligned}
x &= x_0 + at \\
y &= y_0 + bt \\
z &= z_0 + ct
\end{aligned} \qquad (7.4)
$$

The same approach could be used in 2-space to get the equation of a line through two points P and Q or through a point P in the direction of a vector $\mathbf{v}$, except that in this case the points would have only two coordinates and the vectors only two components.

We can tell whether or not two lines are parallel by looking at parametric equations describing the lines. The lines are parallel if and only if direction numbers (a_1, b_1, c_1) for one of the lines and (a_2, b_2, c_2)

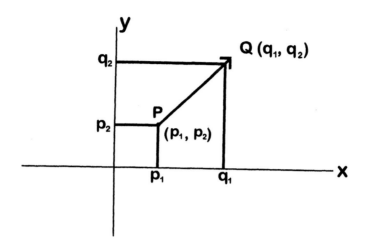

Figure 7.6

for the other satisfy $(a_2, b_2, c_2) = k(a_1, b_1, c_1)$ for some scalar k. If the lines are not parallel, we can try to find a point of intersection. In a plane, two non-parallel lines must intersect. We can find the point of intersection from parametric equations describing the lines.

Example 7.2 *Consider the following two lines in a plane. L_1 is given by*

$$x = 3 + 2t$$
$$y = -1 + t$$

and L_2 is given by

$$x = 2 - s$$
$$y = 3 - 2s$$

Note that we use a different letter as the parameter for the second line. It is very important to do this since the point of intersection may not occur for the same value of the parameter in L_2 as in L_1. At the point of intersection, both lines have the same x value and the same y value. Setting the x values equal we get

$$3 + 2t = 2 - s,$$

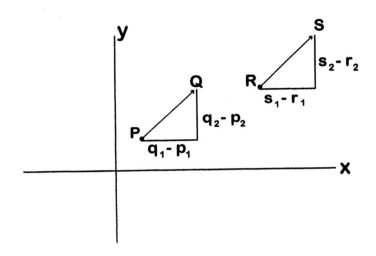

Figure 7.7

and setting the y values equal we get

$$-1 + t = 3 - 2s.$$

Solving simultaneously for s and t, we get $t = -2$ and $s = 3$. Substituting $t = -2$ in the equations for L_1 gives the point of intersection $(-1, -3)$. As a check we can substitute $s = 3$ in the equations for L_2 and see that this gives the same point.

In 3-space two non-parallel lines may fail to intersect. In this case they are called skew. What happens is that when you set equal the x, y, and z values for the two lines, you get three equations involving two unknowns, the two parameters s and t. This system may not have a solution.

Example 7.3 *Example 7.3 Determine whether or not the lines L_1 and L_2 intersect if L_1 is given by*

$$
\begin{aligned}
x &= 1 + t \\
y &= 2 - t \\
z &= -1 + 2t
\end{aligned}
$$

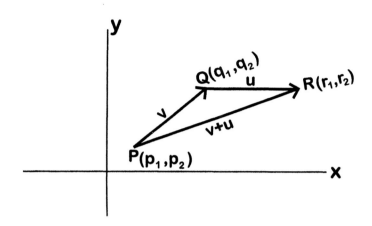

Figure 7.8

and L_2 is given by

$$\begin{aligned} x &= 2 + 3s \\ y &= 1 - 2s \ . \\ z &= 3 + s \end{aligned}$$

Solution: The lines are not parallel since the direction numbers for L_1 are $(1, -1, 2)$ and the direction numbers for L_2 are $(3, -2, 1)$. Setting equal the x values gives

$$2 + 3s = 1 + t,$$

and from the y values we get

$$1 - 2s = 2 - t.$$

Solving simultaneously gives $s = 0$ and $t = 1$. But then this gives different z values for L_1 and L_2. We conclude that the lines are skew.

Exercises

1. Find parametric equations describing a line through the points $P = (1, -3, 2)$ and $Q = (5, 5, -3)$.

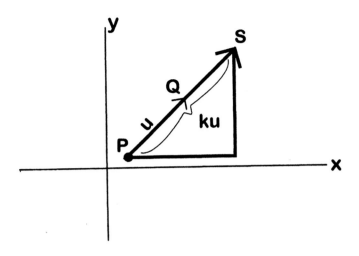

Figure 7.9

2. Find parametric equations describing a line through the point $(1, 4, 1)$ in the direction of $\mathbf{v} = (2, 2, 2)$.

3. Find parametric equations describing a line through the point $(1, 3, 2)$ and parallel to the X axis.

4. Which if any of the following lines are parallel?

$$L_1: \begin{aligned} x &= 3 - t \\ y &= 2 + 3t \\ z &= 4 - 2t \end{aligned} \qquad L_2: \begin{aligned} x &= 5 + 3t \\ y &= 2 - 9t \\ z &= 2 + 6t \end{aligned}$$

$$L_3: \begin{aligned} x &= 10 + 2t \\ y &= 4 - 3t \\ z &= 4 + 2t \end{aligned}.$$

5. Find direction numbers of the following lines:

$$L_1: \begin{aligned} x &= 2 \\ y &= 3 \\ z &= t \end{aligned} \qquad L_2: \begin{aligned} x &= t \\ y &= 4 \\ z &= 5 \end{aligned} \qquad L_3: \begin{aligned} x &= t \\ y &= t \\ z &= t \end{aligned}.$$

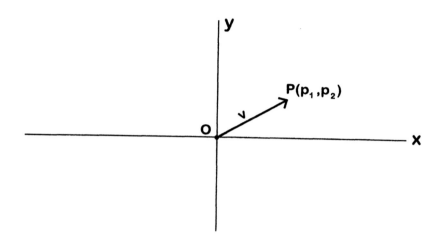

Figure 7.10

6. Describe geometrically each of the lines in problem 5.

7. Find the intersection of the following lines in 2-space:

$$L_1 : \begin{matrix} x & = & 3 & + & 2t \\ y & = & -1 & + & t \end{matrix} \qquad L_2 : \begin{matrix} x & = & 2 & - & s \\ y & = & 3 & + & 2s \end{matrix} .$$

8. Determine whether or not the following pairs of lines in 3-space intersect and find the intersection if it exists:

(a)

$$L_1 : \begin{matrix} x & = & 2 & + & 3t \\ y & = & 2 & - & 3t \\ z & = & -1 & + & t \end{matrix} \qquad L_2 : \begin{matrix} x & = & 6 & + & s \\ y & = & 1 & + & 2s \\ z & = & 1 & + & s \end{matrix} .$$

(b)

$$L_1 : \begin{matrix} x & = & 3 & + & 2t \\ y & = & 2 & - & t \\ z & = & -1 & + & 2t \end{matrix} \qquad L_2 : \begin{matrix} x & = & -1 & + & 8s \\ y & = & -2 & - & 4s \\ z & = & 3 & + & 8s \end{matrix} .$$

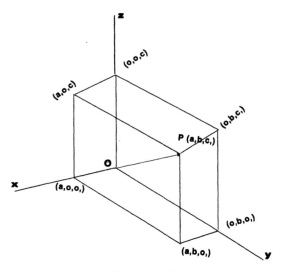

Figure 7.11

(c)

$$L_1 : \begin{array}{rcl} x &=& 5 - 2t \\ y &=& 3 + 2t \\ z &=& -2 - t \end{array} \qquad L_2 : \begin{array}{rcl} x &=& 0 + 3s \\ y &=& 1 + s \\ z &=& 1 + 2s \end{array}.$$

Section 7.2 Dot Product and Planes

The dot product of two vectors can be defined either in terms of their components or in terms of their lengths and directions. Let us use the former approach. If $\mathbf{u} = (x_1, y_1)$ and $\mathbf{v} = (x_2, y_2)$ then

$$\mathbf{u} \cdot \mathbf{v} = x_1 x_2 + y_1 y_2.$$

If we are in 3-space this becomes

$$\mathbf{u} \cdot \mathbf{v} = x_1 x_2 + y_1 y_2 + z_1 z_2.$$

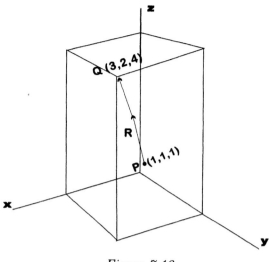

Figure 7.12

We also define the length of a vector **u**, written $|\mathbf{u}|$ to be $\sqrt{\mathbf{u} \cdot \mathbf{u}}$. It is easy to verify that the dot product obeys the following rules:

$$\mathbf{u} \cdot \mathbf{v} = \mathbf{v} \cdot \mathbf{u}$$

$$\mathbf{u} \cdot (k\mathbf{v}) = (k\mathbf{u}) \cdot \mathbf{v} = k(\mathbf{u} \cdot \mathbf{v})$$

$$\mathbf{u} \cdot (\mathbf{v} + \mathbf{w}) = \mathbf{u} \cdot \mathbf{v} + \mathbf{u} \cdot \mathbf{w}$$

$$\mathbf{u} \cdot \mathbf{u} \geq 0, \mathbf{u} \cdot \mathbf{u} = 0 \text{ if and only if } \mathbf{u} = 0.$$

The Law of Cosines says that in a triangle such as those in Figure 7.13 we have $|\mathbf{w}|^2 = |\mathbf{u}|^2 + |\mathbf{v}|^2 - 2|\mathbf{u}||\mathbf{v}| \cos \theta$. Notice that $\mathbf{u} + \mathbf{w} = \mathbf{v}$, so $\mathbf{w} = \mathbf{v} - \mathbf{u}$, thus

$$|\mathbf{v} - \mathbf{u}|^2 = |\mathbf{u}|^2 + |\mathbf{v}|^2 - 2|\mathbf{u}||\mathbf{v}| \cos \theta. \tag{7.5}$$

If $\mathbf{u} = (x_1, y_1, z_1)$ and $\mathbf{v} = (x_2, y_2, z_2)$ then

$$\mathbf{v} - \mathbf{u} = (x_2 - x_1, y_2 - y_1, z_2 - z_1),$$

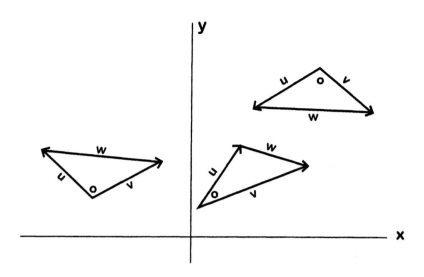

Figure 7.13

and 7.5 becomes

$$(x_2-x_1)^2+(y_2-y_1)^2+(z_2-z_1)^2 = x_1^2+y_1^2+z_1^2+x_2^2+y_2^2+z_2^2-2|\mathbf{u}||\mathbf{v}|\cos\theta.$$

Expanding and canceling gives

$$x_1x_2 + y_1y_2 + z_1z_2 = |\mathbf{u}||\mathbf{v}|\cos\theta,$$

or

$$\mathbf{u}\cdot\mathbf{v} = |\mathbf{u}||\mathbf{v}|\cos\theta. \tag{7.6}$$

Note that the angle between $\mathbf{u}$ and $\mathbf{v}$ is always $< 180°$, and $\mathbf{u}\cdot\mathbf{v} < 0$ if and only if $90° < \theta < 180°$. If $\theta = 90°$, we have $\mathbf{u}\cdot\mathbf{v} = 0$. Conversely, if $\mathbf{u}\cdot\mathbf{v} = 0$, either $\theta = 90°$, $|\mathbf{u}| = 0$, or $|\mathbf{v}| = 0$. Since $|\mathbf{u}| = 0$ implies $\mathbf{u} = (0,0,0)$, $\mathbf{u}\cdot\mathbf{v} = 0$ only if $\mathbf{u}$ and $\mathbf{v}$ are perpendicular or one or both of them is the zero vector. If we agree to regard the zero vector $\mathbf{O}$, (which really has no direction) as perpendicular to every vector, we can say $\mathbf{u}\cdot\mathbf{v} = 0 \iff \mathbf{u}$ is perpendicular to $\mathbf{v}$. We use the symbol $\mathbf{u}\perp\mathbf{v}$. We also use the word orthogonal as a synonym for perpendicular.

Let $\mathbf{n} = (a,b)$ be a fixed vector in a plane and let $P = (x_0,y_0)$ be a fixed point. Then if $Q = (x,y)$ is another point (Figure 7.14), the vector represented by $\overrightarrow{PQ}$ will be perpendicular to **n** if and only if $(x - x_0, y - y_0) \cdot (a, b) = 0$.

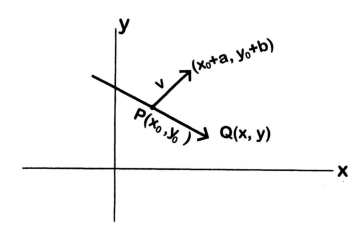

Figure 7.14

This equation describes all points $Q = (x, y)$ such that $\overrightarrow{PQ} \perp$ **n**. These points lie on a line through P and perpendicular to **n**. Thus the equation of a line through the point (x_0, y_0) and perpendicular to the vector **n** $= (a, b)$ is given by

$$(x - x_0, y - y_0) \cdot (a, b) = 0. \qquad (7.7)$$

This can also be written as

$$ax + by + c = 0, \qquad (7.8)$$

where $c = -ax_0 - by_0$.

If we generalize what we have just done to 3-space we get the equation of a plane. If $P = (x_0, y_0, z_0)$ is a fixed point and **n** $= (a, b, c)$ a fixed vector, then the set of points $Q = (x, y, z)$ such that $\overrightarrow{PQ} \perp$ **n** is a plane. For example if P is the origin and **n** is the Z axis, then the set of points Q such that $\overrightarrow{PQ} \perp$ **n** is the XY plane (Figure 7.15).

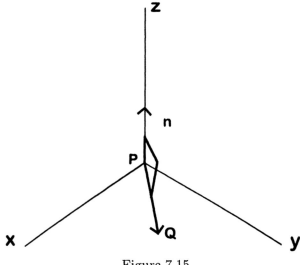

Figure 7.15

The equation of a plane in 3-space which is $\perp$ to the vector $\mathbf{n} = (a, b, c)$ and through the point $P = (x_0, y_0, z_0)$ is

$$(x - x_0, y - y_0, z - z_0) \cdot (a, b, c) = 0. \tag{7.9}$$

A vector $\mathbf{n}$ that is orthogonal to all vectors in a plane is called a **normal** to that plane. Equation 7.9 can be rewritten as

$$ax + by + cz + d = 0, \tag{7.10}$$

where $d = -ax_0 - by_0 - cz_0$.

Note that the coefficients of x, y and z are the components of a normal to the plane. The angle between two planes is defined to be the angle between normals to the planes. The angle between a plane and a line is defined to be 90° minus the angle between a vector in the direction of the line and a normal to the plane. Thus a line is perpendicular to a plane if the line is parallel to a normal to the plane, and a line is parallel to a plane if it is perpendicular to a normal to the plane. A line lies in a plane if the all points on the line satisfy the equation of the plane.

We can find the line of intersection of two non-parallel planes by solving the equations simultaneously.

Example 7.4 *Find the line of intersection of the planes $3x - y + 2z = 2$, and $x - 3y + z = 1$. Solution: Multiply the first equation by 3 and subtract from the second. We get $x = \frac{5}{8} - \frac{5}{8}z$. Substitute this in the second equation of x and get $y = -\frac{1}{8} + \frac{1}{8}z$. Since we only have two equations, the best we can do is solve for two of the unknowns in terms of the third, and this we have done. If we let $z = t$, then we get the parametric equations of a line*

$$
\begin{aligned}
x &= \frac{5}{8} - \frac{5}{8}t \\
y &= -\frac{1}{8} + \frac{1}{8}t \ . \\
z &= t
\end{aligned}
$$

We can check by substituting these values into the equation of the plane and observing that the equation of the plane is satisfied for all values of t.

Although a line in 3-space is usually described by parametric equations, we can always characterize a line in 3-space as the intersection of two planes. If the parametric equations of the line are $x = at + x_0$, $y = bt + y_0$, and $z = ct + z_0$, we can solve one of these for t and substitute in the other two. At least one of the direction numbers a, b and c must be nonzero, for if all three are zero the equations just give a point, not a line. For example if $a \neq 0$, we have $t = \frac{x - x_0}{a}$, so that points on the line must satisfy $y - y_0 = \frac{b}{a}(x - x_0)$ and $z - z_0 = \frac{c}{a}(x - x_0)$. Thus the line is characterized as the intersection of these two planes.

Example 7.5 *Let L be the line given by*

$$
L: \begin{aligned}
x &= 2 + 3t \\
y &= 4 - 2t \ , \\
z &= -1 + t
\end{aligned}
$$

find the equations of two planes whose intersection is L. Solution: $t = \frac{1}{3}(x - 2)$, thus $y = -\frac{2}{3}(x - 2) + 4$ and $z = \frac{1}{3}(x - 2) - 1$. The first can be rewritten as $y + \frac{2}{3}x - \frac{16}{3} = 0$, or $3y + 2x - 16 = 0$, and the second as $x - 3z - 5 = 0$.

Example 7.6 *Let L be the line given by*

$$
\begin{aligned}
x &= 3 \\
y &= 1 + 2t \ , \\
z &= 2 + 3t
\end{aligned}
$$

find the equations of two planes whose intersection is L. Solution: Note that $x = 3$ is the equation of a plane perpendicular to the X axis through the point $(3, 0, 0)$. From the second equation we get $t = \frac{1}{2}(y - 1)$, and substituting this in the equation for z we get $z = \frac{3}{2}(y - 1) + 2$, or $-3y + 2z - 1 = 0$. This is the equation of a plane parallel to the X axis. L is the intersection of the planes $x = 3$ and $-3y + 2z - 1 = 0$.

Exercises

1. Find the equation of the plane through the point $(1, 3, -1)$ and perpendicular to the vector $(1, 1, 1)$.

2. Find a vector perpendicular to the plane given by the equation $3x + 2y - z + 10 = 0$.

3. Describe geometrically the plane $z = 3$.

4. Describe geometrically the plane $3x + 2y = 5$. What does the fact that z does not appear in the equation tell you?

5. Find the point of intersection of the plane $3x + 2y - z = 0$, and the line $x = 3t$, $y = 2 - t$, $z = 1 + 2t$.

6. Show that the line $x = 1 - 3t$, $y = 2 + t$, $z = -2 - t$ is parallel to the plane $2x + 4y - 2z + 3 = 0$. Is the line above or below the plane?

7. Find the line of intersection of the planes $2x + 3y + z = 0$ and $x - 2y - z = 0$.

8. Find the line of intersection of the planes $x + y - 3z = 5$ and $4x + y - z = 4$.

9. Find the equations of two planes that intersect in the line given by the parametric equations $x = -3t + 5$, $y = 2t - 2$, and $z = t + 4$.

10. Find the equations of two planes that intersect in the line given by the parametric equations $x = 2t - 3$, $y = 1$, and $z = t - 6$.

Section 7.3 The Cross Product

A matrix is a rectangular array of numbers which are arranged in rows and columns. If A is a matrix with m rows and n columns we say that A is an $m \times n$ matrix. We denote by a_{ij} the entry in the ith row and j^{th} column of A. Thus $A = \begin{bmatrix} 1 & 2 \\ -3 & 0 \end{bmatrix}$ is a 2×2 matrix with $a_{11} = 1$, $a_{12} = 2$, $a_{21} = -3$, and $a_{22} = 0$. For a 2×2 matrix the determinant of A, written $|A|$, is $a_{11}a_{22} - a_{12}a_{21}$. Thus the determinant of the matrix A given above is 6. If A is a 3×3 matrix then

$$|A| = a_{11}C_{11} + a_{12}C_{12} + a_{13}C_{13}. \qquad (7.11)$$

where

$$C_{11} = \begin{vmatrix} a_{22} & a_{23} \\ a_{32} & a_{33} \end{vmatrix},$$

$$C_{12} = - \begin{vmatrix} a_{21} & a_{23} \\ a_{31} & a_{33} \end{vmatrix},$$

and

$$C_{13} = \begin{vmatrix} a_{21} & a_{22} \\ a_{31} & a_{32} \end{vmatrix}.$$

C_{11}, C_{12}, and C_{13} are called the cofactors of row 1. If A is a 3×3 matrix, the cofactor C_{ij} of a_{ij} is defined by

$$C_{ij} = (-1)^{i+j}|M_{ij}|, \qquad (7.12)$$

where M_{ij} is the 2×2 matrix obtained from A by omitting row i and column j. If you take the entries of any row (or any column), multiply each by its cofactor, and add up these products, you will get the same number each time and that number is $|A|$. We call this "expanding by minors around a row (or column). Notice that equation 7.11 can be rewritten as

$$|A| = (a_{11}, a_{12}, a_{13}) \cdot (C_{11}, C_{12}, C_{13}). \qquad (7.13)$$

A matrix with two rows equal always has determinant zero, thus if the first row of A equals either the second or third row of A, we will have $(a_{11}, a_{12}, a_{13}) \cdot (C_{11}, C_{12}, C_{13}) = 0$. This gives us a way to find a vector that is $\perp$ to two given vectors. Suppose we want a vector $\perp$ to both $\mathbf{u} = (x_1, y_1, z_1)$ and $\mathbf{v} = (x_2, y_2, z_2)$. Then if the matrix A has $\mathbf{u}$ and $\mathbf{v}$ as its second and third rows, the cofactors of the first row of A will form a vector perpendicular to both $\mathbf{u}$ and $\mathbf{v}$. In other words since

$$\begin{vmatrix} \mathbf{u} \\ \mathbf{u} \\ \mathbf{v} \end{vmatrix} = 0 \text{ and } \begin{vmatrix} \mathbf{v} \\ \mathbf{u} \\ \mathbf{v} \end{vmatrix} = 0,$$

we have

$$\mathbf{u} \cdot (C_{11}, C_{12}, C_{13}) = 0 \text{ and } \mathbf{v} \cdot (C_{11}, C_{12}, C_{13}) = 0,$$

where C_{11}, C_{12}, and C_{13} are cofactors of a matrix

$$A = \begin{bmatrix} * \\ \mathbf{u} \\ \mathbf{v} \end{bmatrix} = \begin{bmatrix} * & * & * \\ x_1 & y_1 & z_1 \\ x_2 & y_2 & z_2 \end{bmatrix}.$$

We define the **cross product**,

$$\begin{aligned} \mathbf{u} \times \mathbf{v} &= (C_{11}, C_{12}, C_{13}) \\ &= \left(\begin{vmatrix} y_1 & z_1 \\ y_2 & z_2 \end{vmatrix}, -\begin{vmatrix} x_1 & z_1 \\ x_2 & z_2 \end{vmatrix}, \begin{vmatrix} x_1 & y_1 \\ x_2 & y_2 \end{vmatrix} \right) \\ &= ((y_1 z_2 - z_1 y_2), (z_1 x_2 - x_1 z_2), (x_1 y_2 - y_1 x_2)), \end{aligned}$$

and we have the property that $\mathbf{u} \times \mathbf{v}$ is a vector orthogonal to both $\mathbf{u}$ and $\mathbf{v}$. If $\mathbf{u}$ and $\mathbf{v}$ are equal, or if one is a multiple of the other, $\mathbf{u} \times \mathbf{v}$ will be the zero vector. Unlike the dot product which is defined in the plane, in 3-space and even in n-space for any positive integer n, the cross product, $\mathbf{u} \times \mathbf{v}$, is defined only in 3-space.

Example 7.7 *Let* $\mathbf{u} = (3, 1, 0)$ *and* $\mathbf{v} = (2, 1, 1)$. *Find a vector orthogonal to both* $\mathbf{u}$ *and* $\mathbf{v}$. *Solution: Let*

$$A = \begin{bmatrix} * & * & * \\ 3 & 1 & 0 \\ 2 & 1 & 1 \end{bmatrix}.$$

We have

$$C_{11} = \begin{vmatrix} 1 & 0 \\ 1 & 1 \end{vmatrix} = 1,$$

$$C_{12} = - \begin{vmatrix} 3 & 0 \\ 2 & 1 \end{vmatrix} = -3,$$

and

$$C_{13} = \begin{vmatrix} 3 & 1 \\ 2 & 1 \end{vmatrix} = 1.$$

Thus $\mathbf{u} \times \mathbf{v} = (1, -3, 1)$. *We check that this is orthogonal to both* $\mathbf{u}$ *and* $\mathbf{v}$ *by computing* $(1, -3, 1) \cdot (3, 1, 0)$ *and* $(2, 1, 1) \cdot (1, -3, 1)$, *and indeed both are zero.*

Example 7.8 *Find the equation of the plane through the origin* $(0, 0, 0)$ *and containing the vectors* $\mathbf{u} = (3, 1, 0)$ *and* $\mathbf{v} = (2, 1, 1)$. *Solution: If we know a normal to the plane and a point on the plane, we can write down the equation of the plane using 7.9. We are not given a normal, but we can find one by taking* $\mathbf{u} \times \mathbf{v}$. *We have already done this in the preceding example, so* $(1, -3, 1)$ *is a normal to the plane. The equation of the plane is thus* $(x - 0, y - 0, z - 0) \cdot (1, -3, 1) = 0$ *or* $x - 3y + z = 0$.

Remark **(1)** *We have used here the fact that if* $\mathbf{n}$ *is* $\perp$ *to two non-collinear vectors in a plane, then* $\mathbf{n}$ *is* $\perp$ *to every vector in the plane. This is because if* $\mathbf{u}$ *and* $\mathbf{v}$ *are noncollinear they are linearly independent, and any two linearly independent vectors in the plane are a basis. Thus if* $\mathbf{w}$ *is any other vector in the plane, then* $\mathbf{w} = a\mathbf{u} + b\mathbf{v}$ *and* $\mathbf{n} \cdot \mathbf{w} = \mathbf{n} \cdot (a\mathbf{u} + b\mathbf{v}) = a(\mathbf{n} \cdot \mathbf{u}) + b(\mathbf{n} \cdot \mathbf{v}) = 0$.

Remark **(2)** *Another way to find a vector* $\perp$ *to two given vectors is to solve a linear system of equations. We have* $(x, y, z) \perp (3, 1, 0)$ *if and only if* $(x, y, z) \cdot (3, 1, 0) = 0$ *or* $3x + y = 0$, *and* $(x, y, z) \perp (2, 1, 1)$ *if and only if* $(x, y, z) \cdot (2, 1, 1) = 0$ *or* $2x + y + z = 0$. *Eliminating* y *gives* $x - z = 0$, *or* $x = z$, *while the first equation gives* $y = -3x$. *Thus any triple of the form* $(x, -3x, x)$ *will be a solution. Chosing* $x = 1$, *we get* $(1, -3, 1)$. *It is not surprising that the answer is not unique, since any vector in this direction is* $\perp$ *to both* $\mathbf{u}$ *and* $\mathbf{v}$.

Remark **(3)** *In general if you have a set of vectors and wish to find a vector orthogonal to all of them, put your vectors as the rows of a matrix and find the null space of this matrix. Any vector which satisfies $A\mathbf{x} = \mathbf{O}$ is orthogonal to all the rows of A, in fact $nullsp(A) = (rowsp(A))^{\perp}$ as we have seen in Section 5.2, Lemma 5.2.*

Using the definitions of length, cross product and dot product, one can verify that if $\mathbf{u} = (x_1, y_1, z_1)$ and $\mathbf{v} = (x_2, y_2, z_2)$

$$|\mathbf{u} \times \mathbf{v}|^2 = |\mathbf{u}|^2|\mathbf{v}|^2 - (\mathbf{u} \cdot \mathbf{v})^2. \qquad (7.14)$$

Substituting $\mathbf{u} \cdot \mathbf{v} = |\mathbf{u}||\mathbf{v}| \cos\theta$ in 7.14 gives

$$\begin{aligned} |\mathbf{u} \times \mathbf{v}|^2 &= |\mathbf{u}|^2|\mathbf{v}|^2 - |\mathbf{u}|^2|\mathbf{v}|^2 \cos^2\theta \\ &= |\mathbf{u}|^2|\mathbf{v}|^2(1 - \cos^2\theta) \\ &= |\mathbf{u}|^2|\mathbf{v}|^2 \sin^2\theta. \end{aligned}$$

Thus we get

$$|\mathbf{u}x\mathbf{v}| = |\mathbf{u}||\mathbf{v}||\sin\theta|. \qquad (7.15)$$

We see from 7.15 that $|\mathbf{u} \times \mathbf{v}|$ gives the area of a parallelogram with edges $\mathbf{u}$ and $\mathbf{v}$ (Figure 7.16), since $height = |\mathbf{u}| \sin\theta$, and $base = |\mathbf{v}|$. The area of the triangle formed by $\mathbf{u}$, $\mathbf{v}$, and $\mathbf{v} - \mathbf{u}$ in Figure 7.16 is $\frac{1}{2}|\mathbf{u} \times \mathbf{v}|$.

The triple scalar product of $\mathbf{w}, \mathbf{u}$, and $\mathbf{v}$ is $\mathbf{w} \cdot (\mathbf{u} \times \mathbf{v})$. From our previous discussion it is clear that $\mathbf{w} \cdot (\mathbf{u} \times \mathbf{v})$ is the determinant of a matrix whose rows are $\mathbf{w}$, $\mathbf{u}$, and $\mathbf{v}$.

$$\mathbf{w} \cdot (\mathbf{u} \times \mathbf{v}) = \begin{vmatrix} \mathbf{w} \\ \mathbf{u} \\ \mathbf{v} \end{vmatrix}. \qquad (7.16)$$

From the properties of determinants we know the triple product is zero if any two are equal and exchanging two vectors in the triple product multiplies the result by (-1). For example

$$\mathbf{w} \cdot (\mathbf{u} \times \mathbf{v}) = -\mathbf{u} \cdot (\mathbf{w} \times \mathbf{v}) = \mathbf{u} \cdot (\mathbf{v} \times \mathbf{w}), \text{ etc.}$$

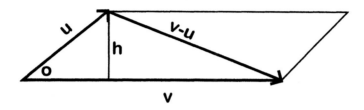

Figure 7.16

We also have

$$\mathbf{w} \cdot (\mathbf{u} \times \mathbf{v}) = |\mathbf{w}||\mathbf{u} \times \mathbf{v}| \cos \theta = |\mathbf{w}||\mathbf{u}||\mathbf{v}|| \sin \Phi| \cos \theta,$$

where θ is the angle between $\mathbf{w}$ and $(\mathbf{u} \times \mathbf{v})$, and Φ is the angle between $\mathbf{u}$ and $\mathbf{v}$. Taking absolute values we get

$$|\mathbf{w} \cdot (\mathbf{u} \times \mathbf{v})| = |\mathbf{w}||\mathbf{u}||\mathbf{v}|| \sin \Phi|| \cos \theta|. \tag{7.17}$$

From 7.17 and Figure 7.17 we see that $|\mathbf{w} \cdot (\mathbf{u} \times \mathbf{v})|$ gives the volume of a parallelapiped with edges $\mathbf{u}$, $\mathbf{v}$ and $\mathbf{w}$. *Height* $= |\mathbf{w}|| \cos \theta|$ and *area of base* $= |\mathbf{u}||\mathbf{v}|| \sin \Phi|$.Thus if A is a 3×3 matrix the determinant of A gives $\pm$ the volume of a parallelapiped with edges the rows of A.

If A is 2×2, the determinant of A gives $\pm$ the area of the parallelogram with edges the rows of A. To see this suppose that $A = \begin{bmatrix} x_1 & y_1 \\ x_2 & y_2 \end{bmatrix}$, where $\mathbf{u} = (x_1, y_1)$ and $\mathbf{v} = (x_2, y_2)$. Then we have

$$|A| = \begin{vmatrix} 0 & 0 & 1 \\ x_1 & y_1 & 0 \\ x_2 & y_2 & 0 \end{vmatrix},$$

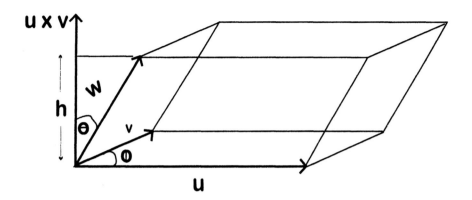

Figure 7.17

and this is± the volume of a parallelapiped with base a parallelogram in the XY plane and height equal to 1. Because of these results in the plane and 3-space, the determinant is sometimes used to extend the idea of volume to n-space.

Exercises

1. Find a vector $\perp$ *to* both $(1, -2, 4)$ and $(2, 5, 3)$.

2. Find the equation of a plane through the origin and containing the two vectors from problem 1.

3. Find the area of the parallelogram with edges $(1, -3, 2)$ and $(1, 1, 4)$.

4. Find the area of the triangle formed by the points $(1, 2, 1)$, $(3, 1, -4)$, and $(1, 0, 1)$.

5. Find the volume of the parallelapiped with edges $(1, 2, 1)$, $(3, 1, -4)$, and $(1, 0, 1)$.

Section 7.4 Projection

We define the projection of a vector **u** on a vector **v** by

$$proj_v \mathbf{u} = \left(\frac{\mathbf{u} \cdot \mathbf{v}}{\mathbf{v} \cdot \mathbf{v}} \right) \mathbf{v}. \tag{7.18}$$

Geometrically this gives the shadow that **u** casts on a line through **v** if the sun's rays are ⊥ to **v** (Figure 7.18). It is apparent from equations 7.18 and 7.6 that $proj_v\mathbf{u}$ is a vector in the direction of **v** (or the opposite direction if **u** · **v** is negative) and that $|proj_v\mathbf{u}| = |\mathbf{u}|| \cos \theta |$.

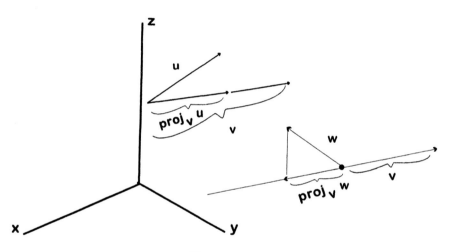

Figure 7.18

Example 7.9 *Find the projection of* **u** $= (1, 3, 2)$ *on* **v** $= (1, 1, 2)$ *and find the cosine of the angle between* **u** *and* **v**. *Solution:*

$$Proj_v \mathbf{u} = \left(\frac{(1, 3, 2) \cdot (1, 1, 2)}{(1, 1, 2) \cdot (1, 1, 2)} \right) (1, 1, 2) = \frac{8}{6}(1, 1, 2) = \left(\frac{4}{3}, \frac{4}{3}, \frac{8}{3} \right).$$

We have

$$\cos \theta = \frac{u \cdot v}{|u| \, |v|} = \frac{8}{\sqrt{14}\sqrt{6}} = \frac{4}{\sqrt{21}}.$$

The idea of projection can be used to find the distance of a point to a plane in 3-space or the distance of a point to a line in 2-space. Consider the plane with normal $\mathbf{n} = (a, b, c)$ and through the point $P_0 = (x_0, y_0, z_0)$. We have seen (equations 7.9 and 7.10) that the equation of this plane is given by $(x - x_0, y - y_0, z - z_0) \cdot (a, b, c) = 0$, and $(x - x_0, y - y_0, z - z_0) \cdot (a, b, c) = ax + by + cz + d$, where $d = -ax_0 - by_0 - cz_0$. The distance D of a point $P_1 = (x_1, y_1, z_1)$ to this plane is given by the length of the projection of $\overrightarrow{P_0 P_1}$ on $\mathbf{n}$ (Figure 7.19).

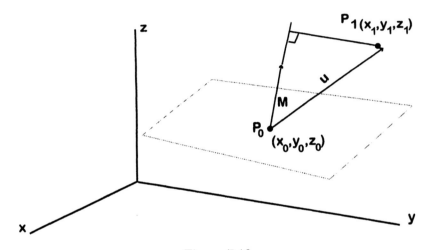

Figure 7.19

Let $\mathbf{u}$ equal the vector represented by $\overrightarrow{P_0 P_1}$. Then $\mathbf{u} = (x_1 - x_0, y_1 - y_0, z_1 - z_0)$. Thus

$$
\begin{aligned}
D &= |proj_{\mathbf{n}} \mathbf{u}| \\
&= \left| \frac{\mathbf{u} \cdot \mathbf{n}}{\mathbf{n} \cdot \mathbf{n}} \right| |\mathbf{n}| \\
&= \left| \frac{(x_1 - x_0, y_1 - y_0, z_1 - z_0) \cdot (a, b, c)}{(a, b, c) \cdot (a, b, c)} \right| |(a, b, c)| \\
&= \frac{|ax_1 + by_1 + cz_1 + d|}{\sqrt{a^2 + b^2 + c^2}},
\end{aligned}
$$

where $d = -ax_0 - by_0 - cz_0$, as before. Thus we see that the distance D from the point $P_1 = (x_1, y_1, z_1)$ to the plane with equation $ax + by + cz + d = 0$ is given by

$$D = \frac{|ax_1 + by_1 + cz_1 + d|}{\sqrt{a^2 + b^2 + c^2}}. \qquad (7.19)$$

The numerator is the value you get if you substitute the coordinates of P_1 for x, y, and z in the equation of the plane, and the denominator is the length of the normal. If $P_1 = (0, 0, 0)$, this gives the distance from the origin to the plane as

$$D = \frac{|d|}{\sqrt{a^2 + b^2 + c^2}}. \qquad (7.20)$$

The distance between two parallel planes can be found by finding the distance from a point on one plane to the other plane. Similarly the distance between a plane and a line parallel to that plane can be found by using any point on the line and finding the distance of that point to the plane.

Using the same methods in 2-space, one can show that the distance D from a point $P_1(x_1, y_1)$ to the line $ax + by + c = 0$ is given by

$$D = \frac{|ax_1 + by_1 + c|}{\sqrt{a^2 + b^2}}. \qquad (7.21)$$

Example 7.10 *Find the distance of the point $(3, 5, -1)$ to the plane $2x - 3y + 5z - 5 = 0$. Solution:*

$$D = \frac{2 \cdot 3 - 3 \cdot 5 + 5 \cdot (-1) - 5}{\sqrt{4 + 9 + 25}} = \frac{19}{\sqrt{38}}.$$

We define the projection of a vector $\mathbf{u}$ on a plane $ax + by + cz + d = 0$ to be $\mathbf{u} - proj_\mathbf{n}\mathbf{u}$, where $\mathbf{n}$ is a vector normal to the plane.

Example 7.11 *Find the projection on the plane given by $2x - 3y + z - 5 = 0$ of the vector $\mathbf{u} = (5, 4, 3)$. Solution: A normal to the plane is $\mathbf{n} = (2, -3, 1)$. The projection of $\mathbf{u}$ on $(2, -3, 1)$ is*

$$proj_\mathbf{n}\mathbf{u} = \left(\frac{(5, 4, 3) \cdot (2, -3, 1)}{(2, -3, 1) \cdot (2, -3, 1)} \right) (2, -3, 1) = \frac{1}{14}(2, -3, 1).$$

Thus the projection of **u** *on the plane is given by*

$$\mathbf{u} - proj_n\mathbf{u} = (5,4,3) - \left(\frac{1}{7}, -\frac{3}{14}, \frac{1}{14}\right) = \left(\frac{34}{7}, \frac{59}{14}, \frac{41}{14}\right).$$

Example 7.12 *Find the projection of* $(5,4,3)$ *on the plane* π *given by* $x + y + z = 0$. *Solution: A normal to the plane* π *is* $\mathbf{n} = (1,1,1)$. *Let* $proj_\pi(5,4,3)$ *denote the projection of* $(5,4,3)$ *on the plane* π. *We have*

$$proj_\pi(5,4,3) = (5,4,3) - \left(\frac{(5,4,3)\cdot(1,1,1)}{(1,1,1)\cdot(1,1,1)}\right)(1,1,1)$$

$$= (5,4,3) - \frac{12}{3}(1,1,1)$$

$$= (5,4,3) - (4,4,4)$$

$$= (1,0,-1).$$

Remark   *The projection of a vector* **u** *on a plane is the same as the projection of* **u** *on any plane parallel to the given plane. We can always take an arrow representing* **u** *starting at a point on the plane we are given.*

We can also use projections to find the point in the plane π given by $ax + by + cz + d = 0$, nearest to a point $P = (x_1, y_1, z_1)$. Suppose $Q = (x_0, y_0, z_0)$ is a point in the plane π. In Figure 7.20, $\overrightarrow{RP}$ is $proj_n\overrightarrow{QP}$, $\overrightarrow{QR}$ is $proj_\pi QP$, and R is the closest point in the plane π to the point P.

We have the vector equation

$$\overrightarrow{OR} = \overrightarrow{OP} + \overrightarrow{PR},$$

or

$$\overrightarrow{OR} = \overrightarrow{OP} - proj_n\overrightarrow{QP}.$$

Thus if $R = (x_R, y_R, z_R)$, we have

$$(x_R, y_R, z_R) = (x_1, y_1, z_1) - \left(\frac{(x_1 - x_0, y_1 - y_0, z_1 - z_0)\cdot(a,b,c)}{(a,b,c)\cdot(a,b,c)}\right)(a,b,c).$$

$$(7.22)$$

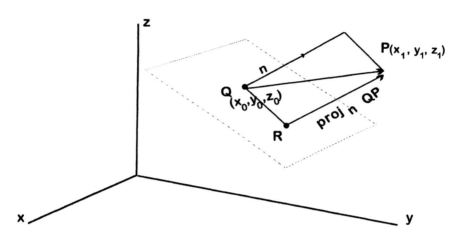

Figure 7.20

Since (x_0, y_0, z_0) is on the plane, we have $d = -ax_0 - by_0 - cz_0$, and thus 7.22 simplifies to

$$(x_R, y_R, z_R) = (x_1, y_1, z_1) - \left(\frac{ax_1 + by_1 + cz_1 + d}{a^2 + b^2 + c^2} \right) (a, b, c). \qquad (7.23)$$

In equation 7.23 x_0, y_0, and z_0 have disappeared and we have a formula in terms of the equation of the plane π and the coordinates of P.

Remark   *The closest point in the plane π to the point (x_1, y_1, z_1) and the projection on the plane π of the vector (x_1, y_1, z_1) are the same triple if and only if the plane contains the origin. To see this, take $(x_0, y_0, z_0) = (0, 0, 0)$ in equation 7.22.*

Example 7.13 *Find the point in the plane π given by $3x + y + 2z + 3 = 0$ that is closest to the point $P = (3, 2, -3)$. Solution: If R is the closest point, we use 7.23 to get*

$$(x_R, y_R, z_R) = (3, 2, -3) - \left(\frac{3 \cdot 3 + 2 \cdot 1 + 2 \cdot (-3) + 3}{9 + 1 + 4} \right) (3, 1, 2)$$

$$= (3,2,-3) - \frac{8}{14}(3,1,2)$$

$$= (3,2,-3) - \frac{4}{7}(3,1,2) = \left(\frac{9}{7},\frac{10}{7},-\frac{29}{7}\right).$$

One can also solve this problem by finding the equation of the line in the direction of a normal to the plane, through the given point and then finding the intersection of this line with the plane. In the above example a normal to the plane is $(3,1,2)$. A line in the direction $(3,1,2)$ through the point $(3,2,-3)$ has parametric equations $x = 3t+3$, $y = t+2$, $z = 2t-3$. Substituting this into the equation of the plane gives $3(3t+t)+(t+2)+2(2t-3)+3 = 0$. Solving for t gives $t = -4/7$. Using this t in the equation of the line gives the point $\left(\frac{9}{7},\frac{10}{7},-\frac{29}{7}\right)$ as the point of intersection of this line and the plane.

Now we consider the distance from a point $P_1 = (x_1,y_1,z_1)$ to a line L with parametric equations $x = at+x_0$, $y = bt+y_0$, $z = ct+z_0$, in other words a line in the direction of (a,b,c) and through the point $Q = (x_0,y_0,z_0)$. First we find the point on the line L closest to P_1. Let R be the point on L closest to P_1.

Looking at Figure 7.21, we have the vector equation

$$\overrightarrow{OR} = \overrightarrow{OQ} + \overrightarrow{QR},$$

which is

$$(x_R,y_R,z_R) = (x_0,y_0,z_0)+\left(\frac{(x_1-x_0,y_1-y_0,z_1-z_0)\cdot(a,b,c)}{(a,b,c)\cdot(a,b,c)}\right)(a,b,c).$$

$$(7.24)$$

Example 7.14 *Find the point on the line L given by $x = -t+2$, $y = t+1$, $z = 2t-1$, closest to the point $P = (3,0,1)$, and find the distance from P to L. Solution: We have $(x_0,y_0,z_0) = (2,1,-1)$, $(a,b,c) = (-1,1,2)$, and $(x_1,y_1,z_1) = (3,0,1)$. Using 7.24 gives*

$$(x_R,y_R,z_R) = (2,1,-1)+\left(\frac{(1,-1,2)\cdot(-1,1,2)}{1+1+4}\right)(-1,1,2)$$

$$= (2,1,-1)+\frac{1}{3}(-1,1,2) = \left(\frac{5}{3},\frac{4}{3},-\frac{1}{3}\right).$$

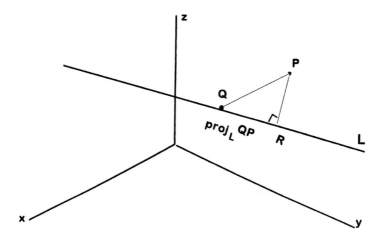

Figure 7.21

This is the point on L closest to the point $(3,0,1)$. *To find the distance of this point to L, we just compute the distance between* $(3,0,1)$ *and* $\left(\frac{5}{3}, \frac{4}{3}, -\frac{1}{3}\right)$, *which is* $\frac{4}{\sqrt{3}}$. *One could also have solved this problem by putting a plane* $\perp$ *to L through the point* (x_1, y_1, z_1) *and intersecting this plane with L. A plane* $\perp$ *to L would have* $(-1,1,2)$ *as a normal, and so a plane* $\perp$ *to L and through* $P_1 = (3,0,1)$ *would have equation* $(-1,1,2) \cdot (x-3, y-0, z-1) = 0$, *or* $-x+y+2z+1 = 0$. *Intersecting this plane with L we get* $-(-t+2) + (t+1) + 2(2t-1) + 1 = 0$, *or* $6t - 2 = 0$, *which gives* $t = \frac{1}{3}$. *Substituting this in the equations for L gives* $x = \frac{5}{3}$, $y = \frac{4}{3}$, $z = -\frac{1}{3}$, *as the closest point on L to* P_1. *This checks with our previous result.*

Remark   *The closest point on the line L to the point* (x_1, y_1, z_1) *and the projection of the vector* (x_1, y_1, z_1) *on L are the same triple if and only L goes through the origin. In this case we can take* $(x_0, y_0, z_0) = (0,0,0)$ *in equation 7.24.*

Exercises

1. Find $\text{proj}_v u$ if u and v are given by:

    (a) $u = (1, 0, 5)$ and $v = (1, 2, 1)$

    (b) $u = (1, 0, 5)$ and $v = (2, 4, 2)$

    (c) $u = (2, 0, 10)$ and $v = (1, 2, 1)$

    (d) $u = (1, 2, 1)$ and $v = (1, 0, 5)$

2. Find the distance from the origin to the plane $3x + 2y - z + 5 = 0$.

3. Find the distance from the plane in problem 2 to the point $(1, 1, 1)$.

4. In 2-space find the distance from the origin to the line $2x - 4y + 5 = 0$

5. In 2-space find the distance of the point $(-3, 7)$ to the line $2x - 4y + 5 = 0$.

6. Find the projection of the vector $(1, -2, 3)$ on the plane $x - y + z = 0$.

7. Find the point on the plane in problem 6 that is closest to the point $(1, -2, 3)$.

8. Find the projection of the vector $(1, -2, 3)$ on the plane $x - y + z + 3 = 0$.

9. Find the closest point in the plane in problem 8 to the point $(1, -2, 3)$.

10. Find the projection of the vector $(2, 1, -3)$ on the plane $2x - 5y + z = 0$.

11. Find the distance of the point $(2, 1, -3)$ to the plane $2x - 5y + z = 0$.

12. Find the distance from the plane $3x - 2y + z - 6 = 0$ to the point $(1, 4, -1)$.

13. Find the point on the line $x = 2t$, $y = -t$, $z = 4t$ that is closest to the point $(-4, 8, 1)$, and find the distance from the line to this point.

14. Find the point on the line $x = 2t + 1$, $y = -t + 3$, $z = 4t$ which is closest to the point $(-4, 8, 1)$, and find the distance of the line to this point.

15. Find the point on the plane $x - y + z + 3 = 0$ that is closest to the origin.

16. Find the point on the line in problem 14 that is closest to the origin.

Section 7.5 More on Planes and Lines

We know how to find the equation of a plane given a normal to the plane and a point on the plane. Now we discuss how to find the equation of a plane containing 3 given noncollinear points $P = (p_1, p_2, p_3)$, $Q = (q_1, q_2, q_3)$, and $R = (r_1, r_2, r_3)$. The points are noncollinear if the vector represented by $\overrightarrow{PQ}$ is not a multiple of $\overrightarrow{PR}$, i.e. if

$$(q_1 - p_1, q_2 - p_2, q_3 - p_3) \neq k(r_1 - p_1, r_2 - p_2, r_3 - p_3).$$

When they are noncollinear three points determine a plane. To find the equation of this plane we could take the cross product of the vectors represented by $\overrightarrow{PQ}$ and $\overrightarrow{PR}$ which would give a normal to the plane, and then using this normal and any one of the three given points we could write the equation of the plane. However we can use the determinant of a 4×4 matrix to write down the equation immediately as

$$\begin{vmatrix} x & y & z & 1 \\ p_1 & p_2 & p_3 & 1 \\ q_1 & q_2 & q_3 & 1 \\ r_1 & r_2 & r_3 & 1 \end{vmatrix} = 0 \qquad (7.25)$$

Expanding by minors around the first row we see that this is the equation of a plane. Substituting $x = p_1$, $y = p_2$, $z = p_3$, we have a matrix with two rows equal, so the determinant is zero. Thus P is on the plane described by equation 7.25. Similarly Q and R can be shown to be on the plane. Thus equation 7.25 is the equation of the plane containing P, Q, and R.

Remark *If the three given points are collinear, the determinant in equation 7.25 is zero for all values of x, y and z, and so equation 7.25 does not give the equation of a plane.*

Example 7.15 *Find the equation of the plane containing the points $(1,1,3)$, $(0,2,1)$, and $(3,1,4)$. Solution:*

$$\begin{vmatrix} x & y & z & 1 \\ 1 & 1 & 3 & 1 \\ 0 & 2 & 1 & 1 \\ 3 & 1 & 4 & 1 \end{vmatrix} = 0.$$

Expanding gives

$$x\begin{vmatrix} 1 & 3 & 1 \\ 2 & 1 & 1 \\ 1 & 4 & 1 \end{vmatrix} - y\begin{vmatrix} 1 & 3 & 1 \\ 0 & 1 & 1 \\ 3 & 4 & 1 \end{vmatrix} + z\begin{vmatrix} 1 & 1 & 1 \\ 0 & 2 & 1 \\ 3 & 1 & 1 \end{vmatrix} - 1\begin{vmatrix} 1 & 1 & 3 \\ 0 & 2 & 1 \\ 3 & 1 & 4 \end{vmatrix} = 0,$$

or

$$x - 3y - 2z + 8 = 0.$$

It is easy to check that the given points satisfy this equation.

Remark *We have used here a 4 × 4 determinant. The formula we have used for expanding it is $|A| = a_{11}C_{11} + a_{12}C_{12} + a_{13}C_{13} + a_{14}C_{14}$, where $C_{1k} = (-1)^{1+k}|M_{1k}|$, and M_{1k} is the 3 × 3 matrix obtained by deleting the first row and k^{th} column of A.*

Remark *An argument similar to that given above can be used to show that the equation of a line in 2-space through the points $P = (p_1, p_2)$ and $Q = (q_1, q_2)$ is given by*

$$\begin{vmatrix} x & y & 1 \\ p_1 & p_2 & 1 \\ q_1 & q_2 & 1 \end{vmatrix} = 0. \tag{7.26}$$

Example 7.16 *Find the equation of the line joining the points* $(3, 5)$ *and* $(-2, 4)$. *Solution:*

$$\begin{vmatrix} x & y & 1 \\ 3 & 5 & 1 \\ -2 & 4 & 1 \end{vmatrix} = 0.$$

Expanding gives

$$x \begin{vmatrix} 5 & 1 \\ 4 & 1 \end{vmatrix} - y \begin{vmatrix} 3 & 1 \\ -2 & 1 \end{vmatrix} + 1 \begin{vmatrix} 3 & 5 \\ -2 & 4 \end{vmatrix} = 0,$$

or

$$x - 5y + 22 = 0.$$

Suppose you want to find the equation of a plane passing through a given line L and a given point $P = (x_0, y_0, z_0)$, with P not on L. As we have seen, a line can be described as the intersection of two planes.

Example 7.17 *Let L be the line of intersection of planes* $3x - 4y + 2z = 8$ *and* $x + 2y + 3z = 6$. *Find the equation of a plane* π *containing L and the point* $P = (1, 0, 1)$. *We could solve the equations of the planes simultaneously, find two points on the line of intersection, and use these two points and* $(1, 0, 1)$ *to write down the equation of the plane by the method of Example 7.15, but this would be the long way around. Instead we observe that*

$$r(3x - 4y + 2z - 8) + s(x + 2y + 3z - 6) = 0 \qquad (7.27)$$

is the equation of a plane for any choice of r *and* s. *All of these planes contain the line L, since points on L satisfy both* $(3x - 4y + 2z - 8) = 0$ *and* $(x + 2y + 3z - 6) = 0$. *Thus equation 7.27 describes a family of planes, all containing L. Now all we have to do is choose* r *and* s *so that the plane goes through the point* $P = (1, 0, 1)$. *We substitute* $x = 1$, $y = 0$, $z = 1$ *in equation 7.27 and get* $-3r - 2s = 0$, *or* $s = -\frac{3}{2}r$. *Consequently for any* r,

$$r(3x - 4y + 2z - 8) - (3/2)r(x + 2y + 3z - 6) = 0 \qquad (7.28)$$

*is the equation of a plane containing L and P. Multiplying equation
7.28 by $\frac{2}{r}$, we get*

$$2(3x - 4y + 2z - 8) - 3(x + 2y + 3z - 6) = 0,$$

or

$$3x - 14y - 5z + 2 = 0.$$

This is the equation of the plane we are looking for.

We saw earlier that to determine whether or not three points P, Q, and R are collinear we simply form the vectors $\overrightarrow{PQ}$ and $\overrightarrow{PR}$ and check whether or not one of these is a multiple of the other. This is the same whether you are in 2-space or 3-space, and it is easy to tell by inspection if one vector is a multiple of another. Of course we could find the equation of the line determined by any two of the points and then check to see if the third is on it, but the first way is quicker. In 3-space we may also desire to know whether or not 4 given points lie in a plane (or even possibly on a line). A set of points is coplanar if there is a plane that contains all the points. If any three of them are collinear, then this line and the fourth point determine a plane. If no three are collinear, we can find the equation of the plane determined by three of them and then check whether or not the fourth point is on it. However there is a shorter way. The points P, Q, R, and S are coplanar if and only if the vectors $\overrightarrow{PQ}$, $\overrightarrow{PR}$, and $\overrightarrow{PS}$ are coplanar, that is if one of these vectors is a linear combination of the other two. In 3-space the vectors $\overrightarrow{PQ}$, $\overrightarrow{PR}$, and $\overrightarrow{PS}$ are triples and they are coplanar if and only if

$$\begin{vmatrix} \overrightarrow{PQ} \\ \overrightarrow{PR} \\ \overrightarrow{PS} \end{vmatrix} = 0.$$

Example 7.18 *Are the points $(1,1,3)$, $(0,2,1)$, $(3,1,4)$, and $(2,4,-1)$ coplanar? Solution: Assume these points are named P, Q, R, and S respectively. Subtract the coordinates of P from each of the other three to get the vectors $\overrightarrow{PQ}$, $\overrightarrow{PR}$, and $\overrightarrow{PS}$. We get $\overrightarrow{PQ} = (-1,1,-2)$, $\overrightarrow{PR} = (2,0,1)$, and $\overrightarrow{PS} = (1,3,-4)$. Putting these vectors as the rows* *of a matrix A and taking the determinant we get*

$$
\begin{vmatrix} \overrightarrow{PQ} \\ \overrightarrow{PR} \\ \overrightarrow{PS} \end{vmatrix} = \begin{vmatrix} -1 & 1 & -2 \\ 2 & 0 & 1 \\ 1 & 3 & -4 \end{vmatrix}
$$

$$
= -1 \begin{vmatrix} 0 & 1 \\ 3 & -4 \end{vmatrix} - 1 \begin{vmatrix} 2 & 1 \\ 1 & -4 \end{vmatrix} + (-2) \begin{vmatrix} 2 & 0 \\ 1 & 3 \end{vmatrix}
$$

$$
= 3 + 9 - 12 = 0,
$$

thus the points are coplanar.

Example 7.19 *Are the points of Example 7.18 collinear? Are any 3 of them collinear: The 4 points are not collinear. The points P, Q, and R do not lie on a line since $\overrightarrow{PQ} = (-1, 1, -2)$ is not a multiple of $\overrightarrow{PR} = (2, 0, 1)$. Thus it is impossible that all 4 points lie on a line. The points P, Q, and S do not lie on a line since $\overrightarrow{PS} = (1, 3, -4)$ is not a multiple of $\overrightarrow{PQ} = (-1, 1, -2)$. It is still possible that the points Q, R, and S lie on a line. To see if this is so we form the vectors $\overrightarrow{QR} = (3, -1, 3)$ and $\overrightarrow{QS} = (2, 2, -2)$. $\overrightarrow{QR}$ is not a multiple of $\overrightarrow{QS}$, so no 3 of the 4 points are collinear.*

Example 7.20 *Determine if the points $(2, 1, 2)$, $(3, 0, 1)$, $(4, 1, 1)$, and $(1, 2, 4)$ are coplanar. Solution: As above let the points be named P, Q, R, and S respectively. Then form the vectors $\overrightarrow{PQ} = (1, -1, -1)$, $\overrightarrow{PR} = (2, 0, -1)$, and $\overrightarrow{PS} = (-1, 1, 2)$. We have*

$$
\begin{vmatrix} \overrightarrow{PQ} \\ \overrightarrow{PR} \\ \overrightarrow{PS} \end{vmatrix} = \begin{vmatrix} 1 & -1 & -1 \\ 2 & 0 & -1 \\ -1 & 1 & 2 \end{vmatrix} = 2,
$$

so the points are not coplanar.

Remark *Since the 4 points in Example 7.20 are not coplanar, no 3 of then can be collinear.*

In an earlier chapter we have seen that the rank of a matrix is the number of nonzero rows in the RREF of the matrix. We can use this to determine whether a set of points in R^3 is collinear, coplanar but not collinear, or not even coplanar. Given n points P_1, P_2, $\cdots$, P_n, form the $n-1$ vectors $\overrightarrow{P_1P_2}$, $\overrightarrow{P_1P_3}$, $\cdots$, $\overrightarrow{P_1P_n}$. Put these as the rows (or columns) of a matrix. The rank of this matrix tells you the dimension of the smallest subspace containing all the vectors. If the rank is 1, the vectors (and hence the points they come from) are collinear. If the rank is 2, the vectors are coplanar but not collinear. If the rank is 3, the vectors are not coplanar.

Exercises

1. Find the equation of the plane through the points $(1, 2, -1)$, $(3, 1, 0)$, and $(1, 1, 5)$.

2. Are the points $(1, 2, -1)$, $(3, 1, 0)$, $(1, 1, 5)$, and $(2, 4, 2)$ coplanar?

3. In each part determine if the 3 points are collinear and if they are, find parametric equations describing the line.

   (a) $(1, 2, 2)$, $(1, 3, 2)$, and $(1, 1, 2)$.
   (b) $(1, 3, 0)$, $(2, 1, 1)$, and $(4.0, 1)$.

4. Are the points $(1, 3, 0)$, $(2, 1, 1)$, $(4, 0, 1)$, and $(-1, 7, 1)$ coplanar?

5. Find the equation of the plane containing the line of intersection of the planes $4x - 2y - 3z + 5 = 0$, and $-3x + 2y - z + 3 = 0$, and the point $(1, 1, 1)$.

6. Find the equation of the plane containing the line L given by $x = 3t + 1$, $y = t - 2$, $z = 2t$ and the point $(-1, 2, 1)$.

Section 7.6 Rotations and Reflections

In 2-space we sometimes call a unit vector in the X direction $\mathbf{i}$, and a unit vector in the Y direction $\mathbf{j}$. Thus $\mathbf{i} = \begin{bmatrix} 1 \\ 0 \end{bmatrix}$ and $\mathbf{j} = \begin{bmatrix} 0 \\ 1 \end{bmatrix}$. If we rotate the plane by an angle θ, **i** moves to **i'** and **j** moves to **j'** (Figure 7.22), and we have $\mathbf{i'} = \begin{bmatrix} \cos\theta \\ \sin\theta \end{bmatrix}$ and $\mathbf{j'} = \begin{bmatrix} -\sin\theta \\ \cos\theta \end{bmatrix}$.

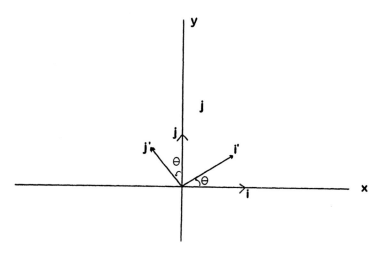

Figure 7.22

If we let $P = \begin{bmatrix} \cos\theta & -\sin\theta \\ \sin\theta & \cos\theta \end{bmatrix}$, then

$$P \begin{bmatrix} a \\ b \end{bmatrix} = a \begin{bmatrix} \cos\theta \\ \sin\theta \end{bmatrix} + b \begin{bmatrix} -\sin\theta \\ \cos\theta \end{bmatrix} = a\mathbf{i'} + b\mathbf{j'}.$$

Multiplying the vector $\begin{bmatrix} a \\ b \end{bmatrix}$ by P rotates $\begin{bmatrix} a \\ b \end{bmatrix}$ through an angle θ. P is called the matrix of the rotation.

If $P = \begin{bmatrix} 0 & 1 \\ 1 & 0 \end{bmatrix}$, then

$$P \begin{bmatrix} a \\ b \end{bmatrix} = \begin{bmatrix} b \\ a \end{bmatrix},$$

so P is the matrix of a reflection about the line $x = y$.

If $P = \begin{bmatrix} -1 & 0 \\ 0 & 1 \end{bmatrix}$, then $P \begin{bmatrix} a \\ b \end{bmatrix} = \begin{bmatrix} -a \\ b \end{bmatrix}$, and P is the matrix of a reflection about the y axis.

The rotation matrix

$$P = \begin{bmatrix} \cos\theta & -\sin\theta \\ \sin\theta & \cos\theta \end{bmatrix}$$

can be viewed in two ways. First we can view it as above as a mapping $\begin{bmatrix} a \\ b \end{bmatrix} \to P \begin{bmatrix} a \\ b \end{bmatrix}$, which rotates each vector by the angle θ. We can also view it as a change of basis. The columns of P are $\mathbf{i}'$ and $\mathbf{j}'$. Suppose that the coordinates of a vector $\mathbf{v} = \begin{bmatrix} x \\ y \end{bmatrix}$ with respect to the basis $\{\mathbf{i}', \mathbf{j}'\}$ are x' and y'. Then $\begin{bmatrix} x \\ y \end{bmatrix} = x'\mathbf{i}' + y'\mathbf{j}'$, which we can write as $\begin{bmatrix} x \\ y \end{bmatrix} = P \begin{bmatrix} x' \\ y' \end{bmatrix}$. Thus

$$\begin{bmatrix} x \\ y \end{bmatrix} = \begin{bmatrix} \mathbf{i}' & \mathbf{j}' \end{bmatrix} \begin{bmatrix} x' \\ y' \end{bmatrix} = \begin{bmatrix} \cos\theta & -\sin\theta \\ \sin\theta & \cos\theta \end{bmatrix} \begin{bmatrix} x' \\ y' \end{bmatrix}. \qquad (7.29)$$

This gives us an easy way to state and remember the rotation of axes equations.

A matrix P is an orthogonal matrix if $PP^T = I$ (Definition 4.14, Section 4.2). It is easy to check that the matrix of a rotation is an orthogonal matrix. The matrices of reflections that we have introduced above are also orthogonal matrices. We now show that in 2-space every orthogonal matrix is either a rotation or a reflection. If P is an orthogonal matrix then $|P| = \pm 1$ (Exercise 5, Section 4.2).

Theorem 7.1 *If P is a 2×2 orthogonal matrix then P is a rotation or a rotation plus a reflection. If $|P| = 1$, P is a pure rotation.*

Proof.   Since P is orthogonal, $PP^T = I$. Assume $|P| = 1$. Let $P = \begin{bmatrix} a & b \\ c & d \end{bmatrix}$, then $P^{-1} = \begin{bmatrix} d & -b \\ -c & a \end{bmatrix}$ (using the formula in Procedure 2.7). Since P is orthogonal we have $P^{-1} = P^T$ which gives

$$\begin{bmatrix} d & -b \\ -c & a \end{bmatrix} = \begin{bmatrix} a & c \\ b & d \end{bmatrix}.$$

Thus $a = d$ and $b = -c$. We also are assuming that $|A| = 1$, so $ad - bc = 1$. This gives $a^2 + b^2 = 1$. Let $a = \cos\theta$. Then $b = \pm\sin\theta$, and we have

$$P = \begin{bmatrix} \cos\theta & \sin\theta \\ -\sin\theta & \cos\theta \end{bmatrix},$$

or

$$P = \begin{bmatrix} \cos\theta & -\sin\theta \\ \sin\theta & \cos\theta \end{bmatrix}.$$

In the second case P is a rotation through an angle θ. In the first case P is a rotation through $-\theta$. If $|P| = -1$, then let

$$P' = \begin{bmatrix} 0 & 1 \\ 1 & 0 \end{bmatrix} P.$$

Then $|P'| = 1$, since P' comes from P by interchanging two rows. But now

$$P = \begin{bmatrix} 0 & 1 \\ 1 & 0 \end{bmatrix}^{-1} P' = \begin{bmatrix} 0 & 1 \\ 1 & 0 \end{bmatrix} P',$$

so P is a pure rotation P' followed by a reflection. ∎

In Chapter 4 we define an eigenvalue of a matrix A to be a number c for which there exists a nonzero vector $\mathbf{v}$ such that $A\mathbf{v} = c\mathbf{v}$. This means that under the mapping $\mathbf{v} \to A\mathbf{v}$, the vector $\mathbf{v}$ goes to a vector in the same direction as itself. It is clear geometrically that a rotation through an angle θ changes the direction of every vector (if $\theta \neq 0$). If $\theta = 180°$, then each vector $\mathbf{v}$ goes to $-\mathbf{v}$, so the rotation by $180°$ has -1 as its only eigenvalue, and every nonzero vector is an eigenvector for -1. If $\theta = 0$, then $\mathbf{v}$ goes to $\mathbf{v}$, the only eigenvalue is 1, and every nonzero vector is an eigenvector for 1. If $\theta \neq 0$, and $\theta \neq 180°$, then the rotation has no real eigenvalues and no real eigenvectors. Now consider the second case in the proof of Theorem 7.1. Here we have a rotation followed by a reflection, $P = \begin{bmatrix} 0 & 1 \\ 1 & 0 \end{bmatrix} P'$, with P' a rotation, so

$$P = \begin{bmatrix} 0 & 1 \\ 1 & 0 \end{bmatrix} \begin{bmatrix} \cos\theta & -\sin\theta \\ \sin\theta & \cos\theta \end{bmatrix} = \begin{bmatrix} \sin\theta & \cos\theta \\ \cos\theta & -\sin\theta \end{bmatrix}.$$

We see that the characteristic equation of P is $k^2 - 1 = 0$, so the eigenvalues of P are 1 and -1. Let $\mathbf{v}$ be an eigenvector for 1 and $\mathbf{u}$

an eigenvector for -1. Since P is symmetric we have $\mathbf{u} \perp \mathbf{v}$ (Lemma 4.7, Section 4.3). For vectors $\mathbf{w}$ in the direction of $\mathbf{v}$, $P\mathbf{w} = \mathbf{w}$. For vectors $\mathbf{t}$ orthogonal to $\mathbf{v}$, $P\mathbf{t} = -\mathbf{t}$. Thus we see that P is a reflection around the line through the origin in the direction of $\mathbf{v}$. So actually a rotation followed by a reflection can also be done by a reflection. We can improve on Theorem 7.1 and say that:

Corollary 1 *Every 2×2 orthogonal matrix represents either a pure rotation or a pure reflection.*

One can tell which it is by the eigenvalues. If there are no real eigenvalues, it is a rotation by an angle θ, $\theta \neq 0$, $\theta \neq 180°$. If the only eigenvalue is 1, it is the rotation by $0°$. If the only eigenvalue is -1, it is the rotation by $180°$. If the eigenvalues are 1 and -1, it is a reflection about a line through the origin, and an eigenvector for 1 gives the direction of this line. This exhausts all possibilities, since the only possible real eigenvalues for an orthogonal matrix are 1 and -1 (Problem 6, Section 4.2).

When we write the general quadratic $ax^2+bxy+cy^2+dx+ey+f = 0$ in matrix form as

$$\begin{bmatrix} x & y \end{bmatrix} \begin{bmatrix} a & \frac{b}{2} \\ \frac{b}{2} & c \end{bmatrix} \begin{bmatrix} x \\ y \end{bmatrix} + \begin{bmatrix} d & e \end{bmatrix} \begin{bmatrix} x \\ y \end{bmatrix} + \begin{bmatrix} f \end{bmatrix} = \begin{bmatrix} 0 \end{bmatrix}$$

and do the substitution $\begin{bmatrix} x \\ y \end{bmatrix} = P \begin{bmatrix} x' \\ y' \end{bmatrix}$, where P is an orthogonal matrix with $|P| = 1$, we are finding the equation of the conic with respect to the rotated coordinate system $\mathbf{i}', \mathbf{j}'$. Recall that if A is symmetric, we can always find an orthogonal P such that $P^{-1}AP$ is diagonal. It is necessary and sufficient that the columns of P be an orthonormal set of eigenvectors of A (Procedure 4.2, Section 4.3). We can always chose P so that $|P| = 1$, since if it is -1, we can exchange the order of the columns. Finding an orthogonal P with $|P| = 1$ which diagoinalizes $\begin{bmatrix} a & \frac{b}{2} \\ \frac{b}{2} & c \end{bmatrix}$ is equivalent to finding an angle through which to rotate the axes that will eliminate the xy term in the equation of the conic.

When we consider quadric surfaces (Section 4.3) we have a 3×3 symmetric matrix A and a 3×3 orthogonal matrix P such that $P^TAP =$

D, where D is a diagonal matrix. How can we describe geometrically the mapping $\mathbf{w} \to P\mathbf{w}$? Since P is 3×3, its characteristic equation is cubic, and so must have at least one real root, which must be ± 1. If it is -1, we can change to the matrix $P' = -P$. Then P' will also be orthogonal and $P'^T A P' = P^T A P = D$. If $\mathbf{v}$ is an eigenvector of P for the eigenvalue -1, $\mathbf{v}$ will be an eigenvector of P' for the eigenvalue 1, since $P'\mathbf{v} = -P\mathbf{v} = \mathbf{v}$. Thus we can diagonalize A with an orthogonal matrix P' which has 1 as an eigenvalue. Now the mapping $\mathbf{w} \to P'\mathbf{w}$ fixes the vector $\mathbf{v}$ and induces a $1-1$ length and angle preserving mapping of the plane orthogonal to $\mathbf{v}$ onto itself. By the corollary to Theorem 7.1, this is a rotation or reflection of this plane. If it is a rotation of the plane, through an angle $\theta \neq 0$ or $180°$, there are no more real eigenvalues, and no real eigenvectors in this plane, so 1 is the only real eigenvector and the mapping is a rotation of all of 3-space around the vector $\mathbf{v}$. This case is characterized by P' having 1 as an eigenvalue of geometric multiplicity 1, and no other real eigenvalues. If the plane orthogonal to $\mathbf{v}$ is rotated by $180°$, then -1 is also an eigenvalue. This is a rotation of 3-space around $\mathbf{v}$ by an angle of $180°$ and is characterized by P' having eigenvalues 1 of multiplicity 1 and -1 of multiplicity 2. If P' induces a reflection in the plane orthogonal to $\mathbf{v}$, then there is another eigenvector $\mathbf{u}$ for 1 in this plane, and P' now fixes the plane spanned by $\mathbf{v}$ and $\mathbf{u}$. In this case P' is a reflection of 3-space through the plane spanned by $\mathbf{v}$ and $\mathbf{u}$. This case is characterized by P' having 1 as an eigenvalue of multiplicity 2 and -1 as an eigenvalue of multiplicity 1.

Example 7.21

$$P = \begin{bmatrix} \frac{8}{9} & -\frac{1}{9} & \frac{4}{9} \\ -\frac{4}{9} & -\frac{4}{9} & \frac{7}{9} \\ \frac{1}{9} & -\frac{8}{9} & -\frac{4}{9} \end{bmatrix}$$

is an orthogonal 3×3 matrix. Describe geometrically the mapping $\mathbf{w} \to P\mathbf{w}$*. Solution: We know that either 1 or -1 must be an eigenvalue of P. To see if 1 is an eigenvalue, we consider*

$$P - I = \begin{bmatrix} \frac{8}{9}-1 & -\frac{1}{9} & \frac{4}{9} \\ -\frac{4}{9} & -\frac{4}{9}-1 & \frac{7}{9} \\ \frac{1}{9} & -\frac{8}{9} & -\frac{4}{9}-1 \end{bmatrix} = \begin{bmatrix} -\frac{1}{9} & -\frac{1}{9} & \frac{4}{9} \\ -\frac{4}{9} & -\frac{13}{9} & \frac{7}{9} \\ \frac{1}{9} & -\frac{8}{9} & -\frac{13}{9} \end{bmatrix}.$$

The RREF of this matrix is

$$RREF(P - I) = \begin{bmatrix} 1 & 0 & -5 \\ 0 & 1 & 1 \\ 0 & 0 & 0 \end{bmatrix}.$$

This matrix has rank two. From this we can see that 1 *is an eigenvalue of geometric multiplicity 1, and* $\begin{bmatrix} 5 \\ -1 \\ 1 \end{bmatrix}$ *is an eigenvector for* 1. *To see if* -1 *is an eigenvalue of* P *we consider*

$$P + I = \begin{bmatrix} \frac{8}{9} + 1 & -\frac{1}{9} & \frac{4}{9} \\ -\frac{4}{9} & -\frac{4}{9} + 1 & \frac{7}{9} \\ \frac{1}{9} & -\frac{8}{9} & -\frac{4}{9} + 1 \end{bmatrix} = \begin{bmatrix} \frac{17}{9} & -\frac{1}{9} & \frac{4}{9} \\ -\frac{4}{9} & \frac{5}{9} & \frac{7}{9} \\ \frac{1}{9} & -\frac{8}{9} & \frac{5}{9} \end{bmatrix}.$$

The RREF of this matrix is the identity matrix, so -1 *is not an eigenvalue. Thus* 1 *is the only real eigenvalue, and the mapping is a rotation of 3-space around a line in the direction of the vector* $(5, -1, 1)$.

Example 7.22

$$P = \begin{bmatrix} \frac{1}{9} & \frac{4}{9} & \frac{8}{9} \\ \frac{4}{9} & \frac{7}{9} & -\frac{4}{9} \\ \frac{8}{9} & -\frac{4}{9} & \frac{1}{9} \end{bmatrix}$$

is an orthogonal 3×3 *matrix. Describe geometrically the mapping* $\mathbf{w} \to P\mathbf{w}$. *Solution: Form*

$$P - I = \begin{bmatrix} -\frac{8}{9} & \frac{4}{9} & \frac{8}{9} \\ \frac{4}{9} & -\frac{2}{9} & -\frac{4}{9} \\ \frac{8}{9} & -\frac{4}{9} & -\frac{8}{9} \end{bmatrix} = \frac{1}{9} \begin{bmatrix} -8 & 4 & 8 \\ 4 & 2 & 4 \\ 8 & -4 & 8 \end{bmatrix}.$$

We see that this is a matrix of rank 1. *Thus* 1 *is an eigenvalue of geometric multiplicity 2. Form*

$$P + I = \begin{bmatrix} \frac{10}{9} & \frac{4}{9} & \frac{8}{9} \\ \frac{4}{9} & \frac{16}{9} & -\frac{4}{9} \\ \frac{8}{9} & -\frac{4}{9} & \frac{10}{9} \end{bmatrix} = \frac{1}{9} \begin{bmatrix} 10 & 4 & 8 \\ 4 & 16 & -4 \\ 8 & -4 & 10 \end{bmatrix}.$$

$$RREF(P + I) = \begin{bmatrix} 1 & 0 & 1 \\ 0 & 1 & -\frac{1}{2} \\ 0 & 0 & 0 \end{bmatrix},$$

showing that -1 is also an eigenvalue of P and showing that $(-2, 1, 2)$ is an eigenvector for the eigenvalue -1. Since P is symmetric, we know that the eigenvector for -1 is orthogonal to all eigenvectors for 1, thus the mapping is a reflection through the plane orthogonal to the vector $(-2, 1, 2)$.

Remark *The matrices of reflections are symmetric (as well as orthogonal). In 2-space a reflection fixes a line and sends vectors orthogonal to this line to their negative. Thus we have two orthogonal eigenvectors, so the matrix is orthogonally similar to a diagonal matrix and hence must be symmetric. In 3-space a reflection fixes a plane and sends vectors orthogonal to the plane to their negative. Thus we have 3 orthogonal eigenvectors (two in the plane for 1, and one orthogonal to the plane for -1). Thus again the matrix is orthogonally similar to a diagonal matrix and so must be symmetric.*

Exercises

In problems 1 through 6 we are speaking of 2-space.

1. What is the matrix of rotation through the angle $\theta = 30°$?

2. What is the matrix of rotation through the angle $\theta = -30°$?

3. What is the matrix of the reflection about the line $y = -x$?

4. Find the matrix of the reflection about the line through the origin making an angle θ with the x axis. (Hint, first rotate the axes through $-\theta$, then reflect about the x axis, then rotate through θ. Multiply together the matrices of these operations. One can also do this problem using geometry and trigonometry.)

5. If we rotate the axes by $30°$, what are the coordinates of the points $(5, -3)$ with respect to the $\mathbf{i'j'}$ axes? (Recall that a rotation matrix is orthogonal, so its inverse is easy to find, it is the transpose.)

6. For each of the following matrices, determine if it is a rotation or a reflection. If it is a rotation, find the angle of rotation. If it is a reflection, find the line of about which the reflection takes place.

(a) $\begin{bmatrix} \frac{1}{\sqrt{2}} & \frac{1}{\sqrt{2}} \\ \frac{1}{\sqrt{2}} & -\frac{1}{\sqrt{2}} \end{bmatrix}$

(b) $\begin{bmatrix} \frac{\sqrt{3}}{2} & -\frac{1}{2} \\ \frac{1}{2} & \frac{\sqrt{3}}{2} \end{bmatrix}$

(c) $\begin{bmatrix} -\frac{\sqrt{3}}{2} & -\frac{1}{2} \\ \frac{1}{2} & -\frac{\sqrt{3}}{2} \end{bmatrix}$

(d) $\begin{bmatrix} \frac{1}{2} & -\frac{\sqrt{3}}{2} \\ -\frac{\sqrt{3}}{2} & -\frac{1}{2} \end{bmatrix}$.

7. Show that if P is a 2×2 or 3×3 orthogonal matrix, the mapping $\mathbf{w} \to P\mathbf{w}$ preserves angles (see Problems 7 and 9, Section 4.2).

8. For the following 3×3 orthogonal matrices P describe geometrically the mapping $\mathbf{w} \to P\mathbf{w}$. If it is a rotation around a line, find the line. If it is a reflection through a plane, find the plane.

(a) $P = \begin{bmatrix} \frac{2}{7} & \frac{3}{7} & \frac{6}{7} \\ -\frac{6}{7} & -\frac{2}{7} & \frac{3}{7} \\ \frac{3}{7} & -\frac{6}{7} & \frac{2}{7} \end{bmatrix}$

(b) $P = \begin{bmatrix} \frac{2}{3} & \frac{2}{3} & \frac{1}{3} \\ -\frac{2}{3} & \frac{1}{3} & \frac{2}{3} \\ \frac{1}{3} & -\frac{2}{3} & \frac{2}{3} \end{bmatrix}$

(c) $P = \begin{bmatrix} 1 & 0 & 0 \\ 0 & 0 & 1 \\ 0 & 1 & 0 \end{bmatrix}$

(d) $P = \begin{bmatrix} \frac{2}{11} & \frac{6}{11} & \frac{9}{11} \\ \frac{9}{11} & -\frac{6}{11} & \frac{2}{11} \\ \frac{6}{11} & \frac{7}{11} & -\frac{6}{11} \end{bmatrix}$.

9. Find the matrix of the mapping of 3-space to itself which sends each vector to its reflection through the plane $x = -y$. Check your answer by finding the eigenvalues and eigenvectors of your matrix.

Appendix A

Linear Transformations

Definition A.1 *We define a linear transformation from R^n to R^m to be a mapping (function) T from R^n to R^m such that for all $\mathbf{u}, \mathbf{v} \in R^n$ and all $c \in R$ we have*

(i) $T(\mathbf{u} + \mathbf{v}) = T(\mathbf{u}) + T(\mathbf{v})$, *and*

(ii) $T(c\mathbf{u}) = cT(\mathbf{u})$.

We can express these properties in words by saying that T "splits over sums" and "scalars come out". Repeated use of (i) and (ii) shows that if $\mathbf{e}_1, \mathbf{e}_2, \cdots, \mathbf{e}_n$ is the natural basis of R^n and if

$$\mathbf{u} = \begin{bmatrix} c_1 \\ c_2 \\ \vdots \\ c_n \end{bmatrix} = c_1\mathbf{e}_1 + c_2\mathbf{e}_2 + \cdots + c_n\mathbf{e}_n,$$

then

$$\begin{aligned} T(\mathbf{u}) &= T(c_1\mathbf{e}_1 + c_2\mathbf{e}_2 + \cdots + c_n\mathbf{e}_n) \\ &= c_1T(\mathbf{e}_1) + c_2T(\mathbf{e}_2) + \cdots + c_nT(\mathbf{e}_n), \end{aligned}$$

so that

$$T(\mathbf{u}) = c_1T(\mathbf{e}_1) + c_2T(\mathbf{e}_2) + \cdots + c_nT(\mathbf{e}_n), \qquad \text{(A.1)}$$

293

or in matrix notation

$$T(\mathbf{u}) = \begin{bmatrix} T(\mathbf{e}_1) & T(\mathbf{e}_2) & \cdots & T(\mathbf{e}_n) \end{bmatrix} \begin{bmatrix} c_1 \\ c_2 \\ \vdots \\ c_n \end{bmatrix}. \qquad \text{(A.2)}$$

Thus

$$T(\mathbf{u}) = A\mathbf{u},$$

where A is the matrix whose columns are the images of the natural basis vectors,

$$A = \begin{bmatrix} T(\mathbf{e}_1) & T(\mathbf{e}_2) & \cdots & T(\mathbf{e}_n) \end{bmatrix}.$$

Definition A.2 *The matrix* $A = \begin{bmatrix} T(\mathbf{e}_1) & T(\mathbf{e}_2) & \cdots & T(\mathbf{e}_n) \end{bmatrix}$ *is called the standard matrix of* T.

Remark   *A is* $m \times n$, *and* A *is the only matrix with the property that* $T(\mathbf{u}) = A\mathbf{u}$, *since if* C *is another matrix with that property we have column i of* $A = T(\mathbf{e}_i) = C\mathbf{e}_i = $ *column i of* C. *Thus* $C = A$.

Definition A.3 *The range of* T *is the set of all images under* T, $\{T(\mathbf{u}) | \mathbf{u} \in R^n\}$.

Definition A.4 *The kernel of* T *is the set of all vectors that are mapped to* $\mathbf{O}$ *by* T, $\{\mathbf{u} | \mathbf{u} \in R^n \text{ and } T(\mathbf{u}) = \mathbf{O}\}$.

Since $T(\mathbf{u}) = A\mathbf{u}$ we see that:

Theorem A.1 *The range of* T *is the column space of* A, *a subspace of* R^m, *and the kernel of* T *is the null space of* A, *a subspace of* R^n.

The domain of T is R^n, so $n = \dim(\text{Domain } T)$. Since for the matrix A we know that

$$rank(A) + nullity(A) = n, \qquad \text{(A.3)}$$

we have proved:

Theorem A.2 *If T is a linear transformation from R^n to R^m, then*

$$\dim(Range\ T) + \dim(Kernel T) = \dim(Domain\ T) = n, \qquad (A.4)$$

and thus

$$dim(Range\ T) \leq \dim(Domain\ T). \qquad (A.5)$$

Definition A.5 T *is called one to one or* $1:1$ *if* $T(\mathbf{u}) = T(\mathbf{v})$ *implies* $\mathbf{u} = \mathbf{v}$.

Definition A.6 T *is called onto if* $Range(T) = R^m$.

Theorem A.3 T *is* $1:1$ *if and only if* $Kernel(T) = \{\mathbf{O}\}$.

Proof. Since $T(\mathbf{O}) = A\mathbf{O} = \mathbf{O}$, we conclude that if T is $1:1$ then $Kernel(T) = \{\mathbf{O}\}$. Conversely, assume that $K = Kernal(T) = \{\mathbf{O}\}$. Then $T(\mathbf{u}) = T(\mathbf{v})$ implies that $T(\mathbf{u}) - T(\mathbf{v}) = \mathbf{O}$, so that $T(\mathbf{u}-\mathbf{v}) = \mathbf{O}$. Thus $\mathbf{u} - \mathbf{v} \in K$, so $\mathbf{u} - \mathbf{v} = \mathbf{O}$, and $\mathbf{u} = \mathbf{v}$ as desired. ∎

Theorem A.4 *If* $m = n$, *then* T *is* $1:1$ *if and only if* T *is onto.*

Proof. If T is $1:1$ then $Kernel(T) = \{\mathbf{O}\}$, whence using (A.4) $\dim(Range\ T) = n$, but $Range(T) \subset R^n$, so $Range(T) = R^n$, and T is onto. Conversely, if T is onto, $\dim(Range\ T) = n$, whence using (A.4) again we have $\dim(Kernel\ T) = 0$, so $Kernel(T) = \{\mathbf{O}\}$. ∎

Theorem A.5 T *is* $1:1$ *if and only if* T *preserves linear independence, that is to say* T *is* $1:1$ *if and only if* $\mathbf{u}_1, \mathbf{u}_2, \cdots, \mathbf{u}_k$ *a linearly independent set in* $Domain(T)$ *implies* $T(\mathbf{u}_1), T(\mathbf{u}_2), \cdots T(\mathbf{u}_k)$ *is a linearly independent set in* $Range(T)$.

Proof. First assume T is $1:1$ and let $\mathbf{u}_1, \mathbf{u}_2, \cdots, \mathbf{u}_k$ be a linearly independent set in R^n. If $c_1 T(\mathbf{u}_1) + c_2 T(\mathbf{u}_2) + \cdots + c_k T(\mathbf{u}_k) = \mathbf{O}$, then $T(c_1 \mathbf{u}_1 + c_2 \mathbf{u}_2 + \cdots + c_k \mathbf{u}_k) = \mathbf{O}$, and thus $c_1 \mathbf{u}_1 + c_2 \mathbf{u}_2 + \cdots + c_k \mathbf{u}_k = \mathbf{O}$. Now this implies that $c_i = 0$ for all $i = 1, 2, \cdots, k$ and thus $T(\mathbf{u}_1), T(\mathbf{u}_2), \cdots T(\mathbf{u}_k)$ is a linearly independent set. Conversely, assuming T preserves linear independence a nonzero vector $\mathbf{u}$ (which alone is a linearly independent set) cannot have $\mathbf{O}$ as its image ($\mathbf{O}$ alone is a dependent set). Thus $Kernel(T) = \{\mathbf{O}\}$ and T is $1:1$ by Theorem A.3. ∎

Corollary 1 T *is* $1:1$ *if and only if the set of images of a basis of* R^n *is a basis of* $Range(T)$.

Proof.   First assume T is $1:1$. Since the images of a basis of R^n obviously span $Range(T)$ and since they are linearly independent by Theorem A.5, they are a basis of $Range(T)$. Conversely, if the set of images of a basis of R^n is a basis of $Range(T)$, then $\dim(Range\ T) = n = \dim(Domain T)$, and we can conclude from (A.4) that $\dim(Kernel\ T) = 0$, so $Kernel(T) = \{O\}$ and T is $1:1$ by Theorem A.3 ∎.

We now restrict ourselves to the case of a linear transformation T from R^n to R^n. As before let A be the matrix whose columns are the images of the natural basis vectors, $A = \begin{bmatrix} T(\mathbf{e}_1) & T(\mathbf{e}_2) & \cdots & T(\mathbf{e}_n) \end{bmatrix}$. In this case A is square, $n \times n$. Let $\mathbf{v}_1, \mathbf{v}_2, \cdots, \mathbf{v}_n$ be another basis of R^n, and let B be the matrix whose columns are these basis vectors, in other words let $B = \begin{bmatrix} \mathbf{v}_1 & \mathbf{v}_2 & \cdots & \mathbf{v}_n \end{bmatrix}$. Now $T(\mathbf{v}_i) = A\mathbf{v}_i$, so we have the matrix equation

$$\begin{bmatrix} T(\mathbf{v}_1) & T(\mathbf{v}_2) & \cdots & T(\mathbf{v}_n) \end{bmatrix} = A \begin{bmatrix} \mathbf{v}_1 & \mathbf{v}_2 & \cdots & \mathbf{v}_n \end{bmatrix}. \quad (A.6)$$

Since $BB^{-1} = I$, we can write

$$\begin{bmatrix} T(\mathbf{v}_1) & T(\mathbf{v}_2) & \cdots & T(\mathbf{v}_n) \end{bmatrix} = BB^{-1}A \begin{bmatrix} \mathbf{v}_1 & \mathbf{v}_2 & \cdots & \mathbf{v}_n \end{bmatrix}. \quad (A.7)$$

Using $B = \begin{bmatrix} \mathbf{v}_1 & \mathbf{v}_2 & \cdots & \mathbf{v}_n \end{bmatrix}$, (A.7) can be rewritten as

$$\begin{bmatrix} T(\mathbf{v}_1) & T(\mathbf{v}_2) & \cdots & T(\mathbf{v}_n) \end{bmatrix} = \begin{bmatrix} \mathbf{v}_1 & \mathbf{v}_2 & \cdots & \mathbf{v}_n \end{bmatrix} B^{-1}AB. \quad (A.8)$$

Definition A.7 *The matrix* $B^{-1}AB$ *is called the matrix of* T *with respect to the basis given by the columns of* B.

Remark   *Note that from equation (A.8) the* i^{th} *column of* $B^{-1}AB$ *contains the coordinates of* $T(\mathbf{v}_i)$ *with respect to the basis* $\mathbf{v}_1, \mathbf{v}_2, \cdots, \mathbf{v}_n$.

If $\mathbf{u} = d_1\mathbf{v}_1 + d_2\mathbf{v}_2 + \cdots + d_n\mathbf{v}_n$, we can write this in matrix notation as

$$\mathbf{u} = \begin{bmatrix} \mathbf{v}_1 & \mathbf{v}_2 & \cdots & \mathbf{v}_n \end{bmatrix} \begin{bmatrix} d_1 \\ d_2 \\ \vdots \\ d_n \end{bmatrix} = B \begin{bmatrix} d_1 \\ d_2 \\ \vdots \\ d_n \end{bmatrix}. \quad (A.9)$$

Definition A.8 *We define* $[\mathbf{u}]_B = \begin{bmatrix} d_1 \\ d_2 \\ \vdots \\ d_n \end{bmatrix}$, *the coordinate vector of* $\mathbf{u}$ *with respect to the basis* $\{\mathbf{v}_1, \mathbf{v}_2, \cdots, \mathbf{v}_n\}$.

Then (A.9) becomes

$$\mathbf{u} = B[\mathbf{u}]_B, \tag{A.10}$$

and we also have

$$[\mathbf{u}]_B = B^{-1}\mathbf{u}. \tag{A.11}$$

Theorem A.6 *There is a unique matrix* $[T]_B$ *such that* $[T(\mathbf{u})]_B = [T]_B[\mathbf{u}]_B$ *for all* $\mathbf{u}$ *in* R^n. *Moreover* $[T]_B = B^{-1}AB$.

Proof.   If $[T(\mathbf{u})]_B = [T]_B[\mathbf{u}]_B$ for all $\mathbf{u} \in R^n$ then

$$[T(\mathbf{v}_i)]_B = [T]_B[\mathbf{v}_i]_B = [T]_B\mathbf{e}_i = column\ i\ of\ [T]_B.$$

Thus the column i of $[T]_B$ contains the coordinates of $T(\mathbf{v}_i)$. This proves uniqueness and also because of (A.8) gives $[T]_B = B^{-1}AB$. We will now show that $B^{-1}AB[\mathbf{u}]_B = [T(\mathbf{u})]_B$. To see this note that $B^{-1}AB[\mathbf{u}]_B = B^{-1}A\mathbf{u} = B^{-1}T(\mathbf{u}) = [T(\mathbf{u})]_B$. ∎

Definition A.9 *From now on we will use* $[T]$ *for the standard matrix of* T *which we have called* A. $[T]$ *is the matrix of* T *with respect to the natural basis and has the property that*

$$T(\mathbf{u}) = [T]\mathbf{u}. \tag{A.12}$$

Definition A.10 *The matrix* $[T]_B$ *is the matrix of* T *with respect to the basis* B, *and multiplication by* $[T]_B$ *sends coordinates of* $\mathbf{u}$ *to the coordinates of* $T(\mathbf{u})$,

$$[T(\mathbf{u})]_B = [T]_B[\mathbf{u}]_B. \tag{A.13}$$

According to Theorem A.6, $[T]$ and $[T]_B$ are related by the equation

$$[T]_B = B^{-1}[T]B \tag{A.14}$$

where B is the matrix whose columns are the basis vectors $\mathbf{v}_1, \mathbf{v}_2, \cdots, \mathbf{v}_n$.

Remark   *In a slight perversion of notation we sometimes use B to stand for both the basis $\{v_1, v_2, \cdots, v_n\}$ and the matrix whose columns are the vectors $v_1, v_2, \cdots, v_n$. This occurs in the proof of the following theorem.*

Theorem A.7 *Two matrices C and D represent the same linear transformation T with respect to two different bases B_1 and B_2 of R^n if and only if there exists an invertible matrix P such that $P^{-1}CP = D$.*

Proof.   First assume C and D represent the same linear transformation T with respect to the bases B_1 and B_2 respectively. Since C represents T with respect to B_1, we have $C = B_1^{-1}[T]B_1$. Also $D = B_2^{-1}[T]B_2$. Solving the first equation for $[T]$ and substituting in the second equation we have $D = B_2^{-1}B_1CB_1^{-1}B_2$. Letting $P = B_1^{-1}B_2$, we have $D = P^{-1}CP$. Conversely, if $D = P^{-1}CP$, let S be the linear transformation defined by $S(\mathbf{u}) = C\mathbf{u}$. Then the matrix of S with respect to the natural basis is C, while the matrix of S with respect to the basis given by the columns of P is $P^{-1}CP$, which is D. ∎

Example A.1 *Let T be defined by $T(x,y,z) = (2x+y, 2x-z, x+3y+2z)$. T is a linear transformation from R^3 to R^3. $T(1,0,0) = (2,2,1)$, $T(0,1,0) = (1,0,3)$, and $T(0,0,1) = (0,-1,2)$. Thus we have*

$$[T] = \begin{bmatrix} 2 & 1 & 0 \\ 2 & 0 & -1 \\ 1 & 3 & 2 \end{bmatrix}.$$

Note that if $\mathbf{u} = (x,y,z)$ and if $\mathbf{u}$ and $T(\mathbf{u})$ are written as columns, we have

$$[T]\mathbf{u} = \begin{bmatrix} 2 & 1 & 0 \\ 2 & 0 & -1 \\ 1 & 3 & 2 \end{bmatrix} \begin{bmatrix} x \\ y \\ z \end{bmatrix} = \begin{bmatrix} 2x+y \\ 2x-z \\ x+3y+z \end{bmatrix} = T(\mathbf{u}).$$

Since $\det[T] = 1 \neq 0$, we know that $Kernel(T) = \{O\}$, T is $1:1$, and T is onto. We now compute the matrix of T with respect to the basis

$$\mathbf{v}_1 = \begin{bmatrix} 1 \\ -1 \\ 2 \end{bmatrix}, \mathbf{v}_2 = \begin{bmatrix} 2 \\ 1 \\ 1 \end{bmatrix}, \mathbf{v}_3 = \begin{bmatrix} 3 \\ 1 \\ 1 \end{bmatrix}. \text{ Let } B = \begin{bmatrix} 1 & 2 & 3 \\ -1 & 1 & 1 \\ 2 & 1 & 1 \end{bmatrix}. \text{ Then}$$

$$[T]_B = B^{-1}[T]B,$$

so we have

$$[T]_B = \begin{bmatrix} 0 & -\frac{1}{3} & \frac{1}{3} \\ -1 & \frac{5}{3} & \frac{4}{3} \\ 1 & -1 & -1 \end{bmatrix} \begin{bmatrix} 2 & 1 & 0 \\ 2 & 0 & -1 \\ 1 & 3 & 2 \end{bmatrix} \begin{bmatrix} 1 & 2 & 3 \\ -1 & 1 & 1 \\ 2 & 1 & 1 \end{bmatrix}$$

$$= \begin{bmatrix} \frac{2}{3} & \frac{4}{3} & 1 \\ \frac{5}{3} & \frac{28}{3} & 12 \\ -1 & -5 & -6 \end{bmatrix}.$$

To check, note that the first column of $[T]_B$ *should be the coordinates of* $T(1, -1, 2)$. *Now*

$$T\begin{bmatrix} 1 \\ -1 \\ 2 \end{bmatrix} = \begin{bmatrix} 2 & 1 & 0 \\ 2 & 0 & -1 \\ 1 & 3 & 2 \end{bmatrix}\begin{bmatrix} 1 \\ -1 \\ 2 \end{bmatrix} = \begin{bmatrix} 1 \\ 0 \\ 2 \end{bmatrix}.$$

Thus $2/3\begin{bmatrix} 1 \\ -1 \\ 2 \end{bmatrix} + 5/3\begin{bmatrix} 2 \\ 1 \\ 1 \end{bmatrix} - 1\begin{bmatrix} 3 \\ 1 \\ 1 \end{bmatrix}$ *should equal* $\begin{bmatrix} 1 \\ 0 \\ 2 \end{bmatrix}$. *It checks.*

We note that since $T(\mathbf{u}) = [T]\mathbf{u}$, the linear transformation T and its standard matrix $[T]$ will have the same eigenvalues and eigenvectors. If we change basis, the eigenvalues will stay the same, since $[T]_B = B^{-1}[T]B$, and similar matrices have the same eigenvalues (Theorem 4.4). However the eigenvectors are different. The eigenvectors of $[T]_B$ are the coordinate vectors with respect to the basis B of the eigenvectors of T. In other words if $\mathbf{u}$ is an eigenvector of T for the eigenvalue k, then $B^{-1}\mathbf{u}$ is an eigenvector of $[T]_B$ for the eigenvector k. This can be easily verified since

$$\begin{aligned} [T]_B\left(B^{-1}\mathbf{u}\right) &= B^{-1}[T]BB^{-1}\mathbf{u} \\ &= B^{-1}[T]\mathbf{u} \\ &= B^{-1}k\mathbf{u} \\ &= k\left(B^{-1}\mathbf{u}\right). \end{aligned}$$

If $B = \begin{bmatrix} \mathbf{v}_1 & \mathbf{v}_2 & \cdots & \mathbf{v}_n \end{bmatrix}$ is a matrix whose columns are a basis of eigenvectors of T, then the matrix of T with respect to this basis is diagonal. This is easy to see since

$$
\begin{aligned}
[T]_B &= \left[\ [T(\mathbf{v}_1)]_B \quad [T(\mathbf{v}_2)]_B \quad \cdots \quad [T(\mathbf{v}_n)]_B \ \right] \\
&= \left[k_1\mathbf{v}_1]_B \quad [k_2\mathbf{v}_2]_B \quad \cdots \quad [k_n\mathbf{v}_n]_B \right. \\
&= \begin{bmatrix}
k_1 & 0 & \cdots & 0 \\
0 & k_2 & & \vdots \\
\vdots & & \ddots & 0 \\
0 & \cdots & 0 & k_n
\end{bmatrix}.
\end{aligned}
$$

Remark   *This result is also a consequence of Theorem 4.6, Section 4.1, since $[T]_B = B^{-1}[T]B$ and we have proved in Theorem 4.6 that $P^{-1}AP$ is diagonal if the columns of P are eigenvectors of A.*

There is a relation between composition of linear transformations and matrix multiplication. The matrix of the composite transformation is the product of the matrices of the transformations. If S and T are two linear transformations from R^n to R^n, we define the composite $T \circ S$ as usual by $(T \circ S)(\mathbf{u}) = T(S(\mathbf{u}))$, and the sum $T + S$ by $(T + S)(\mathbf{u}) = T(\mathbf{u}) + S(\mathbf{u})$. Then using any basis, the matrix of the composite is the product of the matrix of T and the matrix of S, and the matrix of the sum is the sum of the matrix of T and the matrix of S. We state this formally and prove it for the natural basis. The proof would work for any basis, but we would have to write a lot of subscripts.

Theorem A.8 *If T and S are linear transformations from R^n to R^n and if $[T]$ and $[S]$ are the standard matrices of T and S respectively then $[T][S] = [T \circ S]$ and $[T] + [S] = [T + S]$.*

Proof. $[T \circ S]$ is the unique matrix such that $(T \circ S)(\mathbf{u}) = [T \circ S]\mathbf{u}$. Now $(T \circ S)(\mathbf{u}) = T(S(\mathbf{u})) = [T](S(\mathbf{u})) = [T]([S]\mathbf{u}) = ([T][S])\mathbf{u}$, since matrix multiplication is associative. Thus $[T \circ S] = [T][S]$ as desired. The matrix $[T+S]$ is the unique matrix such that $(T+S)(\mathbf{u}) = [T+S]\mathbf{u}$. We have $(T + S)(\mathbf{u}) = T(\mathbf{u}) + S(\mathbf{u}) = [T]\mathbf{u} + [S]\mathbf{u} = ([T] + [S])\mathbf{u}$, by the distributive law of matrix multiplication. Thus $[T + S] = [T] + [S]$ as desired. ∎

Definition A.11 *A linear transformation from an abstract vector space V over the real numbers to another abstract vector space W over the real numbers is a mapping (function) T from V to W which satisfies (i) and (ii) for all* $\mathbf{u}, \mathbf{v} \in V$ *and all* $c \in R$.

Definition A.12 *V is isomorphic to W if there is a* 1 : 1 *linear transformation from V onto W.*

Theorem A.9 *Any vector space V of dimension n over the real numbers is isomorphic to* R^n.

Proof. Let $B = \{\mathbf{v}_1, \mathbf{v}_2, \cdots, \mathbf{v}_n\}$ be a basis of V. For each vector $\mathbf{u}$ in V there exist unique real numbers $c_1, c_2, \cdots c_n$ such that $\mathbf{u} = c_1\mathbf{v}_1 + c_2\mathbf{v}_2 + \cdots + c_n\mathbf{v}_n$. Let T be defined by

$$T(\mathbf{u}) = \begin{bmatrix} c_1 \\ c_2 \\ \vdots \\ c_n \end{bmatrix} = [\mathbf{u}]_B,$$

then T sends each vector to its n-tuple of coordinates with respect to the basis B. It is easy to see that T is 1 : 1 linear transformation from V onto R^n. ∎

Remark *Theorem A.9 is a more precise way (using the concept of isomorphism) of saying what we have stated and proved in Theorem 1.9, Section 1.7.*

Appendix B

Answers to selected questions

Section 1.1

1.

    (a) $\mathbf{u} + \mathbf{v} = (1, 7)$, $c\mathbf{u} = (6, 9)$, $|\mathbf{u}| = \sqrt{13}$

    (b) $\mathbf{u} + \mathbf{v} = (-6, 1)$, $c\mathbf{u} = -\frac{1}{2}(-5, 2)$, $|\mathbf{u}| = \sqrt{29}$

2. $k(1, 1)$ for all real k.

3.

    (a) **O**.

    (b) all vectors which end on the line containing the vector $\mathbf{v}$.

4. All vectors which end on the line $y = 2$.

Section 1.2

1.

    (a) $c = -\frac{1}{3}$, $d = \frac{7}{3}$.

    (b) no solution exists.

2. $c = 5$, $d = -2$.

3.

    (a) Linearly independent

    (b) linearly dependent

    (c) linearly dependent.

4. (a) only.

Section 1.3

1. $(2, 5, 3) = -8(0, 0, 1) + 3(0, 1, 1) + 2(1, 1, 1)$.

2. $(a, b, c) = (c - b)(0, 0, 1) + (b - a)(0, 1, 1) + a(1, 1, 1)$.

3. Yes, every triple (a, b, c) can be written as a linear combination of these three vectors as in problem 2.

4. Yes. The solution to problem 2 shows that the only way to get $(0, 0, 0)$ as a linear combination of $(0, 0, 1)$, $(0, 1, 1)$, and $(1, 1, 1)$ is to use all zero coefficients.

Section 1.5

3. If $\mathbf{u} \neq \mathbf{O}$, then $c\mathbf{u}$ is a different vector for all real numbers c.

5. $(1, 2, 1)$ spans this subspace.

6. Yes. The solution in example 1.15 is unique.

7. Every basis is a spanning set, but not every spanning set is a basis. For a basis every vector must be expressible uniquely as a linear combination of the basis vectors. For a spanning set uniqueness is not required.

Section 1.6

1. No.

2. $c - a - 2b \neq 0$.

3. 2.

Section 1.7

3. $\{(1,1,-1,1),\ (1,0,1,1),\ (1,0,0,0),\ (0,0,0,1)\}$ is a basis of R^4.

4. $\{1+x-x^2+x^3,\ 1+x^2+x^3,\ 1,\ x^3\}$ is a basis of P_3.

Section 1.8

1. $Dim(S)=1,\ Dim(T)=2,\ Dim(S+T)=3,\ Dim(S\sqcap T)=0$.

2. $Dim(S)=2,\ Dim(T)=2,\ Dim(S+T)=3,\ Dim(S\sqcap T)=1$.
$(1,-2,1)$ is a basis of $S\sqcap T$.

Section 2.1

1.

(a) $\begin{bmatrix} -7 & 0 & -1 \\ 0 & 0 & 0 \\ 10 & 2 & 11 \end{bmatrix}$

(b) $\begin{bmatrix} 5 & 7 & 11 \end{bmatrix}$

(c) $\begin{bmatrix} 0 \\ 27 \\ 0 \end{bmatrix}$

(d) impossible

(e) $[19]$

(f) $\begin{bmatrix} 2 & 4 & 1 \\ 6 & 12 & 3 \\ 10 & 20 & 5 \end{bmatrix}$.

2.

(a) two

(b) $\mathbf{u}_1 - \mathbf{u}_4$.

Section 2.2

1. $\begin{bmatrix} -1 & 1 & 4 \\ 0 & 0 & 0 \\ 0 & 0 & 0 \end{bmatrix}, \begin{bmatrix} 0 & 3 & 0 \\ 0 & -1 & 0 \\ 0 & 4 & 0 \end{bmatrix}.$

2.

(a) Yes.

(b) $C = \begin{bmatrix} 3 & 1 & 2 \\ -1 & 4 & 1 \end{bmatrix} \begin{bmatrix} 2 & 0 \\ 2 & 1 \\ 4 & 3 \end{bmatrix} + \begin{bmatrix} 0 & 1 \\ 2 & -1 \end{bmatrix} \begin{bmatrix} 1 & 1 \\ 3 & 2 \end{bmatrix} = \begin{bmatrix} 19 & 9 \\ 9 & 7 \end{bmatrix}$

(c) $F = \begin{bmatrix} 1 & 3 & 0 \end{bmatrix} \begin{bmatrix} -1 & 3 \\ 0 & 0 \\ 2 & 0 \end{bmatrix} + \begin{bmatrix} 1 & 1 \end{bmatrix} \begin{bmatrix} 1 & 1 \\ 0 & 1 \end{bmatrix} = \begin{bmatrix} 0 & 5 \end{bmatrix}$

(d) E is 1×2.

Section 2.3

1.

(a) (Ex.2.21) $\begin{bmatrix} 0 & 1 & 0 & 0 & \frac{1}{2} \\ 0 & 0 & 1 & 0 & \frac{5}{4} \\ 0 & 0 & 0 & 1 & 0 \\ 0 & 0 & 0 & 0 & 0 \end{bmatrix}$

(b) (Ex.2.22) $\begin{bmatrix} 0 & 1 & 2 & 0 & -11 \\ 0 & 0 & 0 & 1 & 5 \\ 0 & 0 & 0 & 0 & 0 \end{bmatrix}$

(c) (Ex.2.23) $\begin{bmatrix} 0 & 1 & 2 & 0 & 4 \\ 0 & 0 & 0 & 1 & 5 \\ 0 & 0 & 0 & 0 & 0 \\ 0 & 0 & 0 & 0 & 0 \\ 0 & 0 & 0 & 0 & 0 \end{bmatrix}$

(d) (Ex.2.24) $\begin{bmatrix} 0 & 1 & 2 & 0 & 4 \\ 0 & 0 & 0 & 1 & 5 \end{bmatrix}.$

2.

(a) $\{(5 - 2s + 3t, s, 7 - 4t, t)|s$ and t are any real numbers$\}$

(b) no solution.

(c) $\{(4 + 2t, 0, t)|t$ is any real number$\}$.

3. $(7, 0, 2) = 1(1, 2, -1) + 2(3, 1, 2) - 1(0, 4, 1)$.

4. $(3, -5, 1) = -6(1, 2, 4) + 3(3, 1, 7) + 1(0, 4, 4)$.

5. not possible.

Section 2.4

1. A and C have the same row space, but no other pair.

2. All have row rank 3.

3. All have column rank 3.

Section 2.5

1.

(a) $\{u_1, u_3\}$.

(b) $u_2 = 2u_1$, $u_4 = \frac{1}{2}u_1 - u_3$, $u_5 = \frac{2}{3}u_1 + 4u_3$.

(c) $rank(A) = 2$.

2.

(a) $\{u_1, u_3, u_6\}$.

(b) $u_2 = 2u_1$, $u_4 = 3u_1 - 2u_3$, $u_5 = u_1 + 2u_3$, $u_7 = 3u_1 - u_3 + 2u_6$.

(c) $rank(A) = 3$.

3.

(a) $\{u_1, u_2, u_3, u_5\}$.

(b) $u_4 = -u_1 + 2u_2 + u_3$, $u_6 = 4u_1 + 2u_2 - u_3 + 2u_5$.

(c) $rank(A) = 4$.

4. $rank(C) = 2$.

5.

   (a) $(1, 0, -5, -5, 0), (0, 1, 2, 1, -1)$.

   (b) $\begin{bmatrix} 1 \\ 1 \\ 2 \\ 3 \end{bmatrix}, \begin{bmatrix} 3 \\ 4 \\ 3 \\ 8 \end{bmatrix}$.

   (c) $rank(A) = 2$.

   (d) $5x_1 - 2x_2 + x_3 = 0, 5x_1 + x_2 + x_4 = 0, x_2 - x_5 = 0$.

6. $x_2 - 2x_3 + x_4 = 0$. $(1, 3, 0, 2) \notin rowsp(A)$, $(2, -1, 1, 3) \in rowsp(A)$, $(1, 1, 2, 1) \notin rowsp(A)$, $(-1, 2, 1, 0) \in rowsp(A)$.

7.

   (a) $\dim(W) = 3$.

   (b) basis of $W = \{(3, -1, 2, 2), (0, -4, -1, -2), (-1, -4, -2, -1)\}$,
     and $(-6, 2, -4, -4) = -2(3, -1, 2, 2)$, $(7, -7, 3, 6) = 3(3, -1, 2, 2)$
     $1(0, -4, -1, -2) + 2(-1, -4, -2, -1)$.

Section 2.6

1. $\begin{bmatrix} 1 & 0 & -1 & 0 \\ 0 & 1 & 4 & 0 \\ 0 & 0 & 0 & 0 \\ 0 & 0 & 0 & 1 \end{bmatrix}$. Basis of null space: $\begin{bmatrix} -1 \\ 4 \\ -1 \\ 0 \end{bmatrix}$.

2. $\begin{bmatrix} 1 & 2 & 0 \\ 0 & 0 & 0 \\ 0 & 0 & 1 \end{bmatrix}$. Basis of null space: $\begin{bmatrix} 2 \\ -1 \\ 0 \end{bmatrix}$.

3. $\begin{bmatrix} 1 & 0 & 0 & -4 \\ 0 & 1 & 0 & 2 \\ 0 & 0 & 1 & 1 \\ 0 & 0 & 0 & 0 \end{bmatrix}$. Basis of null space: $\begin{bmatrix} -4 \\ 2 \\ 1 \\ -1 \end{bmatrix}$.

4. $\begin{bmatrix} 1 & 0 & 0 \\ 0 & 1 & 0 \\ 0 & 0 & 1 \end{bmatrix}$, the null space consists of the zero vector only, so has no basis.

6. Basis of null space $\begin{bmatrix} 2 \\ -1 \\ 0 \\ 0 \\ 0 \end{bmatrix}, \begin{bmatrix} \frac{1}{2} \\ 0 \\ -1 \\ -1 \\ 0 \end{bmatrix}, \begin{bmatrix} \frac{2}{3} \\ 0 \\ 4 \\ 0 \\ -1 \end{bmatrix}$.

7. Basis of null space $\begin{bmatrix} 2 \\ -1 \\ 0 \\ 0 \\ 0 \\ 0 \\ 0 \end{bmatrix}, \begin{bmatrix} 3 \\ 0 \\ -2 \\ -1 \\ 0 \\ 0 \\ 0 \end{bmatrix}, \begin{bmatrix} 1 \\ 0 \\ 2 \\ 0 \\ -1 \\ 0 \\ 0 \end{bmatrix}, \begin{bmatrix} 3 \\ 0 \\ -1 \\ 0 \\ 0 \\ 2 \\ -1 \end{bmatrix}$.

8. $\begin{bmatrix} -1 \\ 2 \\ 1 \\ -1 \\ 0 \\ 0 \end{bmatrix}, \begin{bmatrix} 4 \\ 2 \\ -1 \\ 0 \\ 2 \\ -1 \end{bmatrix}$.

9.

(a) $(1, 0, -5, -5, 0), (0, 1, 2, 1, -1)$

(b) $\begin{bmatrix} 1 \\ 1 \\ 2 \\ 3 \end{bmatrix}, \begin{bmatrix} 3 \\ 4 \\ 3 \\ 8 \end{bmatrix}$.

(c) $\begin{bmatrix} -5 \\ 2 \\ -1 \\ 0 \\ 0 \end{bmatrix}, \begin{bmatrix} -5 \\ 1 \\ 0 \\ -1 \\ 0 \end{bmatrix}, \begin{bmatrix} 0 \\ -1 \\ 0 \\ 0 \\ -1 \end{bmatrix}$.

10. $\begin{bmatrix} 2 \\ 4 \\ -5 \end{bmatrix}$.

Section 2.7

1. Rank $= 2$.

2.

(a) $A^{-1} = \begin{bmatrix} -1 & 0 & 1 \\ \frac{5}{3} & \frac{1}{3} & -1 \\ \frac{1}{3} & -\frac{1}{3} & 0 \end{bmatrix}$.

(b) $B^{-1} = \begin{bmatrix} \frac{4}{5} & \frac{1}{5} & 0 \\ -\frac{1}{5} & \frac{1}{5} & 0 \\ \frac{1}{5} & -\frac{1}{5} & \frac{1}{2} \end{bmatrix}$.

(c) $C^{-1} = \begin{bmatrix} 15 & -3 & -4 \\ -4 & 1 & 1 \\ 10 & -2 & -3 \end{bmatrix}$.

(d) $D^{-1} = \begin{bmatrix} \frac{19}{10} & -\frac{17}{10} & -\frac{13}{10} \\ -\frac{4}{5} & \frac{2}{5} & \frac{3}{5} \\ \frac{3}{10} & \frac{1}{10} & -\frac{1}{10} \end{bmatrix}$.

(e) $F^{-1} = \begin{bmatrix} -\frac{1}{7} & \frac{2}{7} \\ \frac{5}{7} & -\frac{3}{7} \end{bmatrix}$.

3. Is a basis.

4. SEND MONEY

5. $\begin{bmatrix} 3 \\ -2 \\ 5 \end{bmatrix}$.

6. $w = (1, 6, 11)$. $[w]_{B'} = \begin{bmatrix} 4 \\ -\frac{2}{3} \\ \frac{7}{3} \end{bmatrix}$.

Section 2.8

1. You can pick $P = A^{-1}$ and $Q = B$.

2. They are equivalent if and only if they have the same rank.
$\begin{bmatrix} a & 1 \\ 1 & a \end{bmatrix}$ has rank 1 if $a = 1$ or $a = -1$, otherwise it has rank 2.
Thus the matrices are equivalent if and only if either both a and b are from the set $\{1, -1\}$ or neither a nor b is from this set.

3. If $B = PAQ$ then $B = (2P)A(\frac{1}{2})Q$.

4. Recall $colsp(AB) \subset colsp(A)$.

Section 3.1

1. $P = (1\,3\,2\,4)(5\,6)$. $N(P) = 4$. P is even. $P = (1\,3)(1\,2)(1\,4)(5\,6)$.

2. $P^{-1} = (1\,4\,2\,3)(5\,6)$.

3. $(1\,2\,3)$.

4. $P * Q = (1\,5\,8\,3\,2\,7\,4)$, $Q * P = (2\,3\,5\,8\,7\,4\,6)$. $P, Q, P * Q$, $Q * P$ are all even.

5. The least common multiple or r and s. For example if $r = 4$ and $s = 6$, the answer is 12.

6. S_n is commutative for $n = 1, 2$ and S_n is not commutative for $n \geq 3$. A_3 is commutative, but A_n is not commutative for $n \geq 4$.

7. In S_4 the cycles of length 4 and the cycles of length 2 are odd. The remaining 12 permutations make up A_4. A_4 consists of eight cycles of length 3 , three permutations that are products of two disjoint cycles of length 2, and the identity permutation.

8. $P^{-1} = (1\,6\,4\,7\,2)(3\,5\,8)$. $Q^{-1} = (1\,6)(3\,5\,2\,8)$.

9. $n - 1$.

10. $n!$

11. $(2n-1)(2n-3)(2n-5) \cdots (1)$.

13.

(a) $f(x) = x + 1$

(b) $g(1) = 1$, $g(x) = x - 1$ for $x > 1$.

Section 3.2

1. $a = -\frac{9}{17}$

2.

(a) 12

(b) -11

3.

(a) 512

(b) 2160

4. $2e^{3x}$

5.

(a) Yes

(b) Yes.

7.

(a) $\begin{bmatrix} 2 & -1 \\ -3 & 2 \end{bmatrix}$,

(b) $\begin{bmatrix} \frac{1}{2} & \frac{1}{2} & -\frac{1}{2} \\ -1 & 0 & 1 \\ \frac{3}{2} & -\frac{1}{2} & -\frac{1}{2} \end{bmatrix}$.

8. $(a + (n-1)b)(a - b)^{n-1}$.

9.

(a) $\frac{1}{|A|}$.

(b) $c|A|$,

(c) $(|A|)^k$

(d) $(|A|)^{n-1}$

10. -66

11. $x = -\frac{271}{58}$, $y = \frac{67}{29}$, $z = \frac{171}{58}$.

12. $w = 2$, $x = \frac{2}{3}$, $y = 0$, $z = \frac{4}{3}$.

13. $x = \frac{dr-bs}{ad-bc}$, $y = \frac{as-cr}{ad-bc}$.

14. -9240

15.

(b) $96 \le |A| \le 3120$,

(c) $|A| = 816$.

Section 3.3

1. 4

2. 128

3. 600

4. 27

Section 4.1

1.

(a) 1,

(b) $-3, -3, -3$,

(c) 10,

(d) $4, -4, -4$,

(e) 6.

2.

    (a) $k^2 - 5k - 14 = 0$; 7, -2

    (b) $k^2 + k + 1 = 0$; $\frac{-1\pm\sqrt{-3}}{2}$

    (c) $k^2 - 4k + 4 = 0$; 2, 2

    (d) $k^3 - 18k^2 + 81k - 108 = 0$; 3, 3, 12

    (e) $k^3 - k^2 - 18k + 24 = 0$; 4, $\frac{-3\pm\sqrt{33}}{2}$

    (f) $k^3 - 3k^2 - 9k + 27 = 0$; 3, 3, -3

    (g) $k^3 - 5k^2 + 6k = 0$; 0, 2, 3

    (h) $k^3 - 4k^2 - 7k + 10 = 0$; 1, 5, -2

    (i) $k^4 - 9k^3 + 30k^2 - 44k + 24 = 0$; 2, 2, 2, 3

    (j) $k^4 - 9k^3 + 30k^2 - 44k + 24 = 0$; 2, 2, 2, 3

    (k) $k^4 - 9k^3 + 30k^2 - 44k + 24 = 0$; 2, 2, 2, 3

    (l) $k^4 - 8k^3 + 22k^2 - 24k + 9 = 0$; 1, 1, 3, 3.

3. (a), (d), (e), (f), (g), and (h) are similar to diagonal matrices. The others are not.

4. Yes for (b).

7. Use the fact that $(A - kI)^T = A^T - kI$.

8. 1 has geometric multiplicity 1, 2 has geometric multiplicity 3.

9. 2 and 9 have geometric multiplicity 1, 18 has geometric multiplicity 2.

10. $A: |z + 4| \leq 1$, $|z + 5| \leq 2$, $|z - 6| \leq 2$.
    $B: |z - 5| \leq 2$, $|z - 6| \leq 2$, $|z - 7| \leq 3$.
    $C: |z + 8| \leq 4$, $|z - 9| \leq 1$, $|z + 9| \leq 2$.
    $D: |z - 7| \leq 3$, $|z - 6| \leq 3$, $|z + 8| \leq 2$, $|z + 9| \leq 7$.

Section 4.2

1. $\frac{1}{\sqrt{2}}(1,0,1)$, $\frac{1}{\sqrt{38}}(1,-6,1)$.

2. $\left(-\frac{1}{2},-1,0\right)$, $\left(-1,\frac{1}{2},-5\right)$.

3. $\frac{1}{\sqrt{2}}(1,0,-1,0)$, $\frac{1}{\sqrt{18}}(1,-4,1,0)$, $(0,0,0,1)$.

4. $(2,-1,0,0)$, $(3,6,-5,0)$, $(1,2,3,14)$.

Section 4.3

1.

(a) $k=3,5$. For $k=3$, basis of eigenspace is $(1,-1)$. For $k=5$, basis of eigenspace is $(1,1)$.

$$P=\begin{bmatrix} \frac{1}{\sqrt{2}} & \frac{1}{\sqrt{2}} \\ -\frac{1}{\sqrt{2}} & \frac{1}{\sqrt{2}} \end{bmatrix}, D=\begin{bmatrix} 3 & 0 \\ 0 & 5 \end{bmatrix}.$$

(e) $k=0,0,6$. For $k=0$, basis of the eigenspace is $(1,0,-1)$, $(1,-1,1)$. For $k=6$, basis for the eigenspace is $(1,2,1)$.

$$P=\begin{bmatrix} \frac{1}{\sqrt{2}} & \frac{1}{\sqrt{3}} & \frac{1}{\sqrt{6}} \\ 0 & -\frac{1}{\sqrt{3}} & \frac{2}{\sqrt{6}} \\ -\frac{1}{\sqrt{2}} & \frac{1}{\sqrt{3}} & \frac{1}{\sqrt{6}} \end{bmatrix}, D=\begin{bmatrix} 0 & 0 & 0 \\ 0 & 0 & 0 \\ 0 & 0 & 6 \end{bmatrix}.$$

(f) $k=-3,6,6$. For $k=-3$, basis of eigenvectors is $(1,2,-2)$. For $k=6$, basis of eigenvectors is $(0,1,1)$, $(4,-1,1)$.

$$P=\begin{bmatrix} \frac{1}{3} & 0 & \frac{4}{\sqrt{18}} \\ \frac{2}{3} & \frac{1}{\sqrt{2}} & -\frac{1}{\sqrt{18}} \\ -\frac{2}{3} & \frac{1}{\sqrt{2}} & \frac{1}{\sqrt{18}} \end{bmatrix}, D=\begin{bmatrix} -3 & 0 & 0 \\ 0 & 6 & 0 \\ 0 & 0 & 6 \end{bmatrix}.$$

(i) $k=1,1,1,5$.

2.

(a) ellipse,

(b) hyperbola,

(c) parabola.

3.

(a) eigenvalues: -1, -2, 3. Yes.

(b) No.

4.

(a) characteristic polynomial: $(x-2)(x^2-4x+2)$. Yes.

(b) No.

5.

(a) eigenvalues: 2, 2, 3. No.

(b) No.

6.

(a) eigenvalues: 1, -1, 0, 4. Yes.

(b) No.

Section 4.4

2. $\begin{bmatrix} 3 & 0 \\ 0 & 3 \end{bmatrix}$.

3. $\begin{bmatrix} 3 & 0 \\ 1 & 3 \end{bmatrix}$.

5. Characteristic polynomial is $x^2 - 4x - 1$.

$$A^2 - 4A - I = \begin{bmatrix} 0 & 0 \\ 0 & 0 \end{bmatrix}.$$

Thus

$$A - 4I - A^{-1} = \begin{bmatrix} 0 & 0 \\ 0 & 0 \end{bmatrix},$$

or

$$A^{-1} = A - 4I = \begin{bmatrix} -3 & 2 \\ 2 & -1 \end{bmatrix}.$$

Section 4.5

1. Eigenvalues: 1, 1, 3. Let $P = \begin{bmatrix} 1 & 0 & 1 \\ 0 & -2 & 0 \\ 0 & 1 & 1 \end{bmatrix}$. Then

$$P^{-1}AP = \begin{bmatrix} 1 & 0 & 0 \\ 0 & 1 & 0 \\ 0 & 0 & 3 \end{bmatrix},$$

$$P^{-1}A^k P = \begin{bmatrix} 1 & 0 & 0 \\ 0 & 1 & 0 \\ 0 & 0 & 3^k \end{bmatrix},$$

and

$$A^k = P \begin{bmatrix} 1 & 0 & 0 \\ 0 & 1 & 0 \\ 0 & 0 & 3^k \end{bmatrix} P^{-1}$$

$$= \begin{bmatrix} 1 & \left(\frac{1}{2}3^k - \frac{1}{2}\right) & \left(3^k - 1\right) \\ 0 & 1 & 0 \\ 0 & \left(\frac{1}{2}3^k - \frac{1}{2}\right) & 3^k \end{bmatrix}.$$

2. Solve $\begin{bmatrix} \frac{1}{2} & \frac{1}{3} \\ \frac{1}{2} & \frac{2}{3} \end{bmatrix} \begin{bmatrix} p \\ q \end{bmatrix} = \begin{bmatrix} p \\ q \end{bmatrix}$, answer is 40% wet, 60% dry.

7. $\begin{aligned} x_1 &= e^t + e^{2t} + e^{4t} \\ x_2 &= e^t + 2e^{2t} + 4e^{4t} \\ x_3 &= e^t + 4e^{2t} + 16e^{4t} \end{aligned}$.

8. $\begin{aligned} x_1 &= c_1 e^{-t} + 2c_2 e^{-2t} + c_3 e^{3t} \\ x_2 &= \phantom{c_1 e^{-t}} - c_2 e^{-2t} + 2c_3 e^{3t} \\ x_3 &= -c_1 e^{-t} + 2c_2 e^{-2t} + c_3 e^{3t} \end{aligned}$.

Section 5.1

1. $\dim(S) = 2$, $\dim(T) = 2$, $\dim(S + T) = 3$, $\dim(S \sqcap T) = 1$.
   Basis of S : $(2, 1, -1), (-1, 3, 0)$,

Basis of T : $(-1, 0, 1), (2, 2, -1)$,
Basis of $S + T$: $(2, 1, -1), (-1, 3, 0), (-1, 0, 1)$,
Basis of $S \sqcap T$: $(9, 8, -5)$.

2. $\dim(S) = 2$, $\dim(T) = 2$, $\dim(S + T) = 3$, $\dim(S \sqcap T) = 1$.
   Basis of S : $(1, 0, 1), (0, 1, -1)$,
   Basis of T : $(1, -1, 0), (0, 0, 1)$,
   Basis of $S + T$: $(1, 0, 1), (0, 1, -1), (1, -1, 0)$,
   Basis of $S \sqcap T$: $(9, 8, -5)$.

3. $\dim(S) = 2$, $\dim(T) = 2$, $\dim(S + T) = 3$, $\dim(S \sqcap T) = 1$.
   Basis of S : $(0, 0, 1, 0), (1, 1, 0, 0)$,
   Basis of T : $(1, 1, 0, 0), (0, 0, 0, 1)$,
   Basis of $S + T$: $(0, 0, 1, 0), (1, 1, 0, 0), (0, 0, 0, 1)$,
   Basis of $S \sqcap T$: $(1, 1, 0, 0)$.

4. $\dim(S) = 2$, $\dim(T) = 2$, $\dim(S + T) = 3$, $\dim(S \sqcap T) = 1$.
   Basis of S : $(2, 1, 2, 1), (1, 2, 1, 1)$,
   Basis of T : $(7, 5, 7, 4), (1, 0, 2, 0)$,
   Basis of $S + T$: $(2, 1, 2, 1), (1, 2, 1, 1), (1, 0, 2, 0)$,
   Basis of $S \sqcap T$: $(7, 5, 7, 4)$.

5. $\dim(S) = 3$, $\dim(T) = 2$, $\dim(S + T) = 3$, $\dim(S \sqcap T) = 2$.
   Basis of S : $(1, 1, 2, 1), (1, -1, 0, 2), (1, 2, 0, 1)$,
   Basis of T : $(2, 0, 2, 3), (2, 3, 2, 2)$,
   Basis of $S + T$: Same as basis of S,
   Basis of $S \sqcap T$: Same as basis of T.

Section 5.2

1. $(1, 0, -1, 0)$ and $(0, 1, 0, -2)$ will extend the given set to a basis of R^4.

2. $\left(-\frac{1}{2}, 2, -\frac{1}{2}, 1\right)$.

3. $\begin{bmatrix} 1 & 0 & -1 & 0 \\ 0 & 1 & 0 & -2 \end{bmatrix}$.

4. $sp\{(5, 4, -3)\}$.

5. $\left(0, \frac{6}{5}, \frac{8}{5}\right)$.

6. Plane $= \{(x, y, z)|5x + 4y - 3z = 0\}$.

7. $S = \{(w, x, y, z)|w + x - 2y - z = 0.\}$

8. $\left(\frac{37}{13}, \frac{31}{13}, \frac{46}{13}\right)$.

9. $W = \{(w, x, y, z)|w + x - y = 0 \text{ and } w - z = 0\}$.

10. A solution exists if and only if $a + b - c = 0$ and $a - d = 0$.

Section 5.3

1. The vectors $(1, -1, 0, 0)$ and $(0, 0, 1, -2)$ extend the given set to an orthogonal basis of R^4.

2. The vectors $(5, -4, -3, 0)$ and $(-5, 4.0, -3)$ extend the given set to an orthogonal basis of R^4.

3. $(0, 1, 0, 1)$, $(2, 0, 1, 0)$, $(-2, 5, 4, -5)$.

4. Extend by $(1, 1, 0, 0)$, $(3, 0, -1, 0)$, $(1, 0, 0, -1)$.

5. $(2, 1, 0)$, $(-1, 2, 15)$.

6. $(0, 1, -3)$, $(5, 3, 1)$.

Section 5.4

1. $\left(\frac{29}{13}, \frac{23}{13}, \frac{34}{13}, \frac{23}{13}\right)$.

2. $y = x^2 - 3x + 2$.

3.

(b) $y = \frac{1}{3}x^3 - \frac{1}{2}x^2 + \frac{13}{6}x + 1$.

4. $y = \frac{1}{2}x^2 + \frac{3}{2}x + 1$.

5. $y = 3x + \frac{1}{2}$.

Section 7.1

1. $\begin{aligned} x &= 4t + 1 \\ y &= 8t - 3 \\ z &= -5t + 2 \end{aligned}$.

2. $\begin{aligned} x &= 2t + 1 \\ y &= 2t + 4 \\ z &= 2t + 1 \end{aligned}$.

3. $\begin{aligned} x &= t \\ y &= 3 \\ z &= 2 \end{aligned}$.

4. L_1 and L_2 are parallel.

5. Direction numbers for L_1 : $(0, 0, 1)$. Direction numbers for L_2 : $(1, 0, 0)$. Direction numbers for L_3 : $(1, 1, 1)$.

6. L_1 is parallel to the Z axis, through the point $(2, 3, 0)$. L_2 is parallel to the X axis, through the point $(0, 4, 5)$. L_3 is a line through the origin and the point $(1, 1, 1)$.

7. $x = \frac{19}{5}$, $y = -\frac{3}{5}$.

8.

   (a) The lines intersect at the point $(5, -1, 0)$.

   (b) The lines are parallel, do not intersect.

   (c) The lines are skew, do not intersect.

Section 7.2

1. $x + y + z - 3 = 0$.

2. $(3, 2, -1)$.

3. The plane parallel to the XY plane at the height $z = 3$.

4. The plane parallel to the Z axis intersecting the XY plane in the line described by $3x + 2y = 5$. Since z does not appear, it can take any value.

5. $\left(-\frac{9}{5}, \frac{13}{5}, -\frac{1}{5}\right)$.

6. The line is parallel to the plane because the direction numbers of the line, $(-3, 1, -1)$ and a normal to the plane, $(2, 4, -2)$ have dot product equal to 0. The line is below the plane. To see this note the point $(7, 0, 0)$ is on the line, but $x = 7$, $y = 0$ gives $z = \frac{17}{2}$ on the plane.

7. The line of intersection of the planes is given by:
$$\begin{aligned} x &= \tfrac{1}{7}t \\ y &= -\tfrac{3}{7}t \\ z &= t \end{aligned}.$$

8.
$$\begin{aligned} x &= -\tfrac{1}{3} + \tfrac{2}{3}t \\ y &= \tfrac{16}{3} + \tfrac{11}{3}t \\ z &= t \end{aligned}.$$

9. $\frac{5-x}{3} = \frac{y+2}{2}$ and $\frac{5-x}{3} = z - 4$.

10. $\frac{x+3}{2} = z + 6$ and $y = 1$.

Section 7.3

1. $(-26, 5, 9)$.

2. $-26x + 5y + 9z = 0$.

3. $\sqrt{34}$.

4. $\sqrt{29}$.

5. 14.

Section 7.4

1.

    (a) $(1, 2, 1)$

    (b) $(1, 2, 1)$

    (c) $(2, 4, 2)$

    (d) $\frac{3}{13}(1, 0, 5)$.

2. $\frac{5}{\sqrt{14}}$.

3. $\frac{9}{\sqrt{14}}$.

4. $\frac{5}{\sqrt{20}}$.

5. $\frac{29}{\sqrt{20}}$.

6. $(-1, 0, 1)$.

7. $(-1, 0, 1)$.

8. $(-1, 0, 1)$.

9. $(-2, 1, 0)$.

10. $\left(\frac{34}{15}, \frac{5}{15}, -\frac{43}{15}\right)$.

11. $\frac{4}{\sqrt{30}}$.

12. $\frac{12}{\sqrt{74}}$.

13. $\left(\frac{8}{7}, -\frac{4}{7}, \frac{16}{7}\right)$. Distance to the line is $\frac{\sqrt{336}}{7}$.

14. $\left(\frac{13}{7}, \frac{18}{7}, \frac{12}{7}\right)$. Distance to the line is $\frac{\sqrt{3150}}{7}$.

15. $(-1, 1, -1)$.

16. $\left(\frac{23}{21}, \frac{62}{21}, \frac{4}{21}\right)$.

Section 7.5

1. $-5x - 12y - 2z + 27 = 0$.

2. No.

3.

    (a) Yes, collinear. $x = 1$, $y = t$, $z = 2$.

    (b) No, the points are not collinear.

4. No.

5. $16x - 10y + z - 7 = 0$.

6. $x + y - 2z + 1 = 0$.

Section 7.6

1. $\begin{bmatrix} \frac{\sqrt{3}}{2} & -\frac{1}{2} \\ \frac{1}{2} & \frac{\sqrt{3}}{2} \end{bmatrix}$.

2. $\begin{bmatrix} \frac{\sqrt{3}}{2} & \frac{1}{2} \\ -\frac{1}{2} & \frac{\sqrt{3}}{2} \end{bmatrix}$.

3. $\begin{bmatrix} 0 & 1 \\ 1 & 0 \end{bmatrix}$.

4. $\begin{bmatrix} \cos(2\theta) & \sin(2\theta) \\ \sin(2\theta) & -\cos(2\theta) \end{bmatrix}$.

5. $\left(\frac{5\sqrt{3}+3}{2}, \frac{3\sqrt{3}-5}{2} \right)$.

6.

(a) Reflection. Reflection is around the line through the origin containing the eigenvector $\left(\sqrt{2}+1, 1 \right)$, which is an eigenvector for $k = 1$. The line could also be described as the line through the origin making an angle $22.5°$ with the X axis.

(b) Rotation. $30°$.

(c) Rotation. $120°$.

(d) Reflection. Reflection is around the line through the origin containing the vector $\left(\sqrt{3}, -1 \right)$, which is an eigenvector for $k = 1$. The line could also be described as the line through the origin making an angle of $-30°$ with the X axis.

8.

(a) Rotation around the line through the origin containing the vector $\left(1, -\frac{1}{3}, 1 \right)$.

(b) Rotation around the line through the origin containing the vector $(1, 0, 1)$.

(c) Reflection through the plane $y = z$.

(d) Rotation around the line through the origin containing the vector $(5, 3, 3)$.

9. $\begin{bmatrix} 0 & -1 & 0 \\ -1 & 0 & 0 \\ 0 & 0 & 1 \end{bmatrix}$.

Appendix C

MATS

About MATS

MATS is a computer program which will perform many matrix operations. It was written by Prof. Eugene Johnson, an experienced teacher of linear algebra. It was written especially for use by students taking a first course in linear algebra, and a couple of commands were added specifically to fit with this book. The big advantages of MATS are ease of use and the way it fits with the course. It allows the student to use the algorithms being taught while being relieved of the computational drudgery.The disk included with this book contains a copy of the MATS program designed to run on IBM compatible computers. If your computer has a hard drive, you can copy the file mats.exe from the disk onto your hard drive and run it from there. Also it is a good idea to make a copy of the disk supplied with this book.

Starting up MATS

To start the MATS program, put the disk in one of the drives (A or B) on your computer and change to the prompt for that drive (A:> or B:>). Now type *mats* and hit the Enter key. The program will load and you will see the title page with copyright information and a prompt > at the bottom left corner of the screen. You can now begin to use MATS. MATS is very easy to use. There is no complicated syntax to learn. To enter a matrix named A just type $mat(A)$ and hit the Enter key. After that the screen will prompt you what to do. By

typing ? and hitting the Enter key, you will get a list of commands that
MATS recognizes, and a brief discription of what they do. To exit the
MATS program type *quit* and hit the Enter key. This is really all the
instruction you need to start using MATS, but we include the following
discussion of some of the commands and features which may be helpful.

Entering a matrix

To enter a matrix named A type $mat(A)$ and hit the Enter key. (Since it
is necessary to hit the Enter key after every command we will not repeat
this in what follows). The program will ask you for the number of rows.
Enter an integer from 1 to 5. The program will then ask for the number
of columns. Enter an integer from 1 to 10. The program will display
a matrix of the designated size with zeros in all positions and with the
cursor on the upper left position. Type in the number you want in this
position and hit the Enter key. The number will appear in the first row
first column position, and the cursor will move to the next entry in the
first row. In this way you enter the matrix. You may enter fractions
by typing 5/4, for example. If you type 10/8, MATS will respond 5/4.
(MATS also has a decimal mode, but when the program starts up, it
starts in fraction mode.) If you make a mistake in typing an entry for
your matrix, you may correct with the backspace key before hitting
the Enter key. Continue to enter the rest of the matrix. When the
process is complete MATS will tell you "matrix stored as A". Now you
can correct the matrix if necessary using the *edit* command. You can
enter more matrices with different names by typing $mat(B)$, $mat(M)$,
$mat(Q)$, $mat(ID)$, $mat(EXT)$, and so on. The name of a matrix may
be one or more letters.

Remark   *mat(B) and mat B are the same to MATS. The parentheses
are optional.*

Remark   *mat(B) and mat(b) are not the same to MATS. MATS is
case sensitive. The commands are written in lower case letters. It
is recommended that you use upper case letters for naming matrices,
especially if the name has more than one letter, to avoid confusion of
names with commands.*

You can check on which names you have already used by typing *names*. MATS will return the names of all matrices currently defined. You can get rid of previously defined matrices by typing either *del(A)*, which deletes the matrix *A*, or *clr*, which deletes all previously defined matrices and frees up all the names.

Editing a Matrix

You can change individual entries in a matrix named *A* by typing *edit(A)*. MATS will display the matrix *A*, and ask which entry you wish to change. Enter the number of the row and then the number of the column containing the entry you wish to change. Then enter the number you wish to put in that position. When all your changes have been made, type *q* to quit editing. MATS will tell you "matrix *A* overwritten".

Some Useful Commands

Once you have entered your matrix, you can find its reduced row echelon form. Type *rref(A)*. MATS will return the RREF of the matrix *A*. Type *herm(A)*, and MATS will return the Hermite form of *A*. Type *ns(A)* and MATS will return a matrix whose nonzero columns are a basis of the null space of *A*.

If *A* is a square matrix, *det(A)* will return the determinant of *A*, *inv(A)* will return the inverse of *A*, if *A* is invertible. The command *chp(A)* will return the characteristic polynomial of *A*, and MATS will ask if you desire a rational root search (Y/N). If you select Y, MATS will find the rational roots and factor the characteristic polynomial. These rational roots are the rational eigenvalues. MATS will not find irrational or complex roots.

EROS

MATS has a subroutine called *eros* (standing for elementary row operations). This is very useful as and educational aid, because it allows you to do a step by step Gaussian elimination. Instead of just using the *rref* command, you can find the RREF by a step by step Gaussian elimination. If you have done the row reduction by hand, you can check each step on the computer. You can observe the algorithm at work without having to do all the arithmetic. If you type $eros(A)$, MATS will display the matrix A and six operations you can perform.

1. $< p > piv$ If you type p, MATS will tell you it is going to perform "$k * Row\ I + Row\ J$", and ask you to enter first k, then I and J. (The piv stands for pivot.)

2. $< f > fpiv$ If you type f, MATS will tell you it is going to "$Pivot\ on\ the\ (I, J)-entry$", and will ask you to enter I and then to enter J. MATS will add multiples of row I to all other rows so that column J will be zero except in row I. This operation is called a full pivot. Of course it can't be done if the (I, J) entry of your matrix is 0.

3. $< s > swap$ If you type s, MATS will reply "$Row\ I \longleftrightarrow Row\ J$", and ask you to enter I and then J. MATS will exchange row I and row J.

4. $< * > rowp$ If you type $*$, MATS will reply "$k*Row\ I, Enter\ k$". Then it will ask you to enter I. MATS will multiply row I by k.

5. $< / > rowd$ If you type $/$, MATS will reply "$1/k * Row\ I$", and ask you to enter first k and then I. MATS will divide row I by k.

6. $< q > quit$ If you type q , MATS will exit the $eros$ subroutine.

Remark   *If you wish to name the last matrix that has appeared B, just type = B at the prompt. MATS will reply "matrix stored as B". Next time you type B, MATS will display the matrix you have named B.*

The Record and Remark Commands

If you type *record* (followed by Enter), MATS will tell you "the recorder is now off, would you like to change it?(Y/N)". If you choose Y, MATS will ask for a file name. Enter a name for the file. Everything that appears on the screen after that (until the recorder is turned off) will be recorded in this file. After you have exited the MATS program, you can print out this file with the DOS print command. The file can also be edited and printed with any standard word processing program.

If you type *rem*, this tells MATS that you want to insert something that is not a command, but that will appear when the file is typed out. This command is useful if you want to insert a remark into some computations that you are recording and plan to print out.

List of Commands

Remark   *Parentheses are optional and can be replaced by a space.*

$mat(A)$: calls matrix collection routine for matrix named A.

$edit(A)$: calls matrix edit routine to edit matrix A.

$det(A)$: returns the determinant of A.

$inv(A)$: returns the inverse of A if A is invertible.

$rref(A)$: returns the reduced row echelon form of A.

$herm(A)$: returns the Hermite form of A.

$ns(A)$: computes $I - Herm(A)$, the nonzero columns of which are a basis for the null space of A.

$chp(A)$: returns the characteristic polynomial of A.

$adj(A)$: returns the adjoint of A.

$rowmat(A)$: calls collection routine for a row matrix named A.

$rmat(A)$: abbreviation for rowmat(A).

$colmat(A)$: calls collection routine for a column matrix named A.

$cmat(A)$: abbreviation for colmat(A).

$imat(A)$: generates an identity matrix named A (prompts for size.)

$aug(A, B)$: returns the augmented matrix $[A \; B]$.

$eros(A)$: calls the elementary row operations routine on A.

$piv(A)$: calls the pivot routine on A.

$fpiv(A)$: calls the full pivot routine on A.

$rowp(A)$: calls the row multiplication routine on A.

$rowd(A)$: calls the row division routine on A.

$swap(A)$: calls the row swap routine on A.

$A + B$: adds matrices A and B.

$A - B$: subtracts B from A.

$A * B$: multiplies A and B.

$< \mathbf{u}, \mathbf{v} >$: returns the dot product (Euclidean Inner Product) of $\mathbf{u}$ and $\mathbf{v}$.

$\mathbf{u} \times \mathbf{v}$: returns the cross product of $\mathbf{u}$ and $\mathbf{v}$.

$ca(m, n, A)$: returns columns m through n of A as the new active matrix.

$row(i, A)$: returns row i of A.

$col(i, A)$: returns column i of A.

$names:$ lists names of matrices defined.

$= A$: names the current contents of the Xregister as A.

$clr:$ clears all previously used matrices and frees memory.

$del(A)$: deletes the matrix A and frees memory.

$mode:$ reports mode (fract/dec) and allows toggle.

$cof(i, j, A)$: returns the (i, j)-cofactor of A.

$cofmat(A)$: returns the matrix of cofactors of A.

$glinv(A)$: returns an invertible matrix P such that $P * A = RREF(A)$.

A/B: returns A*glinv(B).

$B\backslash A$: returns glinv(B)*A.

$xreg$: returns the current contents of the Xregister.

dec: switches to decimal mode.

$fract$: switches to fraction mode.

fix: sets number of decimal places dispalyed in decimal mode.

$record$: reports recorder status (on/off) and allows for toggle.

rem: tells MATS to ignore current line except to record it if recorder is on.

? : brings up a list of commands and brief discription of what they do.

$quit$: exits MATS, returns you to DOS prompt.

Remark *If you type a command without an argument, the command will be applied to the last matrix the program has seen (current contents of the Xregister).*

If you want to install MATS on a network, contact the author for permission and further instructions.

Subject Index